PHYSIOLOGY - LABORATORY AND CLINICAL RESEARCH

HYPERTHERMIA

RECOGNITION, PREVENTION AND TREATMENT

PHYSIOLOGY - LABORATORY AND CLINICAL RESEARCH

Additional books in this series can be found on Nova's website under the Series tab.

Additional E-books in this series can be found on Nova's website under the E-book tab.

PHYSIOLOGY - LABORATORY AND CLINICAL RESEARCH

HYPERTHERMIA

RECOGNITION, PREVENTION AND TREATMENT

XING-HUA GAO
AND
HONG-DUO CHEN
EDITORS

nova publishers
New York

Library of Congress Cataloging-in-Publication Data

ISBN 978-1-61942-275-9

Published by Nova Science Publishers, Inc. † New York

CONTENTS

PREFACE

By definition, hyperthermia is abnormally high body temperature, usually resulting from infection, medication, or head injury, and sometimes brought about intentionally to treat diseases, especially certain cancers. In this chapter, a variety of hyperthermia related physiological and pathological phenomena were described. Professor Ivanovna described the influence of hyperthermia on liver physiology in different species with different susceptibility to hyperthermia, and stressed the importance of hyperthermia in evolution. Périard described the mechanism of exercised related fatigue in view of hyperthermia induced systemic regulation. Shiodaa et al investigated the role of psychostimulant in inducing hyperthermia and its management. There are a variety of needs for hyperthermia technology, in terms of source of energy, transduction, control and manipulation of energy output. Trujillo and Gutierrez et al overviewed the electromagnetic and ultrasonic hyperthermia, respectively, and they further described the application of their respective techniques. Endogeneous hyperthermia or hyperthermia induced biological changes may be biomarkers in some defined conditions. Patricia et al revealed that plasma heat shock protein may be a marker for heatstroke. Bunai et al described the hyperthermia related pathological findings in forensic autopsy. Ayestaray stressed the importance of hyperthermia in a rare skin condition—pyoderma gangrenosum. Lastly, a couple of clinical uses of hyperthermia were overviewed. Gao et al summarized the application of local hyperthermia on treating human papillomavirus infection of skin and/or mucosa, and discussed the possible mechanisms of actions, especially in immuno-modulatory function. Portela et al introduced the focused hyperthermia technique in treating solid tumors. He described the ferrimagnetic materials aided, focused local hyperthermia in clinical management of solid tumors. The method is less invasive or destructive to normal tissues, making it a relatively safe option in treating tomors. In all, the present chapter will help the readers to get a comprehensive view on the basic and practical knowledge of hyperthermia.

Chapter 1 - Oncology hyperthermia is a therapy to treat cancer in which body tissues are exposed to high temperatures (42°C - 45°C) for a long time (30 min – 60min). Carcinogenic cells do not support specific combinations of both temperature and time while normal cells do. Nowadays, although there are several options for cancer treatment based on different physical principles and protocols, the RF hyperthermia is one of the most promising therapies. The RF hyperthermia is based on power absorption, which produces a temperature increase in tumors. The hyperthermia makes progress in researching diverse biological, technological, physical and clinical aspects; however, for its application in hospitals, it is still required to improve techniques, applicators and also it is needed to know the behavior of all the biological involved aspects.

Chapter 2 - The thermotherapy in which ultrasonic energy is used to increase temperature in a localized region in order to produce a desired effect is called ultrasonic hyperthermia. Although hyperthermia means the elevation of temperature above the human normal value, conventionalisms refer the classical hyperthermia as a temperature increment above 40°C and below 45°C. Recent developments apply high temperature hyperthermia (more than 50°C) to produce instantaneous tissue coagulation. The systems for ultrasonic hyperthermia have been created in relation to the application. The ultrasonic hyperthermia used to produce muscular or skeletal healing requires small temperature increments and hence medium intensity ultrasonic transducers (0.1 – 3.0 W/cm^2) are used. These transducers are normally planar and the application of therapy does not need rigorous control because that intensity of ultrasound rarely produces undesired effects, *e.g.* cavitation and ablation. The oncological hyperthermia is used to kill cancerigenic cells, then acoustical intensities for this therapy are normally high (from 20 W/cm^2 to 20,000 W/cm^2) and focalization is required to concentrate energy. The control in this therapy is very important because temperature should be kept at a constant value during a specific time of treatment in order to warrant cancerigenic cell death. Because this control requires temperature measurement in the region, which is practically impossible nowadays, some authors modified the technique by increasing the ultrasound intensity in order to produce ablation in the tumor in few seconds. In this situation, the temperature measurement is not necessary, but it is required a technique to guide the therapy. The frequency and power of both physiotherapeutic and oncological hyperthermia are quite similar, and the devices are made up generally by the same two parts: the generator and the transducer. The generator produces oscillating signals capable of driving the ultrasonic transducer at its resonant frequency at high power levels. The equipment is composed of an oscillator, an amplifier, the electrical matching, the feedbacks of variables, and control. The transducer is a piezoelectric material with different configurations: one piece planar or curved, and an array of small elements. For physiotherapeutic hyperthermia, it is common to use a planar transducer without backing and with front layer. The plane shape permits to produce a wide beam to be distributed in a large region of tissue; the lack of backing helps in increasing the efficiency and the front layer in matching the acoustic impedances between the piezoelectric and the radiated media. In oncological applications, the transducers are focused either from the shape (spherical transducers) or from the driver signal for transducers planar arrays. Wide-band transducers are lightly backed, while narrow-band transducers are not. Although there are many kinds of transducers for these applications, the design requirements for the ultrasonic hyperthermia systems are equivalent: high efficiency in amplification and high stability in frequency. In this book chapter, the configurations of practical hyperthermia systems and the problems to solve in order to apply them are discussed.

Chapter 3 - Human aerobic performance in hot climatic conditions is markedly influenced by adjustments in circulatory, thermoregulatory, metabolic and neuromuscular function. Characteristically, exercise performed in the heat increases core and skin temperature, cardiovascular strain and perceived exertion. As a result, exercise intensity and duration are compromised. Currently, three mechanisms are suggested to explain the potential pathway(s) by which performance may be impaired. These include i) an increase in cardiovascular strain limiting oxygen delivery and uptake in exercising muscles and possibly the brain, ii) the development of hyperthermia-induced central fatigue reducing neural drive to the contracting musculature, and iii) anticipatory regulation of muscle recruitment ensuring the maintenance of whole-body homeostasis. These hypotheses have led to a diversity of experimental models investigating systemic hemodynamic responses during exercise in the heat, to evaluating force production capacity and voluntary muscle activation during both passive and exercise-induced hyperthermia. Research has shown that voluntary activation in previously active muscles is reduced and force production decreased when core temperature increases to ~40°C. Maximum heart rates have also been reported at these high temperatures on reaching exhaustion, indicating that the cardiovascular system may have attained functional limitations. Hence, the etiology of the impairment to aerobic exercise performance in hot climatic conditions remains contentious. This chapter explores the current state of knowledge regarding the aforementioned mechanisms, and contextualizes fatigue during constant rate and self-paced exercise, as well as identifies avenues of enquiry that remain unresolved.

Chapter 4 - Rules of the evolutionary formation of responsiveness and flexibility of functionally similar tissues from the parallelism theory standpoint, levels of biological systems' organization (molecular, subcellular, cellular, inter-tissular, organ), metabolism strategies (hypercatabolic - resistant; tolerant – hypometabolic, tolerant hypobiosis), differences in DNA concentration, difference in the organisms' standings on the evolution ladder were studied on in situ model in the normal state and after hyperthermia action. Studying this problem in the light of the global warming is becoming especially topical. Hereby the state of organs directly involved in the organism homeostasis maintenance acquires a special importance. Liver is one of such organs.

In particular it was found that liver acinus histotopograohy in the comparative row of intact animals changes from tubular type (fishes, amphibias, reptiles) to tubular-trabecular (birds) and trabecular (mammals) type. Number density of liver acini, volumetric density of mitochondrias, hepatocyte apoptosis/necrosis ratios increase from ectothermal animals to endothermal ones. In the hepatocyte ultrastructure species specificity of the nucleolus structure, volumetric, surface and number indicators of mitochondrias, granular endoplasmic reticula, lysosomes, glycogen and lipids is identified.

In the evolution polyplodization duc to acytolkinetic mitosis bocomes the leading form of hepatocytes' physiological regeneration. And, if in mammals, reptiles and fishes this hepatocyte population dominanation zones are not present, in birds this is portal tract zone and in amphibias – central veins' zone. Hereby the number of PCNA-positive hepatocytes in birds in all acinus zones is by an order of magnitude higher. In other animal groups their number is smaller with the predominant localization in the periporal zone.

The number of desmin-positive Ito cells grows from ectothermal animals to endothermal ones; whereas the maximum number of organ-specific macrophages is found in fishes and mammals.

In intact fishes and amphibias NAD- and NADP-dependent mitochondrial oxidation paths are expressed to an equal degree; in reptiles and birds a higher activity of NADP-dependent dehydrogenases and in mammals – NAD-dependent dehydrogenases is found. The highest activity of hepatocyte apoptosis biochemical markers is found in birds and the lowest activity – in amphibias and fishes.

2. Single hyperthermia detects two animal groups "amphibias/birds" and "fishes/reptiles/mammals" causes species-specific changes of the liver acinus bloodstream morphometrical indicators, increased number of stromal liver cells with the change of the

topography thereof within the acinus; change hepatocytes' physiological regeneration and death pathways. Hyperthermia determines the change of hepatocyte cell cycle structure: in "amphibias/birds" group the number of hepatocytes in the cell cycle G0-G1-stage increases and in the "fishes/reptiles/ mammals" group the increase is observed in S- and G2-M-stages. The number of hypodiploid hepatocytes in all the animal groups increases, the main death pathway is necrosis/autophagy.

Hepatocytes' ultrastructural changes are of mosaic and oppositely-directed nature. Volumetric, number and surface density of the mitochondrias, nucleolus system, granular endoplasmic reticula, lysosomes as well as lipids/glycogen volumetric ratios change. Activation of the liver stromal cells manifested as the increase of their number, transformation of Ito cells into myofibroblasts and increase of autophagolysosomes and cell debris fragments in the Kuppfer cells is marked.

Depending on the development of hypo- or hypermetabolic responses these changes reflect the increase of the number of binucleate or PCNA-positive hepatocytes or both the hepatocyte populations.

In reptiles hyperthermia causes increased activity of NAD- and NADP-containing mitochondric oxidoreductases; in mammals – increased activity of NADP-containing mitochondric oxidoreductases and in fishes - NAD-containing mitochondric oxidoreductases which reflects hypercatabolism development. In birds and amphibias the activity of NAD- and NADP-dependent mitochondric oxidoreductases is inhibited which reflects hypometabolism development. Species differences demonstrate cytosol oxidoreductases' activity, ions' concentration, activities of biochemical markers of hepatocytes' proliferation and death pathways.

Consequently, hypercatabolic metabolism strategy (resistant) after overheating is identified in reptiles and mammals and, to a smaller degree, in fishes. Phylogenetically more ancient hypometabolic strategy (tolerant hypobiosis, tolerant) – in amphibias and birds. Metabolic depression in intact ectothermal animals has natural basis when hypoxic state is changed by hyperoxic state due to reoxygenation. Nevertheless, in birds hypometabolism develops which is manifested at the level of both mitochondrial and cytosol oxidoreductases as well as in volumetric "lipids/glycogen" ratios, cell cycle indicators. It reflects the provision that in the evolution new possibilities/ functions arise on the basis of the already attained level, due to the activation of the evolutionally older hepatocyte functioning and energy-supply mechanisms. In its turn, the identified activation of oxidoreductases of tricarboxylic acids' cycle and cytosol in reptiles is due to the fact that being descendants of the ancient amphibias, reptiles, particularly, tortoises gave birth two major branches of the present-day vertebrates – birds and mammals. Tortoise order descends directly from cotylosaurs and, consequently, their position is the closest to the reptile-like ancestors of mammals. Consequently, it is possible to suggest that in extreme conditions metabolic adaptation strategies' switching is observed: endothermal animals may activate poikilothermal thermal regulation mechanisms and vice versa.

Chapter 5 - The abuse of psychostimulant drugs, including methamphetamine (METH), 3,4-methylenedioxymethamphetamine (MDMA), and cocaine, is popular in many parts of the world, especially among younger generations. These psychostimulant drugs cause various acute and chronic adverse effects, including hyperthermia, arrhythmia, tachycardia, and agitated delirium, of which hyperthermia is one of the most life threatening. Numerous studies have reported that hyperthermia induces multiorgan failure, rhabdomyolysis, disseminated intravascular coagulation, and even mortality. However, pharmacological treatment for psychostimulant-induced hyperthermia has not yet been established. The authors have reported MDMA- and METH-induced dopamine (DA) and serotonin (5-HT) intracellular concentration increases in the hypothalamus, the thermoregulation center, and that selective D1 and 5-HT2A receptor antagonists attenuated psychostimulant-induced hyperthermia. In this review the authors describe the pharmacological mechanisms of these three psychostimulants and present the basic research they have performed thus far. In addition, the authors address pharmacological treatment for hyperthermia induced by these psychostimulants.

Chapter 6 - Heat illnesses range from heat fatigue, heat rashes, disorders resulting from water or salt depletion, including heat exhaustion and heat cramps, to the most serious disorder, heatstroke. Heatstroke is a medical emergency and the resulting physiological changes can be profound and life-threatening. The mortality rate can be as high as 50%. Classic heatstroke tends to affect the very young or old, while exertional heatstroke mainly affects young, fit individuals. Many of the physiological changes found in exertional heatstroke are difficult to differentiate from those associated with intense physical exertion.

The diagnosis of heatstroke is based on an elevated body core temperature (>41°C) and signs of central nervous system disturbance such as disorientation and combative behavior. The presence or absence of sweating is no longer considered diagnostic. It would be useful if there were a laboratory test to confirm a diagnosis of heatstroke, as body core temperature has often returned to normal by the time the patient has been transported to hospital. Heat shock protein 72 (Hsp72) is markedly elevated by heat and/or other stressors to the cell. Hsp72 is also found in the plasma and is elevated immediately after exercise; it is thought to act as a chemokine in the plasma, initiating an inflammatory response.

Plasma Hsp72 is a potentially useful marker of heatstroke, and several studies are discussed that have measured plasma Hsp72 in cases of heatstroke. Further research is needed to examine the mechanism of the increased plasma concentrations of Hsp72 in heatstroke, as passive hyperthermia alone does not appear to lead to the same increase in plasma Hsp72 levels.

Chapter 7 - Heatstroke is a thermal insult induced by an imbalance between accumulation of heat in the body and heat loss from the body. The insult leads to central nervous system disturbances ranging from confusion to deep coma. A multisystem insult follows and leads to a fatal outcome. Heatstroke is associated with certain clinical complications. However, the postmortem findings of heatstroke, including the histopathological findings, are nonspecific, and some fatal heatstroke cases are undoubtedly misdiagnosed as ischemic heart disease at autopsy. If rectal temperature is not measured on admission or there is a long interval between collapse and admission, clinicians are not reminded of the possibility of heatstroke, and pathologists do not receive important information. On the other hand, right-sided cardiac dilation, subendocardial hemorrhage in the left ventricle, severe pulmonary congestion, and venous distention, particularly in the cerebrum, are typical pathological findings associated with heatstroke and remind pathologists of the possibility of the diagnosis. Moreover, the early occurrence and rapid development of postmortem changes brought on by high body temperature at the time of death are also typical findings after death in a state of hyperthermia.

Chapter 8 - Pyoderma gangrenosum is the most common neutrophilic dermatosis. It can appear under different forms : ulcerative, pustular, bullous, vegetative or necrotizing.

Hyperthermia is usually the first revealing sign, in all these variants, before the skin inflammation occurs. Pyoderma gangrenosum may be mistaken for a skin infection. Despite antibiotics associated with a surgical excision, hyperthermia and skin inflammation increase. The only effective treatments are corticoids and immunomodulators.

Chapter 9 - Human papillomavirus (HPV) infection of the skin or mucosa leads to various clinical forms of warts, or even carcinomas. Most of the warts will regress in a matter of years, while some are recalcitrant. Warts are disfiguring or debilitating.

Chapter 10 - Ferrimagnetic materials, exposed to an electromagnetic field, are largely utilized in the experimental treatment of tumours by hyperthermia. Hyperthermia is a widely studied method in the cancer treatment, operating in the tumour cells, tumour neovascularization, and it may induce an immunologic and apoptotic response in the whole organism. The methods currently available to produce hyperthermia are generally limited by the inability to selectively target the tumour cells, with the subsequent risk of affecting adjacent healthy tissues. To overcome this limitation and in the context of the magnetically mediated hyperthermia, the Highly Focalized Thermotherapy (HFT) technique was developed. A new material, a ferrimagnetic cement (FC), composed by a silicate cement and magnetite was created to be used with this technique. FC is injected within the tumour and then exposed to an external high frequency magnetic field. The data obtained in the in vitro and the in vivo studies allows us to conclude that FC can provide a minimally invasive technique to treat solid tumours with HFT.

In: Hyperthermia: Recognition, Prevention and Treatment ISBN 978-1-61942-275-9
Editors: Xing-Hua Gao and Hong-Duo Chen, 1-40

Chapter 1

ELECTROMAGNETIC HYPERTHERMIA TREATMENTS: TECHNOLOGY, MODELING AND EVALUATION OF APPLICATORS FOR CLINICAL USE

*C. Trujillo-Romero, L. Leija and A. Vera**

ABSTRACT

Oncology hyperthermia is a therapy to treat cancer in which body tissues are exposed to high temperatures (42°C - 45°C) for a long time (30 min – 60min). Carcinogenic cells do not support specific combinations of both temperature and time while normal cells do. Nowadays, although there are several options for cancer treatment based on different physical principles and protocols, the RF hyperthermia is one of the most promising therapies. The RF hyperthermia is based on power absorption, which produces a temperature increase in tumors. The hyperthermia makes progress in researching diverse biological, technological, physical and clinical aspects; however, for its application in hospitals, it is still required to improve techniques, applicators and also it is needed to know the behavior of all the biological involved aspects.

INTRODUCTION

Although, there are several applicators for hyperthermia, the regulation of the EM energy deposition over tumors is still required. Several focusing techniques for hyperthermia have been proposed. To develop different kind of applicators in order to know its behavior and interaction with tissue is an expensive task, for this reason the modeling had been an important tool in this process. During the last years, the modeling of EM hyperthermia treatments based on the solution of Maxwell and Bioheat equations by using numerical methods has made progress significantly.

In order to use an applicator in hospitals, it must be evaluated with some tests such as SAR and thermal distributions. To obtain these distributions there are different experimental

* E-mail address: arvera@cinvestav.mx.

setups that include the radiofrequency generation system and different sensors in order to know the behavior of the applicator during the hyperthermia treatment. The applicator behavior can be characterized by means of phantoms which are tissue substitutes. They have dielectric properties similar to those of human tissue in order to obtain a more actual evaluation.

On the other hand, the recent advances in nanotechnology have allowed using nanoparticules in the therapy called magneto hyperthermia. The magneto hyperthermia therapy is just beginning and it requires further research. Although several aspects of hyperthermia are still under research, there are some applicators that are commercialized. In this book chapter, by analyzing the recent researches concerning the most important aspects in hyperthermia treatments, some problems which can be solved in order to reach a major applicability of the therapy around the world and the future challenges in hyperthermia therapy are discussed.

1. Hyperthermia

Oncology hyperthermia refers to a controlled increase of body temperature above 45 °C, in order to damage and kill cancer cells. Oncology hyperthermia is a therapy to treat cancer in which body tissues and tumors are exposed to high temperatures (42°C - 45°C) for a long time (30 min - 60 min). The therapeutic effects of hyperthermia are well-known since XIX century. Carcinogenic cells do not tolerate specific combinations of both temperature and time while healthy cells do. The temperature increase produces an enzymes destabilization and changes in membrane permeability by causing cell death. The damage depends on temperature and time. The success of hyperthermia treatments lies in focusing the heat into the tumor without causing thermal damage to surrounding healthy tissues.

2. Hyperthermia Classification

Nowadays, there are several options to treat cancer based on different physicals principles and applications. The physical analysis of electromagnetic (EM) fields and ultrasound (US) is useful to define systems which are able to deliver power on superficial and deep tumors. In most of the hyperthermia clinic studies, EM and US techniques are used. Around the world, several hyperthermia devices, which work at different frequencies, physical principles, protocols, etc., have been developed in the last years. However, the improvement of treatment protocols, devices, and mathematical models is still required. The EM hyperthermia uses EM fields in a range from 10 MHz to 2.45 GHz [1] (from radiofrequencies (RF) to microwaves (MW)). The main goal of EM hyperthermia is to treat deep tumors; however, there are some problems to be solved to achieve heating at different depths.

Hyperthermia treatments with MW superficial applicators, to treat deep tumors, are impossible because the penetration depth of the waves at these frequencies is only a few millimeters, i.e. the attenuation of EM fields is higher. By using RF is possible to increase the penetration depth of waves, but it is impossible to have control of the affected area and the size of this region. In order to try to solve each of the drawbacks of electromagnetic hyperthermia, there are different heating techniques. Hyperthermia can be classified into different branches: local, regional, whole-body and magnetic fluid hyperthermia [2]. Local hyperthermia is classified as external, interstitial and endocavitary hyperthermia; while regional hyperthermia is classified as hyperthermia of deep tissues, regional perfusion and continuous hyperthermic peritoneal perfusion. Nowadays, although there are several options for cancer treatment based on different physical principles, protocols, and heating sources based on RF, MW, US, etc. the electromagnetic hyperthermia is one of the most promising therapies. The EM hyperthermia is based on power absorption, which produces a temperature increase in tumors. This chapter is focused on EM hyperthermia techniques.

2.1. Electromagnetic Techniques

A) Local Hyperthermia

The main goal in local hyperthermia is to increase tumor temperature (5°C – 7°C) while temperature in surrounding healthy tissue is limited. Oncology treatment is based on achieving a tumor temperature of approximately 42°C and holding it during one hour. If this temperature is reached, cancer cells will be destroyed. Primary malignant tumors have poor blood circulation, which makes them more sensitive to changes in temperature. This elevation of temperature is obtained when the patient is radiated by means of radiofrequencies (RF), microwaves (MW) electromagnetic fields and ultrasound (US). As mentioned above, the local hyperthermia has 3 different approaches:

- *External hyperthermia:* Usually, external hyperthermia is used to heat small areas. External applicators, using high-frequency energy waves, are used when tumors are just below skin. The applicator is placed over the tumor region and the energy is delivered from the applicator to the patient (tumor). Temperature could be measure either into the tumor, by inserting temperature probes inside the tumor, or on the surface of the skin. This treatment could be applied by itself or in combination with radio or chemotherapy. Most tumors treated with this kind of hyperthermia are breast cancer and superficial recurrent melanomas. Examples of applicators for external hyperthermia are shown in Figure 1 a) and b).
- *Interstitial hyperthermia:* Interstitial applicators are used to treat deep tumors. For interstitial techniques, applicators are probes or needles which can be inserted into the tumor. For this reason, temperatures achieved are higher than those obtained with external applicators. To insert this kind of applicators some imaging techniques are used in order to make sure that the applicator is inserted into the tumor. Microwave ablation (MWA) and radiofrequency ablation (RFA) are interstitial techniques that use radio waves to kill cancer cells. Temperature rise also is measured by inserted probes which are aligned with the applicator by means of imaging techniques. The advantage of interstitial hyperthermia is that only the tumor and the tissue in the surroundings are affected. However, this technique has the disadvantage of being invasive and it has the risk of tumor seeding along the insertion path of the antenna

(metastasis) [3, 4]. An example of a needle used for interstitial hyperthermia is depicted in Figure 1 c).

- *Endocavitary hyperthermia:* This kind of hyperthermia is used when the tumor is located inside or near to the body cavities, such as the rectum and esophagus. The applicator or antenna is located inside the cavity and it is inserted into the tumor. Cancer tumors which can be treated with this therapy are located in esophagus, rectum, vagina, cervix, uterus, prostate, bladder, etc.

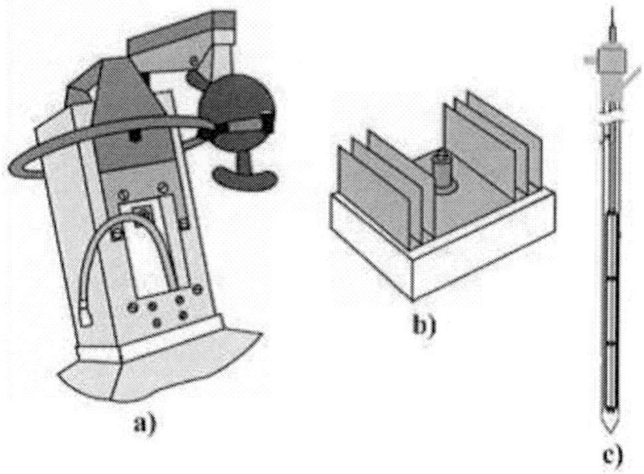

Figure 1. Applicators for local hyperthermia: (a) waveguide applicator; (b) current sheet applicator and (c) interstitial applicator.

B) Regional Hyperthermia

Usually, regional hyperthermia is applied to treat large areas in patients with deep tumors, such as tumors in abdomen, pelvis, organs or limb. Tumors in pelvis, rectal carcinoma, cervical carcinoma, bladder carcinoma and prostate carcinoma are the most indicated tumors to treat with regional hyperthermia. The main goal in regional hyperthermia is to achieve a rise of approximately 5°C, in tumor temperature, above normal physiological temperature. However, because of the tumor location and variations in tissues properties regional hyperthermia is more complex than local hyperthermia. Corry and Barlogie propose an ideal regional hyperthermia system with variable penetration depth (> 12 cm), variable field size and controlled field location; unfortunately, no such a system exist [5].

The regional hyperthermia could be classified as:

- *Hyperthermia of deep tissues:* this kind of hyperthermia is applied to treat cancer such as a bladder or cervical. In this case, external applicators normally based on annular phased arrays (APA) of radiators (multi-antenna applicators) [6], which are arranged in a single ring (2-D applicators) or in three rings (3-D applicators) are used.
- *Regional perfusion:* this kind of hyperthermia is used to treat cancer such as liver or lung. In this procedure, some blood of the patient is removed and heated after the blood is pumped back into the limb or organ.

– *Continuous hyperthermic peritoneal perfusion (CHPP):* it is a technique used to treat cancers within the peritoneal cavity. During surgery, heated anticancer drugs flow from a warming device through the peritoneal cavity. The peritoneal cavity temperature reaches 41°C – 42°C [7].

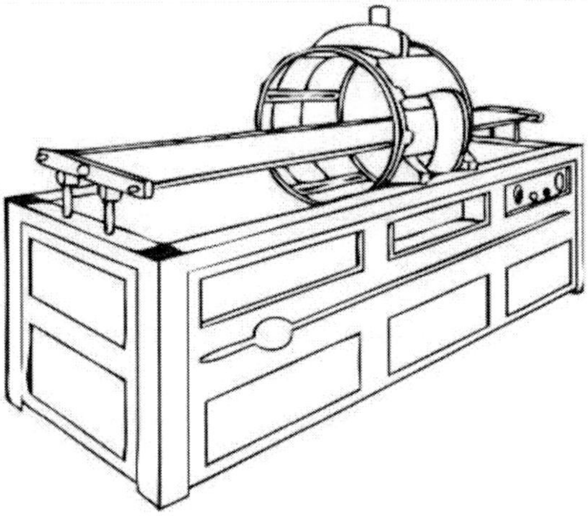

Figure 2. Applicator for regional hyperthermia. Applicator used in hyperthermia of deep tissue (BSD applicator).

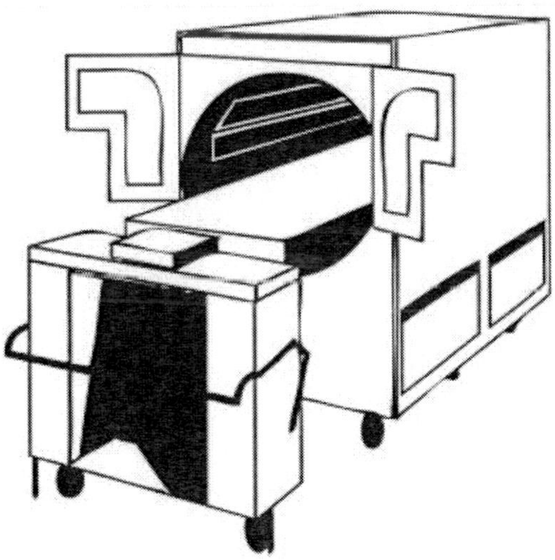

Figure 3. System for whole-body hyperthermia. Aquatherm system for whole-body hyperthermia.

C) Whole Body Hyperthermia

Whole body hyperthermia (WBH) is used to treat metastatic cancer which has been spread throughout the body, i. e. in WBH the whole body is heated in order to treat cancer. Nowadays, the well-known techniques are thermal conduction, extracorporeal induction and EM induction [8]. The temperature increase is usually limited to 41.8°C – 42°C. The systemic heating is usually accomplished through a heat source such as thermal conduction. The most common method of whole body thermotherapy is the heat transfer in and out of the body through skin contact. A variety of techniques, such as warm clothes or blankets, boxes of hot air and fluid therapy have been described by several authors [9, 10]. Some researchs have shown an increased heating rate using additional regional heating devices. The third method of induction heating is through non-contact systems using systems of non-ionizing electromagnetic radiation. Figure 3 shows an example of a system to apply whole-body hyperthermia.

D) Magnetic Fluid Hyperthermia (MFH)

The most recent development in hyperthermia is the magnetic fluid hyperthermia which has had great progress in the last years. This therapy is based on the increase of temperature in tumors by using different superparamagnetic particles which are inserted into the tumor [11-15]. The heating is a consequence of the excitation of magnetic particles exposed to alternating electromagnetic fields that converts the magnetic energy to heat. The relaxation losses are the physical principles that cause heating. This physical phenomenon is divided on Brown and Néel relaxation. The Brown and Néel relaxations depend of the particle size (~ 10 nm) [16]. The Brown relaxation refers to the particles rotation in the carrier fluid, resulting in friction of the particles with the fluid. Néel relaxation refers to the reorientation of the magnetic moment within the particles causing a temperature increase [17]. The heat generated by the nanoparticles has therapeutic effects. The magnetic properties of nanoparticles are important factors to achieve an effective treatment. Maghemite (γ- Fe2O3) and magnetite (Fe3O4) [18] are some particles used to focus the energy over tumors and, as a consequence, to reduce side effects on nearby healthy tissue. Nowadays, several studies have shown the behavior of ferrofluids in hyperthermia treatments by using coils to generate an alternating magnetic field [19, 20] which excite the nanoparticles implanted into the tumor. Magnetic fields at low frequencies (e.g. 153 kHz, 360 kHz, 55 kHz, 800 kHz, etc.) [19-22] are used to characterize nanoparticles in hyperthermia treatments. The nanoparticles are concentrated in the tumor and they remain for at least 6 weeks, allowing hyperthermia treatments. The studies carried out have proven the feasibility of achieving good clinical results by using this technique.

Before starting a hyperthermia treatment, it is necessary to chose one hyperthermia technique. The selection depends primarily on the size and the depth of tumor. Some technical questions to be considered before starting are enlisted here:

a) Which kind of electromagnetic method should be used?
b) What radiofrequencies are the appropriated to use during the treatment?
c) What kind of applicators is the most adequate for the treatment?
d) A non- invasive or an invasive method of hyperthermia is needed?
e) Which kind of thermometry should be used throughout the treatment?

Currently, another way to select an adequate technique of hyperthermia is by means of the hyperthermia treatment planning (HTP) which is discussed later in this chapter.

3. Hyperthermia: Biological Aspects

The effect of the EM waves in biological systems is determined partly by the intensity of the field, partly by the quantity of energy contained in every photon. The non-ionizing radiation, even if it is of high intensity, cannot cause ionization of any biological system; however, it has been demonstrated that these radiations produce other biological effects, such as heating tissue. Electromagnetic waves can produce biological effects which sometimes, but not always, lead to adverse health effects. It is important to understand the difference between both [23]:

- *Biological effect (such as heating tissue during the hyperthermia treatment):* This effect occurs when exposure to EM waves cause some remarkable physiological change in a biological system.
- *Adverse health effect:* this effect occurs when the biological effect is outside of the normal range, which can be compensated by the body, and thus leads to some detrimental health condition.

Table I. Physical and biological effects caused by temperature in biological tissues [1, 24-26]

Temperature range (° C)	Time	Physical Effects	Biological effects
< -50	> 10 min	Freezing	Complete cellular destruction
0-25		Decreased permeability	Decreased blood perfusion, cellular metabolism, hypothermic killing
30-39	No time limit	No changes	Growth
40-46	30-60 min	Changes in the optical properties of tissues	Increased perfusion, permeability, pH, thermotolerance induction, hyperthermic killing
47-50	> 10 min	Necrosis, coagulation	Protein denaturation
> 50	2 min	Necrosis, coagulation	Cell death
60-140	Seconds	Coagulation, ablation	Protein denaturation, membrane rupture
100-300	Seconds	Vaporization	Cell shrinkage and extracellular steam vacuole
> 300	< Second	Carbonization, smoke generation	Carbonization

Thermal effects caused by RF electric fields occur because tissues are electrically conducting. The electric field inside the tissue generates currents and their dissipation leads to energy absorption, therefore, the temperature increase occurs. The interaction of EM waves with biological tissues is related primordially with 3 aspects: 1) the cellular survival that depends on the temperature and the duration of the treatment, 2) the tumor conditions as hypoxia, low pH and poor nutrition, sensitivities to heating of tumors and healthy tissues, etc., which can be a benefit for the therapy, and 3) thermotolerance developed by cells [7].

A) Cellular Survival

Temperatures above 40 °C are cytotoxic, and cells can be killed; however, it depends on the temperature and the duration of the treatment. It is well-known that tumor cells are more sensitive to high temperatures than normal tissue. At certain temperatures and exposure times lethal effects may occur, this means that always it can be find a temperature that applied certain period of time causes cell death. Exposure time is reduced if temperature rises and vice versa. The relationship between temperature and exposure time throughout the hyperthermia treatment has been reported for a variety of biological systems (See Figure 4).

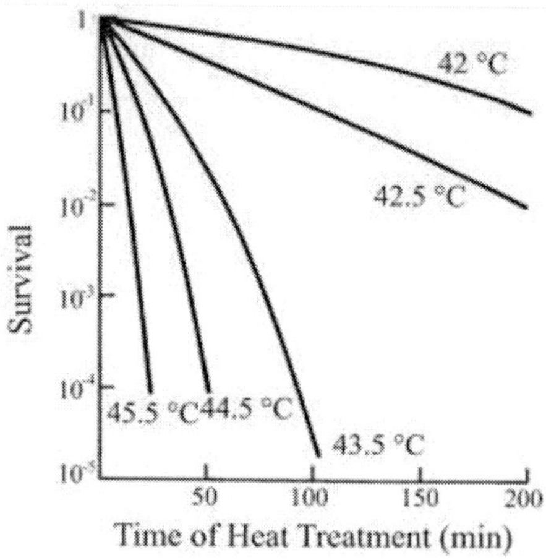

Figure 4. The fractional survival of Chinese hamster cells as function of treatment temperature and treatment time [28].

The cells property that is modified by temperature is the growth rate, which normally increases with increasing temperature; however, there is a maximum temperature in (40 °C – 41°C) which growth is quickly inhibited. Although for several years the phenomenon of how heat kills cancer cells has been studying, the specific mechanisms of damage caused by hyperthermia are still unknown [27]. However, there is a widespread view that cancer cells die because of the thermal denaturalization of critical targets in the cell [24]. In order to causes irreversible damage into cancer cells, hyperthermia therapy relies on a temperature of 42°C to 45°C for periods from 30 to 60 minutes.

B) Tumor Conditions

Usually, tumors have low pH compared with that of normal tissues. The tumors are more acidic than normal tissue, and pH decreases further during hyperthermia. This increased acidity, coupled with a poor nutrition and hypoxia caused by reduced blood flow, enhance the thermal damage done to the cancer cells after hyperthermia. The hypoxic state of tumor cells causes them to be more tolerant to hyperthermia treatments. Increases in blood perfusion, metabolic activity, oxygenation etc. are some of the effects caused by heating tissue. Nevertheless, the blood perfusion is an important parameter due to the different biological changes that occur when there are variations in blood flow after tissue has been exposed to heating [29]. Different studies have demonstrated that if a tumor is heated at 41°C - 43°C the blood flow just change in a factor of less than two; however, blood flow in normal tissue increases in a factor of up to 20 with temperatures between 42°C - 45°C [30]. These factors indicate that throughout the hyperthermia treatment the heat dissipation caused by the blood flow is greater in normal tissue than that in tumors. This behavior is described in Figure 5. As a consequence of this behavior, tumors achieve a higher temperature and they can suffer more damage than healthy tissues.

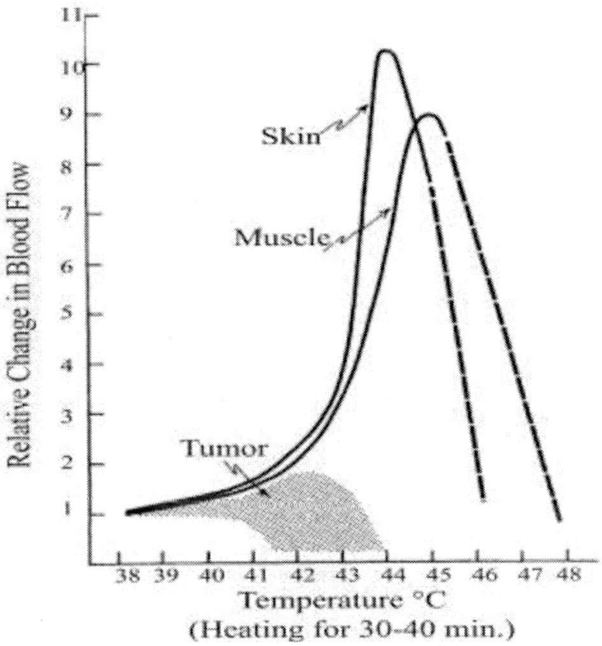

Figure 5. Relative changes in blood flow in the skin, muscle and tumor of rats after heating for 30-40 min at different temperatures [30].

C) Thermotolerance

Thermotolerance is known as the increased thermal resistance developed by cells as a consequence of the previous heat treatment. Thermotolerance is also important because it is considerate as one of the most important factors which influence the sensitivity of cells in hyperthermia treatments. The thermotolerance effect is caused by the heat shock proteins. Heat shock proteins are produced in normal cells as well in neoplastic cells after been exposed to a heat shock. Thermotolerance makes cells thermally treated become more resistant to heat, which involves a problem because thermotolerance protects cells against hyperthermia. However, this phenomenon is temporary and its effect just last a few days. Cells can develop thermotolerance at different times; it depends of the intensity of the first treatment (from just a few hours up to 15 h). Because the thermotolerance phenomenon, hyperthermia treatments must be planned with enough time between each other in order to avoid the thermotolerance effect.

4. Hyperthermia: Physical Aspects

As it was mentioned at the beginning of this chapter, EM fields and US are the two techniques more used in the design of hyperthermia devices. However, this chapter is focused in the EM hyperthermia; therefore, just the physical principles of EM hyperthermia are discussed.

A) Wave Propagation in Tissue

Three basic mechanisms are associated with the interaction of electromagnetic waves with matter:

1) Displacement, drift and diffusion of free electrons and ions,
2) Polarization of atoms and molecules to produce dipole moments, and
3) Orientation of the dipoles in the direction of the applied electric field, E.

Movement of electrons within tissue will result in conduction currents J = σE where σ is the tissue conductivity. In EM hyperthermia treatments the dielectric properties of tissue play an important role in the interaction of tissue with EM waves; principally, permittivity and conductivity of tissues. Both parameters are related by means of the complex permittivity described by Eq. (1):

$$\hat{\varepsilon} = \varepsilon_r \varepsilon_0 - j\frac{\sigma}{\omega} \qquad (1)$$

where ε_r is the relative permittivity, ε_0 is the permittivity of free space, σ is the conductivity and ω is the angular frequency. Permittivity and conductivity are frequency-dependent parameters. There are several studies which describe the behavior of both parameters as function of frequency [31-33].

An understanding of electromagnetic absorption in tissue requires the description of wave propagation in the medium. The propagation of electric and magnetic fields is described in time-harmonic and steady state by means of the Maxwell equations:

$$\nabla \cdot \varepsilon\vec{E} = \rho_v \quad \text{Gauss' law for electric field} \qquad (2)$$

$$\nabla \cdot \vec{B} = 0 \quad \text{Gauss' law for magnetic field} \qquad (3)$$

$$\nabla \times \vec{E} = -j\omega B \quad \text{Faraday' law} \qquad (4)$$

$$\nabla \times \frac{\vec{B}}{\mu} = J + j\omega\varepsilon E \quad \text{Ampere ' law with Maxwell' displasment current} \qquad (5)$$

where E is the complex vector of electric field intensity, B is the complex vector of magnetic flux density, J is the complex vector of the current density, ε is the permittivity, μ is the permeability and ρ_v is the volume charge density. From Maxwell equations and their combination, it is possible to find the wave equation which can be written in terms of either electric *E* or magnetic field *H*. In order to obtain the wave equation, sources are not considerated, i.e. a case with no electric charges (ρ=0) and no currents (J=0) is analized:

$$\nabla^2\vec{E} - \vec{E}(i\mu\sigma\omega - \mu\varepsilon\omega^2) = 0, \tag{6}$$

$$\nabla^2\vec{B} - \vec{B}(i\mu\sigma\omega - \mu\varepsilon\omega^2) = 0. \tag{7}$$

By analizing a general case (conducting medium σ≠0)

$$\vec{E} = E_x(x,y,z)i + E_y(x,y,z)j + E_z(x,y,z)k \tag{8}$$

By sustituying Eq. (8) in Eq. (6)

$$\nabla^2\vec{E} - \vec{E}(i\mu\sigma\omega - \mu\varepsilon\omega^2)$$

$$= \nabla^2(E_x i + E_y j + E_z k) - (E_x i + E_y j + E_z k)(i\mu\sigma\omega - \mu\varepsilon\omega^2) = 0.$$

If the term is defined as γ^2 i.e. $\gamma^2 = i\mu\sigma\omega - \mu\epsilon\omega^2$ and taking into account each component of the eq. (8) then the eq. (6) can be written as:

$$\nabla^2 E_x - \gamma^2 E_x = 0,$$
$$\nabla^2 E_y - \gamma^2 E_y = 0,$$
$$\nabla^2 E_z - \gamma^2 E_z = 0,$$

If $E_x(x,y,z) = X(x)Y(y)Z(z)$, and solving for E_x then

$$\frac{\partial^2(XYZ)}{\partial x^2} + \frac{\partial^2(XYZ)}{\partial y^2} + \frac{\partial^2(XYZ)}{\partial z^2} - \gamma^2 XYZ = 0,$$
$$YZ\frac{\partial^2(X)}{\partial x^2} + XZ\frac{\partial^2(Y)}{\partial y^2} + XY\frac{\partial^2(Z)}{\partial z^2} - \gamma^2 XYZ = 0. \tag{9}$$

By multiplying Eq. (9) by $\frac{1}{XYZ}$, then

$$\frac{1}{X}\frac{\partial^2(X)}{\partial x^2} + \frac{1}{Y}\frac{\partial^2(Y)}{\partial y^2} + \frac{1}{Z}\frac{\partial^2(Z)}{\partial z^2} - \gamma^2 = 0. \tag{10}$$

The solution, for a wave which is propagating in X axe, is obtained as:

$$E_x(x,y,z) = X(x),$$

$$\frac{1}{X}\frac{\partial^2(X)}{\partial x^2} - \gamma^2 = 0 \rightarrow m^2 - \gamma^2 = 0 \therefore m = \mp\gamma,$$

If $X(x) = A_1 e^{\gamma x} + B_1 e^{-\gamma x}$ where $\gamma = \sqrt{i\mu\sigma\omega - \mu\varepsilon\omega^2}$.

Usually, γ is the propagation constant

$$\gamma = \mp(\alpha + i\beta),$$

$$\sqrt{i\mu\sigma\omega - \mu\varepsilon\omega^2} = (\alpha + i\beta) \rightarrow (\alpha + i\beta)^2 = i\mu\sigma\omega - \mu\varepsilon\omega^2,$$

$$\alpha^2 - \beta^2 + 2\alpha\beta i = i\mu\sigma\omega - \mu\varepsilon\omega^2.$$

Then, the real part can be written as:

$$\alpha^2 - \beta^2 = -\mu\varepsilon\omega^2. \tag{11}$$

and the imaginary part can be written as:

$$2\alpha\beta i = i\mu\sigma\omega \rightarrow 4\alpha^2\beta^2 = \omega^2\mu^2\sigma^2,$$

$$\alpha^2 = \frac{\omega^2\mu^2\sigma^2}{4\beta^2}. \tag{12}$$

By sustituying Eq. (12) in Eq. (11)

$$\frac{\omega^2\mu^2\sigma^2}{4\beta^2} - \beta^2 = -\mu\varepsilon\omega^2. \tag{13}$$

By solving Eq. (13)

$$\beta = \omega\sqrt{\frac{\mu\varepsilon}{2}}\left[\sqrt{1+\left(\frac{\sigma}{\omega\varepsilon}\right)^2} + 1\right]^{1/2} \quad \textit{Phase constant} \quad \textit{(Imaginary part)}$$

$$\alpha = \omega\sqrt{\frac{\mu\varepsilon}{2}}\left[\sqrt{1+\left(\frac{\sigma}{\omega\varepsilon}\right)^2} - 1\right]^{1/2} \quad \textit{Attenuation constant} \quad \textit{(Real part)}.$$

The phase and the attenuation constants represent the change in phase per meter and the attenuation of the EM wave along the path traveled by it. These parameters help us to predict the behavior of electromagnetic waves propagation through a medium.

On the other hand, the attenuation constant is also related to the penetration depth; that is the distance in which the amplitude of the EM wave is reduced up to 1/e. The penetration depth can be written as:

$$\delta = \frac{1}{\alpha},$$

where α is the attenuation constant, therefore

$$\alpha = \frac{1}{\omega\sqrt{\frac{\mu\varepsilon}{2}}\left[\sqrt{1+\left(\frac{\sigma}{\omega\varepsilon}\right)^2}-1\right]^{1/2}}. \tag{14}$$

From Eq. (14), it is observed that the penetration depth is a frequency-dependent parameter; therefore, at lower frequencies the penetration depth is higher than the penetration at higher frequencies. However, the penetration in human body is quite difficult to estimate because the EM field interacts with different tissues when it is traveling through the body. Consequently, the EM field distribution into de body is a result of different parameters as frequency, tissue properties (permittivity and conductivity), etc.

In order to know the rate of electromagnetic energy absorbed by tissue, the SAR (Specific absorption rate or absorbed power per unit of mass) can be obtained. SAR (W/kg) can be described by Eq. (15) [34]:

$$SAR = \frac{\sigma}{2\rho}|E|^2, \tag{15}$$

where σ and ρ are the electrical conductivity (S/m) and density (kg/m3) of tissue, respectively, and E is the amplitude of the electric field (V/m). If the EM energy is transferred to the body in a short period of time, thermal conductivity and thermal convection can be neglected, and the SAR can be determined also by using Eq. (16)

$$SAR = c\frac{dT}{dt}, \tag{16}$$

where c is the specific heat of tissue (J/kg°C). From Eq, (16), it is observed that the temperature rise in tissue is related with the specific absorption rate by means of Eq. (17):

$$\Delta T = \frac{1}{c}SAR\Delta t, \tag{17}$$

where Δt is the period of time in which the EM field is applied. Finally, by substituting Eq. (15) in Eq. (16), it is observed that the temperature rise in tissue is related with the electric field applied by means of Eq. (18)

$$\Delta T = \frac{\sigma}{2c\rho}|E|^2\Delta t. \tag{18}$$

By analyzing the Eq. (18), it is observed that the temperature rise obtained throughout the hyperthermia treatment is related with the amplitude of the electric field applied, i.e. if the amplitude of the electric field is higher, it could be achieved a higher temperature throughout the hyperthermia treatment. However, because of the complexity of hyperthermia treatments, during several years, different numerical methods to obtain models of hyperthermia treatments and know the outcome (SAR, E field, and temperature distributions) of each kind of treatment have been developed. By using this numerical methods, it could be analyzed the outcome of a hyperthermia treatment applied to several patients with tumors localized at different depths, locations, sizes, etc.

5. Modeling of Hyperthermia Treatments

Nowadays, there are different modeling methods, whose goal is to generate the models of different multidisciplinary phenomena. By using these multiphysic models, whole hyperthermia system can be modeled. The challenge consists on determining the power absorption in biological tissues, related to temperature increase, which is generated by EM applicators [35]. The modeling of the behavior of hyperthermia applicators throughout a treatment is still required due to its multidisciplinary nature and because of the complexity of this phenomenon.

The hyperthermia modeling has been summarized in four areas where the treatment model should play an interesting role in the power deposition and the evaluation of the resulting dosimetry [36]. The four areas are:

1) *Comparative dosimetry:* it evaluates the performance of different applicators under the same clinical conditions in order to establish guidelines for the use of one system over another in specific circumstances.
2) *Prospective dosimetry:* it includes all the available information about a specific patient and considers the array of heating applicators that can be used to plan the trend of his/her treatment in order to find the most convenient for a specific patient.
3) *Concurrent dosimetry:* it tries to infer the temperature distribution throughout hyperthermia treatment based on temperature measurements and then adjust the input power in order to improve the therapeutic temperature in the tumor.
4) *Retrospective dosimetry:* the complete temperature profile is inferred by means of measurements. However, this infer is carried out after the treatment in order to correlate temperature distributions. This area has the therapeutic benefit of observing and evaluating the ability of an applicator to raise the temperature in the tumor volume.

The approach of hyperthermia treatments by means of modeling is important for planning the dosimetry of each patient. There are two facets in the treatment modeling: the first one is the power absorption while the second one is the heat transfer. The problem lies in calculating the power absorption produced in a biological tissue throughout hyperthermia therapy. The importance of the treatment model as a tool in the research of clinical hyperthermia stems from the impracticability of measure temperature over the region of treatment, because of the depth and the variability among each patient, therefore, there are several ways to generate a model by using partial differential equations (PDE).

The SAR, temperature and electric field distributions are computed by using numerical models which solve the Maxwell and Bioheat equations. The finite element method (FEM) and the finite-difference time-domain method (FDTD) are based in the differential form of the Maxwell equations, while other methods as the conjugate gradient fast Fourier transform are based on the integral form of the Maxwell equations. By using these methods, meshes can be generated which can model accurately tissue boundaries; however, the generation time can be large [37].

Deposed energy or absorbed power density in a point in tissue is proportional to the square of the amplitude of electric field in that point:

$$Q = \sigma |E|^2, \tag{19}$$

where Q is the deposed energy in tissue, σ is the electrical conductivity of tissue and E is the amplitude of electric field. The power deposition can be calculated in steady state, i.e. time dependence is not taken into account. However, time dependence power deposition also can be calculated by means of modeling.

5.1. Modeling of EM Waves

As it is described above, the hyperthermia treatment modeling is governed by the Maxwell equations. For hyperthermia treatments, electric and magnetic field can be written in its differential form as:

$$\nabla \times E = i\omega\mu H, \tag{20}$$

$$\nabla \times H = -i\omega\hat{\varepsilon} E, \tag{21}$$

where E is the electric field amplitude, H is the magnetic field amplitude, μ is the magnetic permeability, $\hat{\varepsilon}$ is the complex permittivity and ω is the angular frequency. The boundary conditions for electric and magnetic fields can be written as:

$$\hat{n} \cdot (\varepsilon_1 * E_1 - \varepsilon_2 * E_2) = 0, \tag{22}$$

$$\hat{n} \times (E_1 - E_2) = 0, \tag{23}$$

$$\hat{n} \times (H_1 - H_2) = 0, \tag{24}$$

$$\hat{n} \cdot (\mu_1 * H_1 - \mu_2 * H_2) = 0. \quad (25)$$

These equations indicate that magnetic field is continuous trough all the boundaries into the body, while the electric field has a normal component that is discontinuous because of the relation of complex permittivity of both tissues that form the boundary. When the EM source is presented over the boundary Eq. 22 and 24 can be written as:

$$\hat{n} \cdot (\varepsilon_1 * E_1 - \varepsilon_2 * E_2) = \rho, \quad (22a)$$

$$\hat{n} \times (H_1 - H_2) = J, \quad (24a)$$

where J and ρ are current and charge densities over the boundary.

Because the main goal is to determine the power deposition and the energy deposited over tissue which is related directly with electric field (See Eq. 19), the next step is to determine the electric field directly or determine the magnetic field and then obtain the electric field. The following set of partial differential equations of second order for E and H field were obtained by means of the Maxwell equations and the boundary conditions:

$$\nabla \times \frac{1}{i\omega\mu} \times \nabla \times E + i\omega\hat{\varepsilon} * E = 0, \quad (26)$$

$$\nabla \times \frac{1}{i\omega\hat{\varepsilon}} \times \nabla \times H + i\omega\mu * H = 0. \quad (27)$$

Finally, Eq. (26) and Eq. (27) can be rewritten as:

$$\nabla \times \nabla \times E - k^2 E = 0, \quad (26a)$$

$$\nabla \times \frac{1}{k^2} \nabla \times H - H = 0, \quad (27a)$$

where $k^2=\omega^2\mu\hat{\varepsilon}$.

By solving these set of PDE, the power deposition over the tissue can be determined if the finite element method (FEM) or the finite-difference time-domain method (FDTD) are used. Several authors describe the generation of solutions by using each of these methods [38-41] as well for the methods that use the Maxwell equations in their integral form [36].

5.2. Modeling of Bioheat Equation

Nowadays, several bioheat equations, which model the heat transfer within tissue, have been proposed. In the bioheat transfer equation the effect of blood perfusion due to tissue vascularity is taking into account. The bioheat equation models can be classified either continuum or vascular models. The continuum models of the bioheat equation use an effective tissue conductivity that incorporates the effects of blood perfusion and the vascular models take into account the detailed vascular system of tissues. The temperature distribution depends on several factors which must be included in the model: tissues thermal properties, geometry of the body, physiological responses, power deposition and power absorption rates of tissues.

A) Continuum Models of the Bioheat Equation

The temperature rise in biological tissues caused by EM waves as a radiation source has been described by means of the bioheat transfer equation (BHTE), which is represented by Eq (28) [42]:

$$\rho c \frac{\partial T}{\partial t} = k\nabla^2 \mathrm{T} - c_b W_b (T - T_a) + Q_p + Q_m. \tag{28}$$

This equation express the temperature rise as function of four factors which are: $\rho c \frac{\partial T}{\partial t}$ is the rate of energy storage where ρ is the tissue density, c is the specific heat of tissue, $k\nabla^2 \mathrm{T}$ is the rate of heat transfer by conduction where k is the thermal conductivity, T is the temperature, $c_b W_b (T - T_a)$ is the rate of heat transport by blood perfusion where c_b is the specific heat of blood, W_b is the blood mass flow rate and T_b is the blood temperature, Q_p (W/m^3) is the rate of energy absorbed and Q_m is the rate of metabolic heat generation. The contribution of the temperature difference due to the metabolism and thermal conduction is negligible being the most important effects the convection (blood flow) and the absorbed power density. The model proposed by Pennes (BHTE) assumes a uniform and isotropic blood perfusion rate W_b, constant blood temperature T_b, isotropic blood perfusion, not heat transfer between venous and arterial blood and thermal equilibrium between the capillary blood and the surrounding tissue. Because of these reasons, the BHTE is considered not completely valid [43]. However, the BHTE still be the most widely used [44].

In ordert to improve the modeling of temperature distributions, several modifications of the BHTE proposed by Pennes have been developed. Weinbaum and Jiji developed a non vascular model which improves the BHTE by means of a thermal conductivity index keff [45]:

$$k_{eff} = k\left(\delta_{ij} + \frac{\pi^4}{4\sigma k^2} n a^2 {k_b}^2 Pe^2 l_i l_j\right), \tag{29}$$

where k is the thermal conductivity, *ij* is the Kronecker delta, σ is the shape factor defined as $\sigma = \frac{\pi}{cos h^{-1}\left[\frac{L}{2a}\right]}$ where a is the artery radius and L is the centre-to-centre spacing between vessels and finally, Pe is the Peclet number defined as $Pe = P_r R_e = \frac{2\rho_b c_b a u}{k_b}$ where the subscript b denotes blood and u is the velocity.

Weinbaum and Jiji further estimated the enhancement of thermal conductivity for countercurrent vessel pairs of various diameters. They found that for vessels smaller than 50 mm in diameter the countercurrent heat exchange is negligible and the direct perfusion in the capillaries is the only significant contributor. For vessels with diameters in the range from 70 mm to 150 mm the two effects are of comparable importance, and for vessels larger than 150 mm in diameter the countercurrent heat exchange is the more significant contributor to the enhancement in conductivity. The Weinbaum and Jiji equation considered the complex vasculature system which includes varying vessel size, density, velocity and direction and it is independent of blood temperature. The model was based on countercurrent vessels that occur mainly in skeletal muscle; hence, it is not valid for tissue in that the countercurrent vascular architecture is not dominant, such as tumors. A later study showed that the simple BHTE model could qualitatively reproduce the major features of the temperature patterns, seen experimentally, better than the Weinbaum and Jiji model could do it, and the simple BHTE model produced better quantitative fits to the experimental data than the Weinbaum and Jiji model did it [46].

In 1990 Charny developed a hybrid model as a result of the comparison of the Weinbaum - Jiji and Pennes equations [47]. The result showed that by applying both equations it could be obtained a more appropriated bioheat transfer equation to describe perfused tissues. However, this model allows choosing one of both equations in accordance with a particular problem. The most frequently used equations, the BHTE of Pennes and the Weinbaum- Jiji equation, uses for instance either a heat sink or an increased thermal conductivity in order to take into account the blood perfusion. Both methods introduce modeling inaccuracies when they are applied to an ordinary tissue with a variety of vessel sizes. Hence, Wren develops a hybrid equation which includes both parameters, the heat sink term from the Pennes equation and the keff term from the Weinbaum - Jiji equation [48]:

$$\rho_c c_t \frac{\partial T_t}{\partial t} = \nabla k_t \nabla T_t + \beta(\nabla k_p \nabla T) - (1-\beta) S \omega_b \rho_b c_b (T_t - T_a) + \dot{Q} \tag{30}$$

The effective conductivity term is into the k_t term, which is the tissue thermal conductivity with no blood perfusion, and k_p which is the thermal conductivity due to the blood perfusion. Therefore, $k_{eff} = k_t + k_f$, the coefficient S depends on the temperature equilibration of the capillarity blood with the surrounding tissue. In the original Pennes equation S=1 and the heat exchange between the blood and the tissue takes place in the capillaries. The coefficient β accounts for the relative importance of the heat sink and effective conductivity term, such that when $0 \leq \beta \leq 1$, the effect of blood perfusion is modeled as a mixture of both terms, for $\beta=1$ the equation coincides with the Weinbaum - Jiji model, and for $\beta=0$ the equation coincides with the Pennes model. β, K_{eff} and ω can be varied independently of each other within the tissue area to produce a model that can describe the different perfusion rates in tissue. The hybrid equation developed by Wren improves the accuracy of the model while retaining the simplicity of the Pennes and Weinbaum- Jiji equations.

B) Vascular Models of the Bioheat Equation

Blood flow is the main factor in the thermoregulation of tissues. The vascular models are considered more accurate models than continuum models. The vascular models need detailed information about the vasculature in the treatment area and need to keep a register of all blood temperatures throughout this system in order to calculate tissue temperature. This high level of complexity results in high computational power in order to perform calculations on such models. In order to predict the effect of blood vessels, it is necessary to know the relation between the geometry of vessels and several parameters, such as surrounding tissue, the thermal properties of blood and tissues, blood flow rate and finally, the presence of other blood vessels in the surroundings. In vascular models, it is necessary to identify the location of blood vessels, which is quite complex, therefore, this method usually is used in combination with a continuum model. The continuum model is used to predict large-scale variations in tissue temperature while the vascular method is used to predict local temperature distribution in the tumor. Any continuum model, described previously, can give local temperature distributions in a tissue volume; however, vascular models make up for the precision in the area of large blood vessels. These kinds of methods are better for hyperthermia treatments.

Baish et al. carried out a parametric comparison of three model types (unidirectional, countercurrent and large/small/large vessel models) and showed that the continuum models are limiting cases of the vascular models [49]. In the unidirectional vessels model are two unknown quantities: tissue temperature and vessel blood temperature, where the heat transfer takes place between blood and tissue around it along the length of the vessel.

The conservation of energy equations for tissue and blood vessels are:

$$k\frac{\partial^2 T_t}{\partial x^2} - nk\sigma_a(T_t - T_a) + \dot{g} = \rho c\frac{\partial T_t}{\partial t}, \tag{31}$$

$$\pi r_o{}^2 \rho_b c_b \bar{u}\frac{dT_a}{dx} = k\sigma_a(T_t - T_a), \tag{32}$$

where σ is defined as:

$$\sigma = \frac{2\pi}{\ln\left(\frac{r_s}{r_0}\right) + \frac{2k}{k_b Nu - \frac{1}{2}}} \tag{33}$$

and the Nusselt number is taken as Nu=4.

The equilibration length of a blood vessel which is the length in that the vessel temperatures are equilibrated with the surrounding tissue can be written as:

$$L_{UD} = \frac{\pi r_0{}^2 \rho_b c_b \bar{u}}{k\sigma_a}. \tag{34}$$

For cases where LUD<<1, the direct perfusion model agrees with results from the unidirectional model. When LUD=1 the results diverge from the continuum model and the UD (unidirectional) model should be used. When LUD>>1, the blood vessels resemble heat sinks with a net strength and the vascular model takes the form of the Pennes equation.

Other vascular model of the bioheat equations is the countercurrent model. The energy balance in the countercurrent vessels model is described for the blood in arteries, veins and blood in the surrounding tissue [44].

The energy in the blood in arteries is described by the next equation:

$$\pi {r_0}^2 \rho_b c_b \bar{u} \frac{dT_a}{dx} = -k\sigma_\Delta (T_a - T_v) + \frac{k\sigma}{2}\left[T_t - \frac{(T_a + T_v)}{2}\right], \tag{35}$$

while the energy in the blood in vein is described by the equation:

$$\pi {r_0}^2 \rho_b c_b \bar{u} \frac{dT_v}{dx} = -k\sigma_\Delta (T_a - T_v) + \frac{k\sigma}{2}\left[T_t - \frac{(T_a + T_v)}{2}\right]. \tag{36}$$

Finally, the energy in the blood in tissue is described by:

$$k\frac{\partial^2 T_t}{\partial x^2} - nk\sigma\left[T_t - \frac{T_a + T_v}{2}\right] + \dot{g} = \rho c \frac{\partial T_t}{\partial t} \tag{37}$$

where σ_Δ and σ_Σ are conduction shape coefficients.

The equilibration length is:

$$L_{cc} = \frac{\pi r^2 \rho_b c_b \bar{u}}{k\sqrt{\sigma_\Delta \sigma}}. \tag{38}$$

For typical blood vessels $\sqrt{\sigma_\Sigma \sigma_\Delta}$ is approximately 2.4. The CC (countercurrent) model approaches the results of the effective conductivity k_{eff} when Lcc<<1. For long equilibration lengths, when, Lcc>>1, the CC equation takes the form of the Pennes equation where the heat sink term is $nk\sigma_\Sigma(T_t - T_a)$.

Finally, in the large/small/large vessels model the tissue contains two large countercurrent vessels with smaller perpendicular vessels connecting them. Only the small vessels are considered to exchange heat with the surrounding tissue. The resulting energy conservation equation is:

$$k\frac{\partial^2 T_t}{\partial x^2} - \rho_b c_b (T_a - T_v)\frac{d}{dx}(n\pi {r_0}^2 \bar{u}) + \dot{g} = \rho c \frac{\partial T_t}{\partial t} \tag{39}$$

This equation takes the form of the Pennes equation when $\omega_b = -\rho_b \frac{d}{dx}(n\pi {r_0}^2 \bar{u})$.

Nowadays, there are several models of the bioheat equation described in detail by different authors [44, 50, 51]. The vascular models need detailed information about the vasculature in treatment area in order to calculate tissue temperature. The vascular models can be reduced to a continuum model, which are the simplest cases that can occur. The continuum models usually are applied in cases where tissues have many small vessels which are nearly at thermal equilibrium with surrounding tissue while the vascular models are applied for complex vasculature. Table II describes a summary of the bioheat equation models.

Table II. Summary of the bioheat equation models

Model	Description	Evaluation
Pennes	Capillary diffusion by source-like perfusion term.	Describe collective thermal effect for tissue not adjacent to vessel which may introduce local convection.
Weinbaum- Jiji	Countercurrent and bleed-off effects between two paired vessels portrayed by effective conductivity.	Neglects some heat retained in local tissue when the countercurrent heat exchange occurs between the paired vessels.
Charny	Applied Pennes and Weinbaum- Jiji equations to obtain a more appropriated equation in perfused tissues.	Allows choosing one of both equations in accordance with a particular problem.
Wren	Includes the heat sink term from the Pennes equation and the keff term from the Weinbaum- Jiji equation.	The parameters β, Keff and ω can be varied independently of each other within the tissue to describe the different perfusion rates in the tissue.
Unidirectional vessels	Two main parameters: tissue temperature and vessel blood temperature.	The heat transfer takes place between blood and tissue around it along the length of the vessel.
Countercurrent	The energy balance is described for the blood in arteries, veins and surrounding tissue.	The CC model approaches the results of the Weinbaum- Jiji or the Pennes equation.
Large/small/large vessels	Tissue contains two large countercurrent vessels with smaller perpendicular vessels connecting them.	Only the small vessels are considered to exchange heat with the surrounding tissue.

6. Clinical Aspects

6.1. Applicators Classification

In the clinical applications there are several kinds of applicators which can be used to produce heating into tumors during a hyperthermia treatment. The hyperthermia applicators can be classified as non-invasive and invasive applicators.

A) Non-Invasive Applicators

The non-invasive applicators are classified into three main groups [52]:

- *Electric field applicators:* this kind of applicators generates a strong electric field which interacts with tissue by causing heating into tissue. A RF *E* field applicator is always accompanied by a magnetic field; however, the magnetic field at small frequencies is weak and it can be neglected. At high frequencies, Maxwell equations show that there is a strong coupling between *E* and *H* field; therefore, the *H* field must be taken into account. Hence, the *E* field applicator classification is limited to low frequencies. The most used frequencies to operate these applicators are 13.56

MHz and 27.12 MHz (ISM frequencies). Capacitor-plate applicators are typical *E* field applicators. Figure 6 depicts a typical set-up in hyperthermia treatments by using an *E* field applicator. Although capacitive applicators are able to heat deep tumors, they produce an overheating in fat. The overheating is caused because there is a difference between muscle and fat permittivity; fat has lower permittivity, therefore, the E field in fat is higher [53]. In the *E* field applicators the energy deposition is sensitive to the applicator position due to the presence of a dielectric medium (tissues) which has a strong effect on the *E* field.

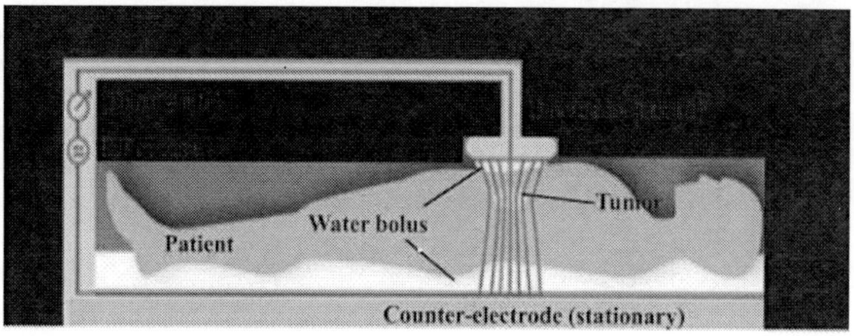

Figure 6. Set-up of hyperthermia therapy by using a capacitor-plate applicator.

– *Magnetic field applicators:* This kind of applicators generate mainly *H* field. As it occurs with the *E* field applicators, any RF magnetic field applicator is accompanied by an *E* field. However, at low frequencies the *E* field can be neglected because of its weakness. The simplest form of an inductive applicator (*H* field applicator) is a coaxial current ring. The current generates an *H* field which is mainly axial and almost uniform in the middle of the ring. The magnetic field, which is varying in time, induces an electric field into the body by causing heating in tissues and tumors. Because the body is essentially nonmagnetic, it not disturbs the magnetic field generated by the applicator, and because the electric field is weak the permittivity of the body not affect the deposition of energy into the tissue. Figure 7 describe the set-up of hyperthermia therapy with a magnetic field applicator.

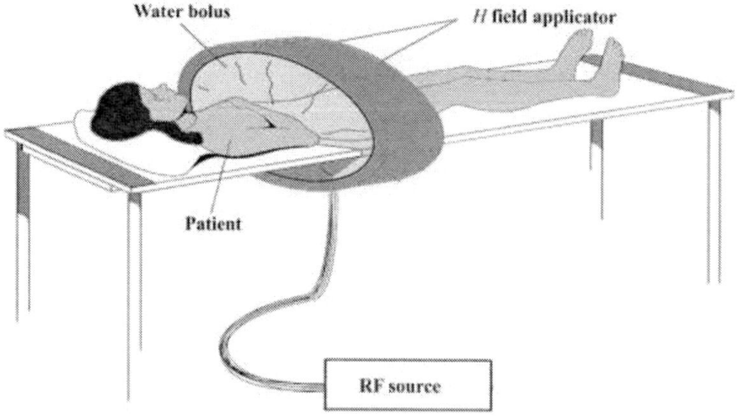

Figure 7. Set-up of hyperthermia therapy by using a coaxial current ring applicator.

- *Radiative applicators:* In the radiative applicators the heating is caused by a wave which is propagated from the applicator to the patient. The radiative applicators can be classified into unfocusing-type applicators and focusing-type applicators. Typical unfocusing applicators for hyperthermia treatment are waveguides. Several kind of open-ended waveguide applicators have been studied over the last years [54-56]. The waveguide aperture depends of the work-frequency, at highest frequencies, smallest applicators can be designed. However, as the penetration of EM energy into the body decreases when frequency increases, it is needed to find a balance between frequency and penetration depth to design an applicator with the correct dimensions to be used in hyperthermia treatments. If the applicator is filled with a dielectric medium (the work frequency is reduced), its dimensions can be reduced while the penetration depth can be kept. The reduction rate is directly proportional to the square root of the relative dielectric constant εr; hence, by filling the empty space inside the applicator with a dielectric medium, the aperture of the waveguide can be reduced in a factor up to 12. On the other hand, focusing applicators are arrays of multiple antennas which irradiated the body. The main goal of these applicators is to converge the EM energy on deep regions of the body and to achieve a higher focusing. Figure 8 represents the main devices for hyperthermia treatments by using radiative applicators.

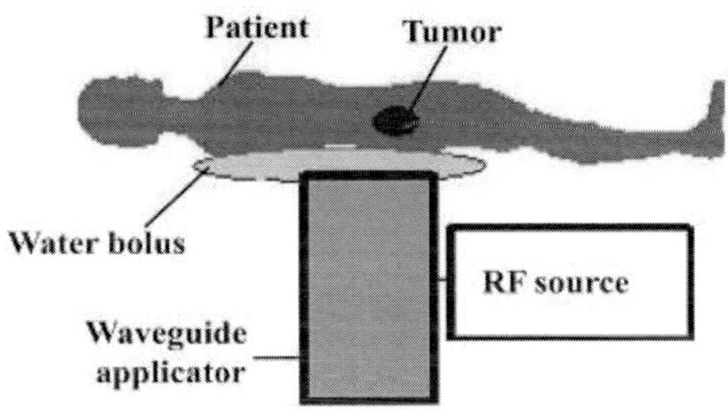

Figure 8. Set-up of hyperthermia therapy by using a waveguide applicator.

B) Invasive Applicators

The invasive applicators produce more uniform temperature distributions. By using these applicators the temperature distribution can be focused over the tumor; however, the applicator need to be inserted or implanted into the tumor which is no feasible for all the tumors. The invasive applicators are classified into three main groups [57]:

- *Needles array:* These applicators produce RF localized currents fields. The RF currents produced by voltages sources generate ohmic heating. The heating pattern is affected by tissue inhomogeneities, since the current tend to follow the paths of least resistance. The thermal distribution can be controlled by switching sources repeatedly between alternate pairs of needles [57, 58].

- *Radiating microwave antennas:* This kind of applicators can be used by itself or in a multi-antenna array. If a multi-antenna array is used, it could be possible to work with coherent and incoherent waves and treat several kinds of tumors [59, 60]. A typical implanted antenna is a coaxial dipole antenna. The coaxial antenna is composed of a thin semi-rigid coaxial cable [57, 61, 62].
- *Inductively heated ferromagnetic seeds:* In this therapy, ferromagnetic seeds are implanted into the tumor [63-65]. In order to heat the tumor, an external magnetic field is used. The heat generated by the interactions of seeds and the external magnetic field depends of the magnetic properties seeds, size, shape, etc.

6.2. Radiofrequency Generation Systems

In order to treat patients with a new hyperthermia device and achieve the desired effect, the device must fulfill certain standards. The hyperthermia equipment must heat the human tissue in a controlled, safe and reproducible manner. There are some parameters that can be monitored successfully; these parameters can show the effect of hyperthermia in humans. The RF hyperthermia devices operate by delivering electromagnetic radiation. Electromagnetic power is supplied by a power source and it is delivered to the patient by the applicator. These applicators must deliver the energy in a small volume of tissue. The minimum requirements for a set-up of hyperthermia treatments are shown in Figure 9.

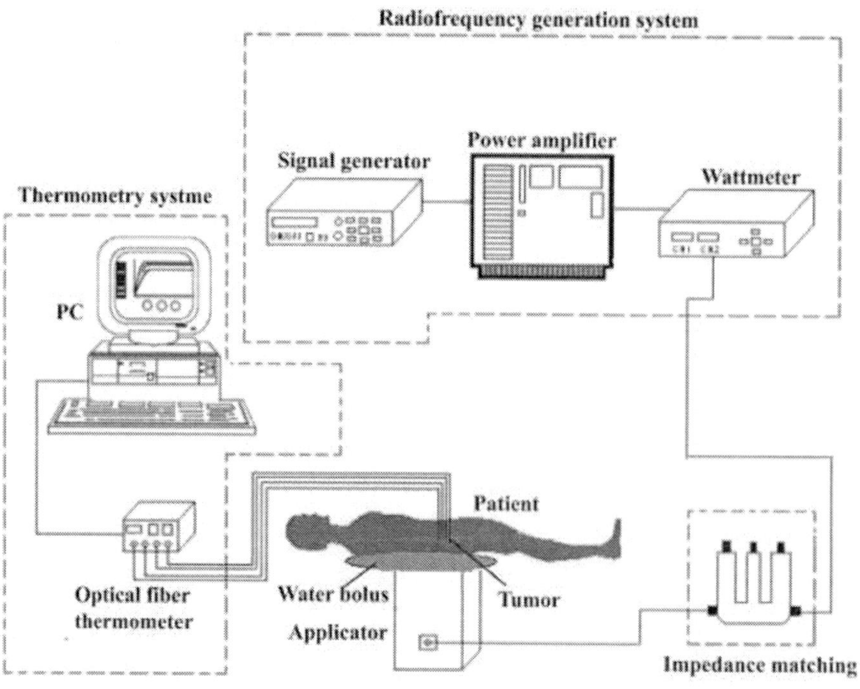

Figure 9. Set-up of hyperthermia treatments in clinical applications.

Figure 9 describes the RF experimental set-up for hyperthermia treatments; it consists of three main systems: radiofrequency generation system, thermometry system and impedance matching system.

- *Radiofrequency generation system:* The radiofrequency system is used as a RF power supply of the applicator.
- *Signal generator:* The signal generator produces a time-dependent signal with specific characteristics as frequency and amplitude.
- *Power amplifier:* The power amplifier is used to amplify the signal power and transmit it to the applicator with a maximum efficiency. The power amplifier is one of the most important devices in the RF generation system because it may work with high powers (approximately from 100 to 200 Watts) at high frequencies (Several frequencies are used in hyperthermia treatments; however, they can varied from 10 MHz to 2.45 GHz) [66, 67]. The power amplifier must be capable of tolerating some impedance mismatch. The long-term stability of the power level should be monitored for 60 minutes (typical duration of the hyperthermia treatment) and the variations in output power should not exceed 10%. Although there are different commercial systems, due to each applicator has its own requirements, some work-groups have designed different power amplifier to be used for a specific hyperthermia applicator [66-68]. Currently, multi-channel power amplifiers for hyperthermia treatments have been designed; they are able to control amplitude and phase of the signals delivered of each applicator, in the case of the multi-array antennas, in order to achieve better focusing of the EM energy over the tumor.
- *Wattmeter:* The wattmeter is used to register the incident and the reflected power. The main goal of the wattmeter is to provide information about the power that has been transmitted to the patient and the quantity of lost power. The power could be reflected and it could cause damage the power amplifier.

A) Thermometry System

It is used to register temperature inside the tumor and healthy tissue around it.

- *Optical fiber sensors:* In most of the RF hyperthermia treatments, optical fiber sensors are chosen because they do not interfere with the electric field generated by the applicators; therefore, do not modify its radiation pattern. The accuracy of the thermometers must be at least of 0.2 °C in a range from 37 °C to 46 °C. The calibration must be made by using at least two calibrated mercury thermometers and a water-bath with a uniform temperature.
- *PC:* Temperature data are recorded by a PC. By the knowledge of the temperature achieved into the tumor (recorded by sensors) the power supplied by the radiofrequency system can be adjusted to improve the heating into the tumor.

B) Impedance Matching System

Finally, so as to match the RF system with the applicator, an impedance matching system is used. Usually, STUB or balun transformers are used as devices to match impedance between the applicator and the RF system. The goal of this system is to ensure the maximum power transmission from the RF generation system to the applicator. If both systems are well matched, almost all the power is transmitted and the reflected power is less that it cannot damage the equipment. The level of the incident and reflected power is monitored by the wattmeter. An important parameter to know the level of mismatch in a hyperthermia is the SWR (standing wave ratio). The SWR is defined as the ratio of the maximum amplitude of the standing wave to the minimum amplitude. It is well-known that the higher ratio of reflected power to forward power, the greater SWR; hence, the ideal SWR is 1. When the SWR is high, the power loss is greater, which means that the RF system and the applicator are not well matched. In order to reduce the mismatch, the impedance matching system is required.

In order to prevent overheating and damage in patients and to ensure the reliability of the data recorded throughout the treatment, each component of the set-up for hyperthermia treatments must have a periodic review of its performance. The recommended tests are shown in Table III.

Table III. Tests recommended on the quality assurance of RF hyperthermia equipment [69]

Test	Acceptability criteria	Frequency
Thermometry		
Calibration probes		Yearly
Accuracy (20-50 °C)	0.1 °C	
Clinical thermometry		
Accuracy (30-50 °C)	0.2 °C	Quarterly
Precision	0.1 °C	Quarterly
Stability	0.1 °C/hr	Quarterly
Response time	< 10 sec	Yearly
Perturbations/artifacts	< 0.1 °C	Yearly
EM interference	< 0.1 °C	Yearly, when some component is changed
Inspection of probe damage	Qualitative	Each treatment
Power source and power meter		
Electrical safety	ANSI-AAMI	Yearly
EM and RF shielding	US CFR part 18	When the equipment is installed
Accuracy of net power into applicator	10 %	Yearly
Applicators and matching devices		
Check temperature distributions	Qualitative	Yearly
Single point SAR reproducibility	10 %	Each treatment

6.3. Quality Assurance Guideline and Protocols for Hyperthermia Treatments

Nowadays, there are different protocols and guidelines to evaluate and characterize hyperthermia applicators and treatments. The most important guidelines were written by the European Society for Hyperthermic Oncology (ESHO) and the American Association of Physicists in Medicine (AAPM). To generate these guidelines the technical and the clinical committees of the ESHO evaluated the experience of the institutes which are active in this kind of treatments by using radiative equipments; while the AAPM guideline is prepared by memberships who work in task groups of hyperthermia. The main subjects evaluated are requirements and characterization of equipment, treatment planning, treatment, treatment documentation, safety aspects, hyperthermia staff requirements and instrumentation for quality assurance [70-72]. The goal of the quality assurance is to ensure detailed documentation about the hyperthermia treatment [69].

6.3.1. Characterization of RF-Hyperthermia Applicators: SAR Distributions

There are different methodologies to evaluate RF applicators for hyperthermia treatments. These methodologies can be different. It depends on the kind of applicator to be evaluated; if it is either external or interstitial, if the evaluation is for one applicator or an array of applicators.

A) External Radiative Applicators

- One applicator

To evaluate the performance of one external applicator the most used method is by means of SAR distributions. The SAR patterns are based on the measurement of temperature rise during a short period of time. To obtain the SAR, Equation 16, described previously, is used:

$$SAR = c\frac{dT}{dt}, \tag{16}$$

$\frac{dT}{dt}$ is a time interval sufficiently small that the conduction effects can be neglected; hence, the temperature increment is proportional to the time interval. The temperature increment can be measured by an array of temperature sensor as well as by a thermographic camera (See Figure 10). By using a thermographic camera, a complete temperature distribution is obtained; however, the accuracy of the measurement is lower than the accuracy obtained by using temperature probes. Regardless of the method used, measurements must be repeated several times in order to obtain an average.

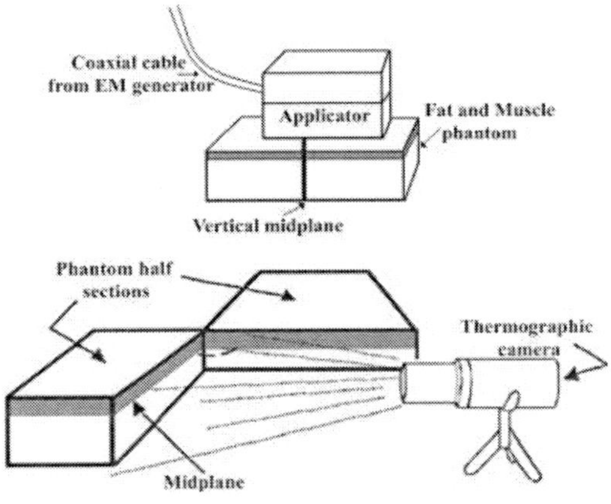

Figure 10. Experimental set-up to measure SAR distributions by using a thermographic camera and the split phantom technique [71].

For the evaluation of an applicator, a spatial distribution of SAR is better; however, it is impractical to obtain the spatial distribution. Hence, to evaluate an applicator at least 3 profiles should be obtained: The central depth profile and two orthogonal profiles at 1 cm depth in muscle in a perpendicular plane to the depth profile. The lateral profiles have to extend several centimeters beyond the limits of the applicator. The main objective to analyze beyond to the limits of the applicator is to determine the regions in where hot spots are generated or if they are generated. It is necessary to keep in mind that the water bolus has a great influence over the power deposition [73].

Phantom to Evaluate the Applicator (SAR Distributions)

In order to evaluate the performance of the applicators before using them in clinical applications, tissue substitute materials (phantoms) are used. At least, a phantom of two layers should be used. The first layer simulates the dielectric properties of fat (permittivity and conductivity); the thickness of this layer should be of approximately 1 cm. The second layer simulates the dielectric properties of muscle. The thickness of the muscle layer must be of 10 cm and it must be extended at least 5 cm beyond the dimensions of the applicator to be evaluated. The fat layer could not be taking into account; however, it is needed to keep in mind that the fat layer affects the effectiveness of matching.

Multi-Applicators

The multi-array applicator must be evaluated as a unit. In this case the lateral SAR profiles must cover the space between the radiating elements. At least two depth profiles should be obtained. Figure 11 depicts the profiles of SAR which can be obtained for an array of rectangular applicators for hyperthermia treatments. The phantom used to evaluate a multi-array of applicators must be following the same requirements as the phantom for the evaluation of a single applicator.

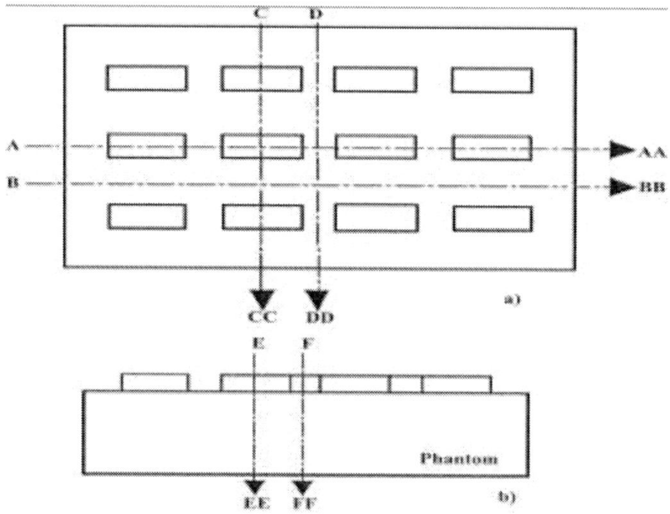

Figure 11. a) Recommended location of lateral profiles for the evaluation of a planar array of radiating-element applicators. All profiles are at 1 cm depth in a phantom. b) Minimum recommended locations of depth profiles for the evaluation of a planar array of radiating-element applicators. Either profile E-EE or F-FF should pass through the center of the array [71].

B) Multi- Applicators for Deep Hyperthermia

This kind of applicators is typically annular rings of radiating apertures. Usually, a standard torso phantom is used to evaluate it. Electric field probes are used to measure the strength of the *E* field near to the patient or into the phantom. These measurements are necessary to evaluate the symmetry of the fields. Temperature probes must be long enough to measure throughout the length of the phantom. In order to obtain SAR distributions, the output power of each applicator must be adjusted to obtain a symmetric power deposition. Three SAR profiles should be done, the first one correspond to the central plane while the others two are parallel planes at each side of the central plane.

C) External RF Applicators

Magnetic Field Applicators

To evaluate these applicators, it is necessary to test separately each component of the hyperthermia system, to ensure that each one is working properly. Finally, the complete system should be evaluated by using a phantom tissue. The objective of these two evaluations is because during the interaction between the components unexpected problems can occur.

Inductive Applicators

Usually, the induction applicators are coils which present high voltages and power levels. Because of these reasons, these applicators must be isolated correctly in order to avoid burns to the patient or the technicians. The phantom, used to evaluate these applicators, is usually a saline solution which can be contained in any electrically non conducting container with the shape of the region to be treated.

Capacitive Heating Systems

As it was described below, the capacitive applicators are usually a set of electrodes while the complete hyperthermia system consists of the RF source, the wattmeter and the matching system. The best way to evaluate these applicators is by analyzing the performance of the whole hyperthermia system. Another requirement for these applicators is that the electrodes must have a temperature control (usually by circulating water). It is necessary to use a temperature sensor to measure the electrodes temperature.

D) Interstitial and Intracavitary Applicators

Interstitial Applicators

In order to test the behavior of the antenna, a wattmeter could be connected between the RF generator and the antenna. The measurement should be carried out with the generator at 25, 50 and 100% of its maximum power, to ensure the linearity of the power distribution. The output should fall within 10 % of the mean value. If the system exceeds this tolerance, it must be repaired or replaced. On the other hand, the coaxial cables and connectors must be inspected frequently for power losses, overheating, etc. Another way of evaluating the cables is by connecting a 50 ohm load instead of the antenna. If the reflected power measured, when the 50 ohms load is connected, exceeds the 20% of the forward power, it means that there is a physical damage either in cables or in connectors. Reflected power lower than 10% of the forward power indicates acceptable matching. In this case, the SAR measurements should be carried out at 5 mm from the antenna axis, that should be the point of maximum SAR. At least one SAR profile must be obtained. Because these antennas are inserted into the tumor, the heating time should be as short as possible (30 seconds is reasonable).

Interstitial RF (Localized Current Field)

In order to evaluate these applicators, it is also needed to use a power meter and a 50 ohm load as it was described above. The linearity and accuracy of the cycle timing can be determined by displaying the output waveform with an oscilloscope. The design of the equipment should minimize the risk of the generation of electric current paths not intended for treatment. Visual inspection procedures can help assure proper operation of the equipment. All cables and connectors should be checked for physical damage.

Interstitial Ferromagnetic Seeds

In the characterization of ferromagnetic seeds, the thermo regulating temperature of curie-point into a phantom must be obtained. A thermometer probe should be measure the temperature at 5 mm radius from the midpoint of the seed. Enough RF power should be applied to reach the thermoregulation mode, and maintained for at least 10 min. It is necessary to compare the temperature at the reference point with the original seed performance; adjust the thermal dosimetry (ferro-seed spacing) accordingly, or discard the seed. Nowadays, research has shown that there are different ways to characterized ferromagnetic seeds, nanoparticules and ferrofluids for its use in hyperthermia treatments.

6.3.2. Characterization of Applicators: Complementary Parameters

Effective Field Size and Penetration Depth

In order to have a complete evaluation of hyperthermia applicators, some other parameters could be obtained. The effective field size (EFS) and the penetration depth (PD) are the most used parameters to characterize hyperthermia applicators. The EFS is defined by the 50% SAR contour measured at a depth of 1 cm from the surface of an homogeneous muscle phantom while the PD is defined as the distance below 1 cm in which the SAR is 50% of that value obtained at 1 cm depth [74]. These parameters must be obtained by using a complete hyperthermia set-up, i.e. the water bolus must be included and its initial temperature must be in agreement with the temperature of muscle phantom. In the case of the evaluation of multi-applicators or capacitive applicators, geometrically realistic phantoms are required due to the EFS depends critically on the particular geometry. In the case of capacitive applicators, the thickness of the phantom must be equal to the tissue between the electrodes in the clinical set-up.

Power Efficiency

The power efficiency in the complete hyperthermia system must be measured; i.e. the power efficiency between the generator, power amplifier, matching system, applicator and bolus. The energy delivered in the muscle phantom should be determined by using a calorimetric technique [34]. The power obtained by means of the calorimetric technique should be compared with those values obtained by the wattmeter.

Leakage Radiation

In order to ensure the safety of patients and technicians, the leakage radiation must be measured by using an isotropic probe. These kinds of measurements must be made throughout the hyperthermia treatment to avoid high electromagnetic fields. For multi-arrays applied in deep hyperthermia, it is necessary to measure the leakage because these antennas usually have leakage problems. The leakage levels can cause a overheating outside the treatment region.

Water Bolus

As it was mentioned above, the water bolus could cause big changes in the effective field size of the applicator; for this reason, it is necessary to take into account the water bolus in each measurement. On the other hand, the technician must ensure that there is a good contact between the bolus and the patient. A special care must be taken in areas in which there are catheters to measure interstitial temperatures. During all the hyperthermia treatment the temperature of the water bolus should be measured.

6.3.3. Measurements Errors

Either during the characterization of applicators or during the treatment some errors can occur, especially in recorded temperature. Temperature measurements are oversensitive to errors because they are taken in the presence of strong electromagnetic fields. This interference can be reduced if the probes are correctly shielded and if the correct filtering is applied to all the input signals and power lines. The presence of EM interference is rapidly detected by a sudden increase in temperature when the RF system is turned on while if the RF system is turned off the temperature changes occur in the opposite side and with the same amplitude. Figure 12 describes the behavior of temperature probes in presence of strong EM fields.

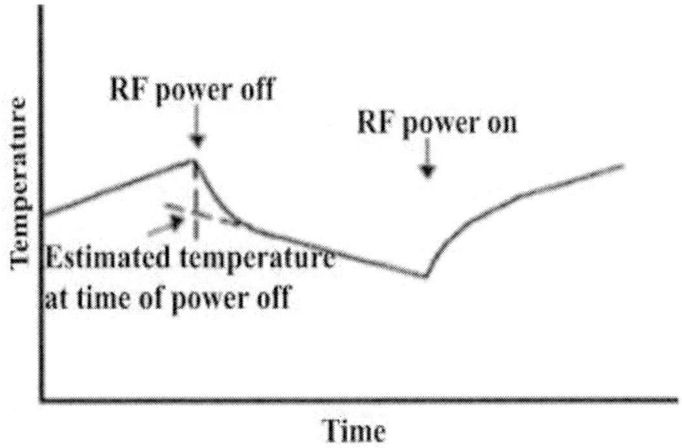

Figure 12. Typical response of a temperature probe with EM interference or probe heating.

Another error to keep in mind is that one that occurs when temperature probes or catheters (in which the probes are inserted) absorb energy faster than the surrounding tissue. As a consequence, probes are more heated than the tissue, therefore, the recorded temperature is not that of tissue; it corresponds to the probe temperature. Finally, there are some temperature probes which perturb the radiation pattern of the antennas. Therefore, they can change temperature distributions in tissue.

In order to ensure the correct behavior of the applicators during the hyperthermia treatment, it is totally necessary to follow and keep in mind each requirement in their characterization. By following these protocols and guidelines not only it is possible to use an applicator in hyperthermia treatments, but also the safety of the patient is assured.

6.4. Hyperthermia Treatment Planning (HTP)

The main objective is to determine the pattern of power deposition required to obtain good results (in a specific patient and tumor), by achieving therapeutic temperatures in tumor while temperature in healthy tissue is kept at the minimum change. Because of the complexity of hyperthermia treatments, the hyperthermia treatment planning can help to obtain 3D distributions of SAR and temperature. The hyperthermia treatment planning is a process of two main steps: 1) Calculation of the absorbed power (EM simulation), and 2) calculation of the temperature distribution (Thermal simulation). However, in order to obtain both distributions, some other steps should be done.

In order to start with the treatment planning, the tumor must be identified. It is necessary to use the most advanced techniques to localize the tumor and star with the treatment planning. The most appropriated techniques to localize the tumor into de human body is by means of X-ray transmission computed tomography [CT] and magnetic resonance imaging (MRI) [75]. The multiple consecutive slice of the CT or MRI contains the most complete information about the tumor and the human body. In the treatment planning, the accurate location of the thermometers, which record the temperature during throughout the hyperthermia treatment, is needed in order to compare the results obtained with the treatment planning and that obtained during treatment. The reason to use the CT or MRI is to clearly discriminate between each kind of tissue (muscle, long, bone, tumor, etc.) and identify and define each tissue with its own dielectric and thermal properties [76]. Nowadays, there are few hospitals which are able to use this information to predict SAR distributions.

The second step in the hyperthermia treatment planning is to choose the correct applicator or multi-applicator. The choice of the applicator depends of the tumor location and size. If the tumor is located within 4 cm from the skin, a superficial applicator should be chosen, otherwise, an applicator for deep hyperthermia must be chosen. The hyperthermia treatment planning has been studied with several kinds of applicators [77-81]. The next stage in the HTP is to obtain SAR and temperature distributions in order to predict hot-spots and other risks during treatment. The modeling of the treatment by means of numerical models of Maxwell and Bioheat equations, described previously, are used in order to obtain SAR and temperature distributions. The final process is to consider the impact of the hyperthermia treatment in the patient. Therefore, possible disadvantages of using a specific applicator must be recognized in the treatment planning, and taken into account in the selection of the best treatment for a specific patient. An assessment of the impact of the hyperthermia treatment on safety and efficiency should be made early on by the HTP.

In order to know the temperature achieved during the treatment there are different protocols which specified the number of probes required for different tumors and locations. For example, for breast cancer at least three sensors must be inserted into the tumor and one probe more must record the temperature at the skin. For neck nodes at least one multi-probe of three sensors must be inserted into the tumor while other one must record the skin temperature and in the case of melanomas, if the tumor does not exceeds 16 cm^2, at least two probes must be inserted into the tumor and 1 more in the skin; however, if the tumor is larger, the probes must be increased at least in one for each additional 16 cm^2 [74]. Figure 13 describe the minimum number of probes to use in hyperthermia treatments of breast and neck nodes. These measurements of temperature must be compared with temperatures obtained with the HTP.

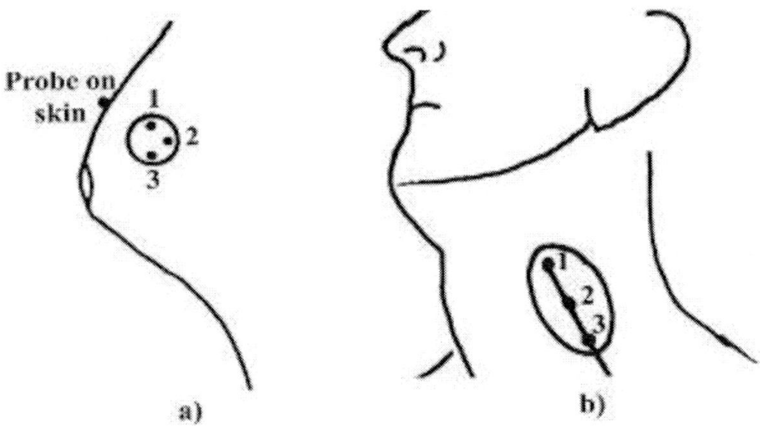

Figure 13. a) Locations of the minimum number of temperature probes for treatment of breast carcinoma, b) Locations of the minimum number of temperature probes for treatment of neck nodes.

Most of the hyperthermia treatments last up to 60 minutes. However, this period of time start when sensors inserted into the tumor reached ~ 43°C (approximately 10 minutes after turning on the hyperthermia equipment). In general, if there are some problems during the treatment, such as patient complains due to pain, and the equipment has to be turned off, the treatment can be extended up to a maximum time of 90 minutes. The temperature rise in tumor is from 0.25 to 1°C per minute. These requirements need to be taken into account during the hyperthermia treatment planning in order to compare results.

In order to compare hyperthermia treatment planning with the results obtained during treatment, it is necessary to report some particularities of the hyperthermia treatment. Document the set-up of each treatment by means of different photographs and diagrams in order to know the number of applicators used and its relative positions as well as relative positions of tumor and temperature probes. The number of thermometers and temperatures must be recorded during each treatment. If there are some problems and the equipment must shut down or the treatment must to be finished, the reasons must be reported. On the other hand, the power (input, output and reflected) should be recorded throughout the treatment.

Currently, the BSD Medical Corporation commercialized the applicators for superficial and deep hyperthermia which are part of the few devices able to apply hyperthermia treatment planning. The hyperthermia treatment planning allows tracing the size and shape of the tumor. Treatment plans can then be developed by simulating the placement of antenna in the tumor. Power and phase of each channel can be adjusted on a screen. Figure 14 described the main steps of the treatment planning for the BSD applicator. Nowadays, there are just few hospitals using hyperthermia treatment planning. Although during the last years, the hyperthermia treatment planning has had great progress, it is still required to improve models, segmentation of tissues, tumor, temperature probes, reconstruction of the array of applicators, etc. These improvements are necessaries to achieve better results and ensure the success of the hyperthermia treatment.

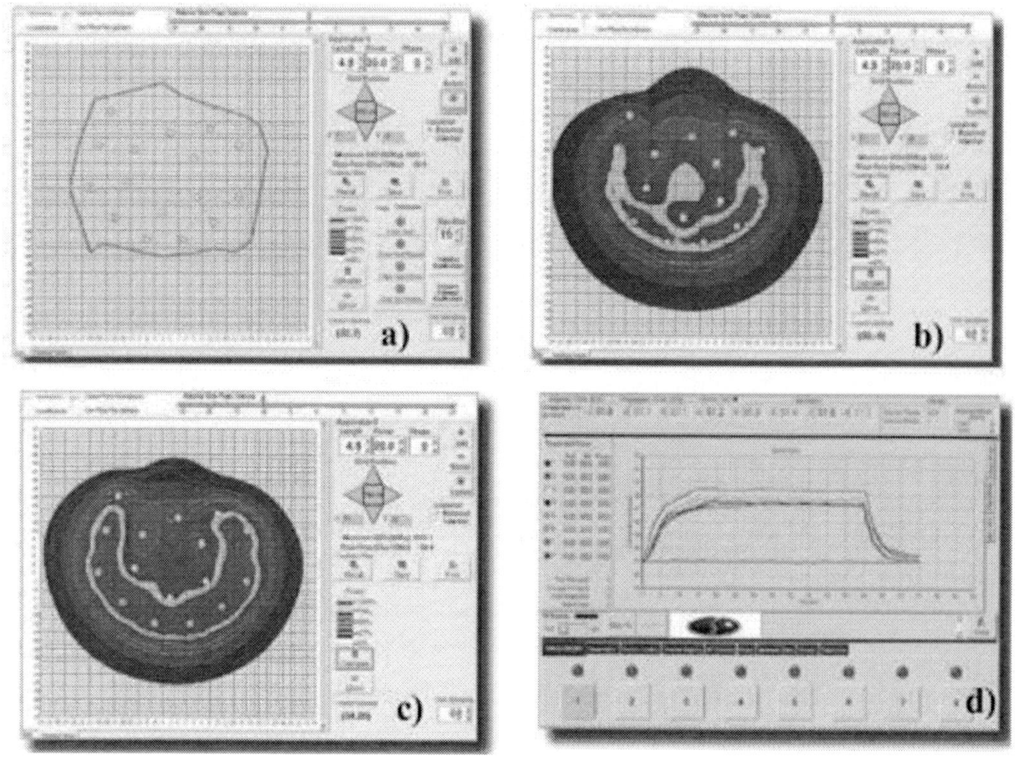

Figure 14. Hyperthermia treatment planning for the BSD applicator. a) Trace tumor and position Applicators, b) Calculate energy distribution pattern, c) Adjust phase and power to achieve desired heating pattern, d) Transfer settings to hyperthermia treatment screen.

CONCLUSION

This chapter book describe the most important aspects to keep in mind during a hyperthermia treatment; as well as the requirements to validated and characterized the applicators and systems used throughout the hyperthermia treatment. Hyperthermia treatments are used in order to treat cancer tumors. The heat generated throughout the treatment has some thermal effects which produce the death of cancer cells while the healthy tissue heating is tried to maintain in a minimum level. In hyperthermia treatments, temperature and SAR distributions are important in order to know if the treatment is successfully applied, i.e. they are needed in order to know if the temperature in tumor is in the range of therapeutic temperatures and how much of the surrounding healthy tissue is damaged due to these temperatures. As it was mentioned above, there are different parameters in the biological aspect of hyperthermia which can change temperature distributions in the treatment region. In order to analyze how the treatment is affected by this kind of parameters, different models have been developed. These models are based in Maxwell and Bioheat equations. Several improvements to the bioheat equations had been developed; however, it is still necessary to use these improvements to generate further studies about 3-D temperature distributions in order to predict the behavior of hyperthermia treatments. Although up to now research had shown the most important biological aspects which are involved in the response of the tumor to heat, more studies are necessaries in order to improve hyperthermia techniques, models and applicators.

On the other hand, in order to be able of developing successful HTP, it is required that the modeling takes into account each parameter and characteristics of the hyperthermia treatment and the patient. As it was described previously, in order to successfully develop an HTP for a specific patient and tumor, it is necessary to take into account all the set-characteristics of the hyperthermia treatment; i.e. the treatment data must be recorded, the antennas set-up must be accurately described, patient and tumor must be segmentated with high accuracy in order to define each tissue with its particular thermal and electric properties and the temperature probes locations must be defined. Although some commercialized applicators are currently using HTP, all the HTP process requires further research and optimization in order to be able of using it in each hospital that works with hyperthermia treatments.

REFERENCES

[1] P. R. Stauffer, "Evolving technology for thermal therapy of cancer," *Int. J. Hyperthermia,* vol. 21, pp. 731-44, Dec 2005.

[2] J. van der Zee, "Heating the patient: a promising approach?," *Ann. Oncol.*, vol. 13, pp. 1173-84, Aug 2002.

[3] T. Shuto, T. Yamamoto, S. Tanaka, A. Kanazawa, S. Takemura, H. Tanaka, S. Kubo, K. Hirohashi, H. Sakaguchi, and S. Seki, "Resection of needle-tract implantation after percutaneous puncture for hepatocellular carcinoma," *Journal of Gastroenterology*, vol. 39, pp. 907-908, 2004.

[4] H. Takahashi, M. Konishi, T. Nakagohri, K. Inoue, S. Takahashi, Y. Tanizawa, M. Monden, and T. Kinoshita, "*Aggressive Multimodal Treatment for Peritoneal Dissemination and Needle Tract Implantation of Hepatocellular Carcinoma: a Case Report.*" vol. 34, 2004, pp. 551-555.

[5] G. H. Nussbaum, *Physical Aspects of Hyperthermia*. Washington: American Institute of Physics, 1982.

[6] P. Wust, R. Beck, J. Berger, H. Fahling, M. Seebass, W. Wlodarczyk, W. Hoffmann, and J. Nadobny, "Electric field distributions in a phased-array applicator with 12 channels: measurements and numerical simulations," *Med. Phys.*, vol. 27, pp. 2565-79, Nov 2000.

[7] R. W. Habash, R. Bansal, D. Krewski, and H. T. Alhafid, "Thermal therapy, part 2: hyperthermia techniques," *Crit. Rev. Biomed. Eng.,* vol. 34, pp. 491-542, 2006.

[8] J. Milligan, "Whole-body hyperthermia induction techniques," *Cancer Res.,* vol. 44, pp. 4869s-4872s, Oct 1984.

[9] J. M. Bull, D. Lees, W. Schuette, J. Whang-Peng, R. Smith, G. Bynum, E. R. Atkinson, J. S. Gottdiener, H. R. Gralnick, T. H. Shawker, and V. T. DeVita, Jr., "Whole body hyperthermia: a phase-I trial of a potential adjuvant to chemotherapy," *Ann. Intern. Med.*, vol. 90, pp. 317-23, Mar 1979.

[10] J. Conway and A. P. Anderson, "Electromagnetic techniques in hyperthermia," *Clin. Phys. Physiol. Meas.*, vol. 7, pp. 287-318, Nov 1986.

[11] R. Hiergeist, W. Andra, N. Buske, R. Hergt, I. Hilger, U. Richter, and W. Kaiser, "Application of magnetite ferrofluids for hyperthermia," *Journal of Magnetism and Magnetic Materials*, vol. 201, pp. 420-422, 1999.

[12] Moroz, P, Jones, K. S, Gray, and N. B, *Magnetically mediated hyperthermia: current status and future directions* vol. 18. London, ROYAUME-UNI: Taylor andamp; Francis, 2002.

[13] H. Ingrid and et al., "*Magnetic nanoparticles for selective heating of magnetically labelled cells in culture: preliminary investigation*." vol. 15, 2004, p. 1027.

[14] H. Rudolf and et al., "*Magnetic particle hyperthermia: nanoparticle magnetism and materials development for cancer therapy*." vol. 18, 2006, p. S2919.

[15] R. Hergt and S. Dutz, "Magnetic particle hyperthermia--biophysical limitations of a visionary tumour therapy," *Journal of Magnetism and Magnetic Materials,* vol. 311, pp. 187-192, 2007.

[16] R. E. Rosensweig, "Heating magnetic fluid with alternating magnetic field," *Journal of Magnetism and Magnetic Materials*, vol. 252, pp. 370-374, 2002.

[17] S. A. Rovers, R. Hoogenboom, M. F. Kemmere, and J. T. F. Keurentjes, "Relaxation Processes of Superparamagnetic Iron Oxide Nanoparticles in Liquid and Incorporated in Poly(methyl methacrylate)," *The Journal of Physical Chemistry C*, vol. 112, pp. 15643-15646, 2008.

[18] S. E. Barry, "*Challenges in the development of magnetic particles for therapeutic applications*." vol. 24, 2008, pp. 451-466.

[19] R. Ivkov, S. J. DeNardo, W. Daum, A. R. Foreman, R. C. Goldstein, V. S. Nemkov, and G. L. DeNardo, "*Application of High Amplitude Alternating Magnetic Fields for Heat Induction of Nanoparticles Localized in Cancer*." vol. 11, 2005, pp. 7093s-7103s.

[20] J. Motoyama, N. Yamashita, T. Morino, M. Tanaka, T. Kobayashi, and H. Honda, "Hyperthermic treatment of DMBA-induced rat mammary cancer using magnetic nanoparticles," *BioMed. Central Ltd.*, 2008.

[21] L.-Y. Zhang, H.-C. Gu, and X.-M. Wang, "Magnetite ferrofluid with high specific absorption rate for application in hyperthermia," *Journal of Magnetism and Magnetic Materials,* vol. 311, pp. 228-233, 2007.

[22] Skumiel, A. Józefczak, M. Timko, P. Kopčanský, F. Herchl, M. Koneracká, and N. Tomašovičová, "Heating Effect in Biocompatible Magnetic Fluid," *International Journal of Thermophysics*, vol. 28, pp. 1461-1469, 2007.

[23] W. H. Organization, "Electromagnetic fields and public health: physical properties and effects on biological systems in Press releases, fact sheets, backgrounders and statements from WHO," *Fact Sheet* N182., 1998.

[24] R. W. Habash, R. Bansal, D. Krewski, and H. T. Alhafid, "Thermal therapy, part 1: an introduction to thermal therapy," *Crit. Rev. Biomed. Eng.,* vol. 34, pp. 459-89, 2006.

[25] M. W. Dewhirst, M. Lora-Michiels, B. L. Viglianti, W. C. Dewey, and M. Repacholi, "Carcinogenic effects of hyperthermia," *Int. J. Hyperthermia,* vol. 19, pp. 236-51, May-Jun 2003.

[26] J. R. Lepock, "How do cells respond to their thermal environment?," International journal of hyperthermia : *the official journal of European Society for Hyperthermic Oncology*, North American Hyperthermia Group, vol. 21, pp. 681-7, 2005.

[27] J. R. Lepock, "Cellular effects of hyperthermia: relevance to the minimum dose for thermal damage," *Int. J. Hyperthermia*, vol. 19, pp. 252-66, May-Jun 2003.

[28] L. E. Gerweck, "Hyperthermia in cancer therapy: the biological basis and unresolved questions," *Cancer Res.,* vol. 45, pp. 3408-14, Aug 1985.

[29] M. R. Horsman, "Tissue physiology and the response to heat," *Int. J. Hyperthermia,* vol. 22, pp. 197-203, May 2006.

[30] C. W. Song, A. Lokshina, J. G. Rhee, M. Patten, and S. H. Levitt, "Implication of blood flow in hyperthermic treatment of tumors," *IEEE Trans. Biomed. Eng.*, vol. 31, pp. 9-16, Jan 1984.

[31] F. R. Andreuccetti D, and Petrucci C, , "*Dielectric Properties of Body Tissues,*" 1997-2002.

[32] S. Gabriel, R. W. Lau, and C. Gabriel, "The dielectric properties of biological tissues: III. Parametric models for the dielectric spectrum of tissues," *Phys. Med. Biol.,* vol. 41, pp. 2271-93, Nov 1996.

[33] C. Gabriel, S. Gabriel, and E. Corthout, "The dielectric properties of biological tissues: I. Literature survey," *Phys. Med. Biol.,* vol. 41, pp. 2231-49, Nov 1996.

[34] C. H. Durney, *Radiofrequency Radiation Dosimetry Handbook*, Fourth Edition ed. Salt Lake: The University of Utah, 2002.

[35] K. D. Paulsen Calculation of power deposition patterns in hyperthermia, clinical thermology: *Thermal modeling and thermal dosimetry* vol. vol. 2: Springer-Verlag, 1988.

[36] G. M. Chato JC, Roemer RB, and Paulsen KDBerlin ; New York :: Springer-Verlag, 1990., *Thermal Dosimetry and Treatment Planning* Springer-Verlag, 1990.

[37] K. D. Paulsen, S. Geimer, J. Tang, and W. E. Boyse, "Optimization of pelvic heating rate distributions with electromagnetic phased arrays," *Int. J. Hyperthermia,* vol. 15, pp. 157-86, May-Jun 1999.

[38] Zienkiewicz and Taylor, *The Finite Element Method*, Sixth ed.: Elsevier Butterworth-Heinemann, 2005.

[39] A. Taflove and S. C. Hagness, *Computational Electrodynamics: The Finite-Difference Time-Domain Method*, Third ed.: Artech House, 2005.

[40] P. Monk, *Finite Element Methods for Maxwell's Equations*. USA Oxford University Press, 2003.

[41] K. S. Kunz and R. J. Luebbers, *The Finite Difference Time Domain Method for Electromagnetics:* Taylor and Francis Group, 1993.

[42] H. H. Pennes, "Analysis of skin, muscle and brachial arterial blood temperatures in the resting normal human forearm," *Am. J. Med. Sci.,* vol. 215, p. 354, Mar 1948.

[43] H. Brinck and J. Werner, "*Use of vascular and non-vascular models for the assessment of temperature distribution during induced hyperthermia.*" vol. 11, 1995, pp. 615-626.

[44] H. Arkin, L. X. Xu, and K. R. Holmes, "Recent developments in modeling heat transfer in blood perfused tissues," *IEEE Trans. Biomed. Eng.,* vol. 41, pp. 97-107, Feb 1994.

[45] S. Weinbaum and L. M. Jiji, "A new simplified bioheat equation for the effect of blood flow on local average tissue temperature," *J. Biomech. Eng.,* vol. 107, pp. 131-9, May 1985.

[46] E. G. Moros, A. W. Dutton, R. B. Roemer, M. Burton, and K. Hynynen, "*Experimental evaluation of two simple thermal models using hyperthermia in muscle in vivo."* vol. 9, 1993, pp. 581-598.

[47] C. K. Charny, S. Weinbaum, and R. L. Levin, "An evaluation of the Weinbaum-Jiji bioheat equation for normal and hyperthermic conditions," *J. Biomech. Eng*., vol. 112, pp. 80-7, Feb 1990.

[48] J. Wren, M. Karlsson, and D. Loyd, "A hybrid equation for simulation of perfused tissue during thermal treatment," *Int. J. Hyperthermia*, vol. 17, pp. 483-98, Nov-Dec 2001.

[49] J. W. Baish, P. S. Ayyaswamy, and K. R. Foster, "Heat transport mechanisms in vascular tissues: a model comparison," *J. Biomech. Eng.,* vol. 108, pp. 324-31, Nov 1986.

[50] M. Stanczyk, G. M. Van Leeuwen, and A. A. Van Steenhoven, "Discrete vessel heat transfer in perfused tissue--model comparison," *Phys. Med. Biol.,* vol. 52, pp. 2379-91, May 7 2007.

[51] J. Crezee, J. Mooibroek, J. J. Lagendijk, and G. M. van Leeuwen, "The theoretical and experimental evaluation of the heat balance in perfused tissue," *Phys. Med. Biol*., vol. 39, pp. 813-32, May 1994.

[52] C. H. Durney, "Antennas and other electromagnetic applicators in biology and medicine," *Proceedings of the IEEE* vol. 80, 1992 pp. 194 - 199

[53] N. M. Reddy, V. Maithreyan, A. Vasanthan, I. S. Balakrishnan, B. K. Bhaskar, R. Jayaraman, V. Shanta, and S. Krishnamurthi, "Local RF capacitive hyperthermia: thermal profiles and tumour response," *Int. J. Hyperthermia,* vol. 3, pp. 379-87, Jul-Aug 1987.

[54] P. J. Rietveld, W. L. van Putten, J. van der Zee, and G. C. van Rhoon, "Comparison of the clinical effectiveness of the 433 MHz Lucite cone applicator with that of a conventional waveguide applicator in applications of superficial hyperthermia," *Int. J. Radiat. Oncol. Biol. Phys.,* vol. 43, pp. 681-7, Feb 1 1999.

[55] R. H. Johnson, A. W. Preece, and J. L. Green, "Theoretical and experimental comparison of three types of electromagnetic hyperthermia applicator," *Phys. Med. Biol*., vol. 35, pp. 761-79, Jun 1990.

[56] C. J. Trujillo, L. Leija, and A. Vera, "Design and Preliminary Evaluation of a superficial applicator for hyperthermia with a new coaxially fed antenna: Theoretical models " in Health Care Exchange (PAHCE), 2010 *Pan. American Lima* 2010 pp. 62 - 67

[57] J. A. Mechling and J. W. Strohbehn, "A theoretical comparison of the temperature distributions produced by three interstitial hyperthermia systems," *Int. J. Radiat. Oncol. Biol. Phys.,* vol. 12, pp. 2137-49, Dec 1986.

[58] J.-S. Huang, D. A. Gervais, and P. R. Mueller, "Radiofrequency Ablation: Review of Mechanism, Indications, Technique, and Results," *Chin. J. Radiol*., pp. 119-134, 2001.

[59] C. J. Simon, D. E. Dupuy, and W. W. Mayo-Smith, "Microwave ablation: principles and applications," *Radiographics*, vol. 25 Suppl 1, pp. S69-83, Oct 2005.

[60] K. Ito, K. Saito, H. Yoshimura, Y. Aoyagi, and H. Horita, "Coaxial-slot antenna for interstitial microwave thermal therapy and its application to clinical trial," *Conf. Proc. IEEE Eng. Med. Biol. Soc.,* vol. 4, pp. 2526-9, 2004.

[61] A. S. Wright, L. A. Sampson, T. F. Warner, D. M. Mahvi, and F. T. Lee, Jr., "Radiofrequency versus microwave ablation in a hepatic porcine model," *Radiology,* vol. 236, pp. 132-9, Jul 2005.

[62] W. Schramm, D. Yang, B. J. Wood, F. Rattay, and D. Haemmerich, "Contribution of direct heating, thermal conduction and perfusion during radiofrequency and microwave ablation," *Open Biomed. Eng. J.,* vol. 1, pp. 47-52, 2007.

[63] R. Hergt, W. Andra, C. G. d'Ambly, I. Hilger, W. A. Kaiser, U. Richter, and H.-G. Schmidt, "Physical limits of hyperthermia using magnetite fine particles," Magnetics, *IEEE Transactions on* vol. 34 pp. 3745 - 3754 1998

[64] H. N. Gleich B, Bridell H, Jurgons R, Seliger C, Alexiou C, Wolf B and Weyh T „ "Design and evaluation of magnetic fields for nanoparticle drug targeting in cancer.," IEEE Trans. on Nanotechnology., vol. 6, pp. 164-170., 2007.

[65] R. Hergt and W. Andrä, *Magnetic Hyperthermia and Thermoablation:* Wiley-VCH Verlag GmbH and Co. KGaA, 2007.

[66] P. Wust, H. Fahling, T. Helzel, M. Kniephoff, W. Wlodarczyk, G. Monich, and R. Felix, "Design and test of a new multi-amplifier system with phase and amplitude control," *Int. J. Hyperthermia*, vol. 14, pp. 459-77, Sep-Oct 1998.

[67] J. F. Bakker, M. M. Paulides, A. H. Westra, H. Schippers, and G. C. Van Rhoon, "*Design and test of a 434 MHz multi-channel amplifier system for targeted hyperthermia applicators*." vol. 26, pp. 158-170.

[68] H. D. Trefna, P. Togni, J. V. Mikael, and Persson;, "*Design of the time-reversal hyperthermia system," in Antennas and Propagation*, 3rd European Conference on Berlin 2009.

[69] P. Shrivastava, K. Luk, J. Oleson, M. Dewhirst, T. Pajak, B. Paliwal, C. Perez, S. Sapareto, T. Saylor, and R. Steeves, "Hyperthermia quality assurance guidelines," *International Journal of radiation oncology, biology, physics*, vol. 16, pp. 571-587, 1989.

[70] Lagendijk JJ, Van Rhoon GC, Hornsleth SN, Wust P, De Leeuw AC, Schneider CJ, Van Dijk JD, Van Der Zee J, Van Heek-Romanowski R, Rahman SA, and G. C., "ESHO quality assurance guidelines for regional hyperthermia.," *Int. J. Hyperthermia*, vol. 14, pp. 125-133.

[71] G. S. lbbott, I. Brezovich, P. Fessenden, Y. Pipman, T. Sandhu, V. Sathiaseelan, P. Stauffer, A. Galdi, and T. Saylor, "Performance Evaluation of Hyperthermia Equipment," *American Association* of Physicists in Medicine by the American institute of Physics 26, 1989.

[72] A. M. D. Nallo and et al., "*Practical experience in electromagnetic hyperthermia quality control procedures within the context of international guidelines*." vol. 11, 1990, p. 231.

[73] M. L. Van der Gaag, M. De Bruijne, T. Samaras, J. Van der Zee, and G. C. Van Rhoon, "Development of a guideline for the water bolus temperature in superficial hyperthermia," *Int. J. Hyperthermia*, vol. 22, pp. 637-56, Dec 2006.

[74] J. W. Hand, J. J. Lagendijk, J. Bach Andersen, and J. C. Bolomey, "Quality assurance guidelines for ESHO protocols," *Int. J. Hyperthermia,* vol. 5, pp. 421-8, Jul-Aug 1989.

[75] J. B. Van de Kamer, A. A. De Leeuw, S. N. Hornsleth, H. Kroeze, A. N. Kotte, and J. J. Lagendijk, "Development of a regional hyperthermia treatment planning system," *Int. J. Hyperthermia*, vol. 17, pp. 207-20, May-Jun 2001.

[76] N. V. W. A. A. C. D. L. J. J. W. L. J. B. Van De Kamer, "*The significance of accurate dielectric tissue data for hyperthermia treatment planning.*" vol. 17, 2001, pp. 123-142.

[77] J. J. Lagendijk, "Hyperthermia treatment planning," *Phys. Med. Biol.,* vol. 45, pp. R61-76, May 2000.

[78] B. J. James and D. M. Sullivan, "Creation of three-dimensional patient models for hyperthermia treatment planning," *IEEE Trans Biomed. Eng.,* vol. 39, pp. 238-42, Mar 1992.

[79] Z. Li, M. Vogel, P. F. Maccarini, V. Stakhursky, B. J. Soher, O. I. Craciunescu, S. Das, O. A. Arabe, W. T. Joines, and P. R. Stauffer, *"Improved hyperthermia treatment control using SAR*/temperature simulation and PRFS magnetic resonance thermal imaging." vol. 27, pp. 86-99.

[80] H. P. Kok, P. M. van Haaren, J. B. van de Kamer, P. J. Zum Vorde Sive Vording, J. Wiersma, M. C. Hulshof, E. D. Geijsen, J. J. van Lanschot, and J. Crezee, "Prospective treatment planning to improve locoregional hyperthermia for oesophageal cancer," *Int. J. Hyperthermia*, vol. 22, pp. 375-89, Aug 2006.

[81] G. Sreenivasa, J. Gellermann, B. Rau, J. Nadobny, P. Schlag, P. Deuflhard, R. Felix, and P. Wust, "Clinical use of the hyperthermia treatment planning system HyperPlan to predict effectiveness and toxicity," *International journal of radiation oncology, biology, physics,* vol. 55, pp. 407-419, 2003.

In: Hyperthermia: Recognition, Prevention and Treatment ISBN 978-1-61942-275-9
Editor: Xing-Hua Gao and Hong-Duo Chen, 41-83

Chapter 2

TECHNOLOGY IN ULTRASONIC HYPERTHERMIA

Mario-Ibrahín Gutiérrez, Raquel Martínez, Arturo Vera, and Lorenzo Leija

ABSTRACT

The thermotherapy in which ultrasonic energy is used to increase temperature in a localized region in order to produce a desired effect is called ultrasonic hyperthermia. Although hyperthermia means the elevation of temperature above the human normal value, conventionalisms refer the classical hyperthermia as a temperature increment above 40°C and below 45°C. Recent developments apply high temperature hyperthermia (more than 50°C) to produce instantaneous tissue coagulation. The systems for ultrasonic hyperthermia have been created in relation to the application. The ultrasonic hyperthermia used to produce muscular or skeletal healing requires small temperature increments and hence medium intensity ultrasonic transducers (0.1 – 3.0 W/cm^2) are used. These transducers are normally planar and the application of therapy does not need rigorous control because that intensity of ultrasound rarely produces undesired effects, *e.g.* cavitation and ablation. The oncological hyperthermia is used to kill cancerigenic cells, then acoustical intensities for this therapy are normally high (from 20 W/cm^2 to 20,000 W/cm^2) and focalization is required to concentrate energy. The control in this therapy is very important because temperature should be kept at a constant value during a specific time of treatment in order to warrant cancerigenic cell death. Because this control requires temperature measurement in the region, which is practically impossible nowadays, some authors modified the technique by increasing the ultrasound intensity in order to produce ablation in the tumor in few seconds. In this situation, the temperature measurement is not necessary, but it is required a technique to guide the therapy. The frequency and power of both physiotherapeutic and oncological hyperthermia are quite similar, and the devices are made up generally by the same two parts: the generator and the transducer. The generator produces oscillating signals capable of driving the ultrasonic transducer at its resonant frequency at high power levels. The equipment is composed of an oscillator, an amplifier, the electrical matching, the feedbacks of variables, and control. The transducer is a piezoelectric material with different configurations: one piece planar or curved, and an array of small elements. For physiotherapeutic hyperthermia, it is common to use a planar transducer without backing and with front layer. The plane shape permits to produce a wide beam to be distributed in a large region of tissue; the lack of backing helps in increasing the efficiency and the front layer in matching the acoustic impedances between the piezoelectric and the radiated media. In oncological applications, the transducers are focused either from the shape (spherical transducers) or from the driver signal for transducers planar arrays. Wide-band transducers are lightly backed, while narrow-band transducers are not. Although there are many kinds of transducers for these applications, the design requirements for the ultrasonic hyperthermia systems are equivalent: high efficiency in amplification and high stability in frequency. In this book chapter, the configurations of practical hyperthermia systems and the problems to solve in order to apply them are discussed.

1. Introduction

Ultrasound is a mechanical wave of acoustic pressure with a frequency greater than the upper limit of human hearing. This mechanical energy propagates in media where the molecules are close enough to transmit vibration among them. At the beginning, the ultrasound in medicine was oriented to applications in therapy rather than diagnosis (Vera et al., 2008). In general, the term ultrasonic therapy could be defined as the use of high-intensity ultrasound in order to induce changes in the tissue by means of their thermal and non-thermal effects. When the main desired effects are the thermal effects, the therapy is referred as ultrasonic hyperthermia.

In accordance with the target, hyperthermia can be classified in physiotherapeutic hyperthermia, oncological hyperthermia, ocular hyperthermia, etcetera (Driller and Lizzi, 1987); the variety of the therapy protocols is the main reason to call them that way. In physiotherapy, the elevation of temperature should not be more than five Celsius degrees to avoid cells death, but not too low to permit the apparition of therapeutic effects (Draper et al., 1993); treatment sessions could last from 5 min to 30 min. In oncology, temperature rises are larger than those in physiotherapy and treatment time could be extended up to 60 min. Oncological hyperthermia is often used in conjunction with chemotherapy or radiotherapy, which increases cancerigenic cells dead more than the two other therapies alone (ter Haar, 1999). Physiotherapeutic hyperthermia has been applied for many years, but protocols are not standard and there are many inconsistencies in the results (Robertson and Baker, 2001). For oncological application, better systems and devices are still required in order to apply the techniques in clinic.

Hyperthermia is a relatively new therapy for cancer treatment. Its effects are obtained by increasing the temperature range in the tumor up to 42-45°C, in classical hyperthermia, and even more than 50°C in high temperature hyperthermia (Diederich and Hynynen, 1999). When the rate of temperature increment exceeds the ability of the body regulation system to dissipate the heat, the cells die. This therapy is effective when the temperature increase is maintained for about 60 minutes per treatment. The clinical results obtained by several clinical researchers are encouraging for the treatment of some kinds of tumors. The heating of cells induces conformational changes of certain proteins that depend on pH values. These conformational changes lead to alteration of multimolecular structures like cytoskeleton, membranes and also some structures in the cell nucleus. Metabolic changes like the increase of metabolic rates, lactate rates and the decrease of pH are also induced by heat (Field and

Bleehen, 1979). These conformational and metabolic changes lead to alteration of the microenvironment in tumors and has an impact on cellular death induced by heat.

The heating of cancerous tissues can be accomplished by several means, like electromagnetic and ultrasonic radiation among others. Each technique has its advantages and disadvantages which determine their application to selective tumor regions. It is important to mention that the areas where ultrasound cannot be successfully applied are those where bone or air regions block or reflect the acoustic field and can cause "hot spots". However, some of the advantages of using ultrasound include the feasibility of constructing applicators of almost any shape and size and the good penetration at frequencies where the wavelengths are on the order of millimeters. Also, these small wavelengths easily permit focalization in order to produce high temperature regions into a small region in short time.

2 Hyperthermia with Ultrasound

Etymologically, hyperthermia derives from the greek terms hyper- that means beyond, and -thermos that means heat (Vera et al., 2001). In medicine, hyperthermia refers to any temperature increase beyond the normal human temperature, but in this chapter, the term hyperthermia is used just for referring to the temperature artificially increased in organic tissues in order to cause a desired effect. In physiotherapy hyperthermia, the temperature rise should be capable to increase blood perfusion, immune response and cell membrane interchange in order to provoke tissue regeneration. In oncological hyperthermia (Field and Bleehen, 1979), temperature increments combined with time should be enough to cause cell dead or at least to sensitize cells to other oncological therapies as radiotherapy or chemotherapy (ter Haar, 1999).

2.1. Mild or "Classical" Hyperthermia

In past years, hyperthermia was defined as temperature increases between 42 to 45°C. Because hyperthermia of more than 50°C has been widely used recently, we refer to the "classical" hyperthermia as the mild hyperthermia to separate it from the high temperature hyperthermia. Into this temperature range, it could be included a non-oncological application of ultrasonic hyperthermia: the physiotherapeutic hyperthermia.

Nowadays, physiotherapeutic hyperthermia requires temperature increases from 40°C up to 45°C to improve tissue regeneration in combination with other rehabilitation therapies (massages, ice pads, medicaments, etc. (Draper et al., 1995b; Merrick et al., 2003)). More development in this area should be made in order to effectively reach the minimum therapeutic threshold of 40°C and warrant a correctly applied therapy (Merrick et al., 2003). Therapy protocols are often ineffective due to inappropriate election of the transducer by therapist (Draper et al., 1995a) or due to the lack of calibrated machines (Hekkenberg et al., 2004; Pye and Milford, 1994). Repeatability between acoustic field patterns produced by different ultrasound devices (Merrick et al., 2003) and even by transducers of same model should be improved (Johns et al., 2007).

The oncological mild hyperthermia is defined in the temperature range of 42°C to 45°C and treatment durations up to 60 minutes. This temperature elevation has shown its effectiveness in sensitized cells to other therapies as chemotherapy (ter Haar, 2007), radiotherapy (Diederich and Hynynen, 1999) or electromagnetic fields (Larkin et al., 2005). Also, hyperthermia is cytotoxic by itself when the tumor environments are low perfused, hypoxic, and at low pH, such conditions commonly exist in regions of malignant tumors (Field and Bleehen, 1979). The sensitization to other therapies produced by hyperthermia is not tumor specific and healthy tissues are equally affected; localized heating is required to just damage tumors. Observations in cell cultures from *in vivo* heated tumors and healthy tissue have indicated that hyperthermia does not significantly affect cell survival for either tumors or normal cells, but it could be implicated in increasing the immune response (Paliwal and Mitragotri, 2008) and in creating an inadequate environment to survival (vascular damage) (Field and Bleehen, 1979). Temperature feedback is required to effectively maintain the temperature above 42°C and avoid tumor thermotolerance.

2.2. High Temperature Hyperthermia and Ablation

For high temperature hyperthermia, the temperature is increased more than 50°C to produce cytotoxic cell effects. The temperature elevations are higher than the classical hyperthermia to potentiate the cytotoxic effects of temperature in cells. At some combinations of temperature and treatment times, an effect called ablation is produced. The term ablation has different meanings in different applications. In medicine, ablation is the surgical removal of an organ, structure or part, but we will refer to ablation as the melting or transformation of tissues by high temperature elevations. Some authors prefer to use the term thermal surgery for the therapy in which temperature provokes tissue coagulation (Diederich and Hynynen, 1999). Thermal surgery or ablation is the hyperthermia modality most used for the treatment of either benign or cancerigenic tumor. During the procedure, high ultrasound intensities are used to produce high temperature elevations in short periods of time to almost literally kill the cells of the tumor. The temperature range are of 50-55°C during 1 to 2 min (Diederich and Hynynen, 1999) and even higher temperature during shorter time for instant ablation, e.g. 60-80°C during 1-2 seconds (ter Haar, 1999). The election of short time to heat tissues at high temperature emerges from the lack of non-expensive non-invasive thermometry systems to supervise the temperature for long therapy time (ter Haar, 1999). Also, the modeling is simplified, because at short heating time, blood perfusion and tissue thermal conduction can be neglected. The use of high intensity focused ultrasound (HIFU) transducer permits high temperature increments in short period of time in a clearly defined region, the focus.

3. Ultrasound Propagation

Ultrasound propagates differently in relation to the media. In solids, the propagation can be transversal, longitudinal, by Rayleigh waves, etc. In liquids, the ultrasound mainly propagates longitudinally. This kind of propagation provokes molecules vibration in the same direction of the wave travel. The propagation in biological tissues is similar to that in liquid media and then the equations used for models are often equivalent. Nonlinear propagation should be considered when high power ultrasound waves are studied (Khokhlova et al., 2006; Wojcik et al., 1999). In this book chapter, emphasis in system development is made, and general aspects about hyperthermia treatments are analyzed.

3.1. Acoustic Wave Equation

The linear wave equation for homogeneous media is

$$\nabla^2 p - \frac{1}{c^2}\frac{\partial^2 p}{\partial t^2} = 0, \tag{1}$$

where p is the acoustic pressure, c is the wave velocity in the media, t is the time and ∇ is the gradient operator. If the wave propagates harmonically, the equation (1) is modified to the homogenous Helmholtz equation for harmonic propagation (continuous wave excitation):

$$\nabla^2 p - k^2 p = 0, \tag{2}$$

where k is the wavenumber ($k = \omega/c$), ω is the angular frequency $\omega = 2\pi f$ and f is the frequency. For axisymmetric problems, the wave equation is simplified to

$$\nabla^2 p + p\left(k^2 - \frac{b^2}{r^2}\right) = 0. \tag{3}$$

Here, r is the radial coordinate, and b is the circumferential wavenumber that represents the unique dependence on the azimuthal coordinate φ; for equation (3), the acoustic pressure is a function of the coordinates as $p(r, \varphi, z) = p(r, z)e^{-jb\varphi}$.

3.2. Attenuation, Absorption and Scattering

When the wave propagates in a biological tissue, the energy attenuation should be considered in the equations. Each biological tissue has its own attenuation, and even into the same kinds of tissue the attenuation could be different in relation to the concentration of cells, tissue micro-layers, water content, tissue degradation (for measurements in-vitro), etc; the attenuation also is highly dependent on the temperature and the frequency of the ultrasound. There are works about measurements of tissues attenuation in-vitro, but a closed value has not been found (López-Haro et al., 2010).

The attenuation is an effect caused principally by both the absorption and the scattering of the tissue. The absorption represents the acoustical energy that is converted into other kind of energy, principally thermal energy. The scattering is produced by the intracellular components (and fat cells) that reflect, in an incoherent way, the acoustical energy. Scattering is not equal to dispersion, because this last refers to the separation of the wave in its frequency components when it passes through a medium. When a wave travels in attenuating media, the acoustic pressure at each depth z is given by

$$p(z) = p_i e^{-\mu z}; \tag{4}$$

here μ is the pressure attenuation coefficient in Np/m (Nepers per meter), and p_i is the initial pressure at $z = 0$. The attenuation coefficient can be separated by

$$\mu = \alpha + \gamma, \tag{5}$$

where α is the pressure absorption coefficient and γ is the pressure scattering coefficient, both in Np/m. Because in biological tissues the absorption is the main source of attenuation, the scattering coefficient is considered zero for most of the problems (Moros et al., 1993). In Table I, the attenuation and absorption coefficients of some biological tissues are shown.

When measuring attenuation, temperature and ultrasound frequency should be registered in conjunction with the attenuation coefficient value. Also, care should be taken when the results of measurements are interpreted. In literature, there are reported values of "attenuation coefficient" for some tissues without emphasis about which attenuation coefficient is, i.e. if it is the attenuation coefficient of pressure, power, energy, amplitude, peak-peak value, etc. In most reports, the methods for calculating the attenuation coefficient of ultrasonic pulses use the amplitudes of the Fourier spectra; hence those results are with reference to intensity (or power) attenuation coefficient. In equations (4) and (5), the used pressure coefficients represent the half of the acoustic intensity coefficients (or power, or energy).

Table I. Acoustic properties of some biological tissues (Bazán-Trujillo, 2005; Gómez-Flores, 2006; Vázquez-Hernández, 2006)

Tissue	Density[a] [kg m^{-3}]	Acoustic wave velocity[a] [m s^{-1}]	Intensity absorption coefficient[a,b] [Np m^{-1}]	Intensity attenuation coefficient[a,b] [Np m^{-1}]
Bone	1380-1810	1500-3700	-	198.8-463.9
Fat	921	1400-1490	-	6.56-11.97
Kidney	1040	1564-1640	3.3	3.91-13.24
Liver	1060	1540-1640	2.3-3.2	4.26-23.83
Muscle	1050-1270	1508-1630	2-11	5.87-19.92
Tendon	1200	1750	14	39.72-92.79

a: depends on temperature and tissue conditions (nutrients, water contents, blood, degradation, etc.).
b: depends on frequency.

3.3. Temperature Elevation by Ultrasound

The time average heat Q produced in media depends on both acoustic intensity I and pressure absorption coefficient α as $Q = 2\alpha I$ (Moros et al., 1993); here, the coefficient α is given by equation (5). The temperature distribution T in a medium can be gotten by the Pennes bioheat equation (Pennes, 1948):

$$\rho C \frac{\partial T}{\partial t} - \kappa \nabla^2 T = \rho_b C_b \omega_b \left(T_b - T\right) + Q_{met} + Q_{ext}, \quad (6)$$

where ρ is the medium density (kg m^{-3}), C is the medium heat capacity (J kg^{-1} K^{-1}), κ is the medium thermal conductivity (W m^{-1} K^{-1}), ρ_b, C_b and ω_b are respectively the density, heat capacity and rate perfusion (1/s) of blood, Q_{met} is the heat due to metabolism (W m^{-3}), and Q_{ext} is the heat produced (W m^{-3}) by an external source. The heat by external source is related to the acoustic pressure (Nyborg, 1981) by

$$Q_{ext}\left(r,z\right) = 2\alpha I\left(r,z\right) = \frac{\alpha\left|p\left(r,z\right)e^{-\mu z}\right|^2}{\rho c}\left[\frac{W}{m^3}\right]; \quad (7)$$

here, c is the ultrasound velocity in the medium (m s^{-1}), and | | means the amplitude of the spatial acoustic pressure. The exponential term multiplying the pressure emulates the effect of reducing the amplitude due to attenuation μ (Moros et al., 1993).

4. Ultrasound Generation

Ultrasound is produced by a piezoelectric transducer that vibrates at a specific frequency when it is electrically excited. The development of high performance ultrasound transducers is very important in order to control the energy delivered to the tissues and warrant an effective therapy. Last advances in this matter have been produced in relation to new composites to improve the mechanical response and then the efficiency in the energy conversion and produced heat.

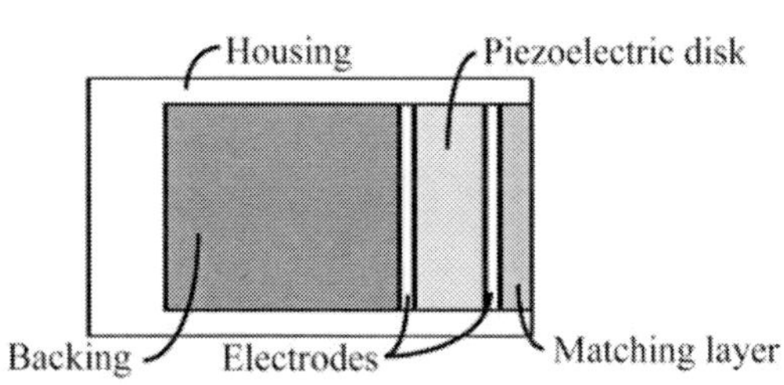

Figure 1. Main components of ultrasound transducers. The use of backing and matching layer depends on the application. High power applications often do not required backing.

4.1. Piezoelectric Materials

Piezoelectric materials are those that present electrical polarization when mechanical stress is provoked at some of their faces, and vice versa, i.e. they present a displacement or strain when an electric potential is generated at some of their faces. Before the discovery of piezoelectric ceramics, quartz crystals were used as ultrasound transducers. However, ceramics have shown better performance than crystals in high-power applications and better sensitivity in imaging. Among others, the most common materials for medical ultrasound are: barium titanate ($BaTiO_3$), lead titanate ($PbTiO_3$), lead zirconate ($PbZrO_3$) and combinations of these last two as lead zirconate titanate (PZT). This later is produced at different quantities of the constitutive components which results in many other piezoelectric materials to list separately: PZT-4, PZT-5A, PZT-5H, etc.

4.2. Types of Transducers

The general representation of an ultrasound transducer is shown in Figure 1. With reference of this figure, the backing material limits the mechanical response of the piezoelectric ceramic which increases the band-width; the matching layer (or layers) is used to improve the transmission of ultrasound to a medium with different acoustic impedance than the piezoelectric ceramic. The presence of each component depends on the application, e.g., in high power systems, the transducer usually does not have backing and even matching layer.

4.2.1. Plane Transducers

When the energy is applied to a wide region, the most used transducers are commonly planar. The radiation produced by a planar transducer has two main zones: near-field or Fresnel zone, and far-field or Fraunhofer zone both shown in Figure 2.

The Fresnel zone of the stationary acoustic field of planar transducers is composed by some overlaps due to constructive and destructive interferences. This zone is more irregular than the Fraunhofer zone, but it is here where the thermal effects are mainly produced. In the transition between Fresnel and Fraunhofer zones, it is located the last maximum peak that could be the natural focus of planar transducers.

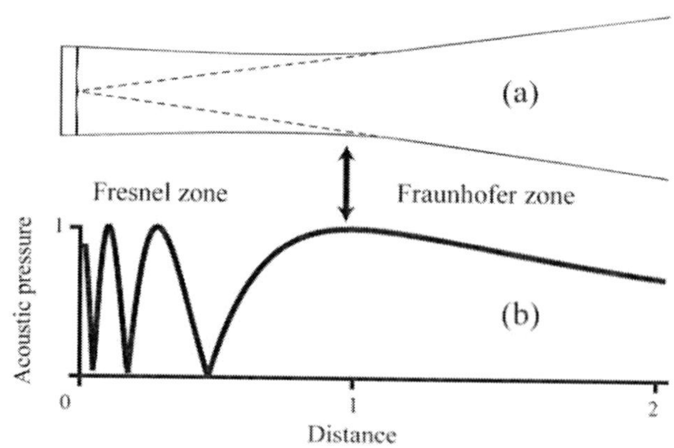

Figure 2. Fresnel and Fraunhofer zones in acoustic field of planar transducer. Notice in (a) the diffraction of acoustic field: in Fresnel zone there is a little convergence and in Fraunhofer zone a markedly divergence. (b) Acoustic pressure along the propagation axis of an ideal circular plate radiator.

4.2.2. Arrays

These transducers are constructed using many elements of small transducers aligned to form a matrix of transducers. They can be aligned on a planar or on a curve surface to get different shapes for a wide variety of applications. The acoustic field produced depends on the excitation of each element which can be in phase but with different intensity at each element or by varying both the phase and intensity. In Figure 3, some classical shapes of arrays are shown.

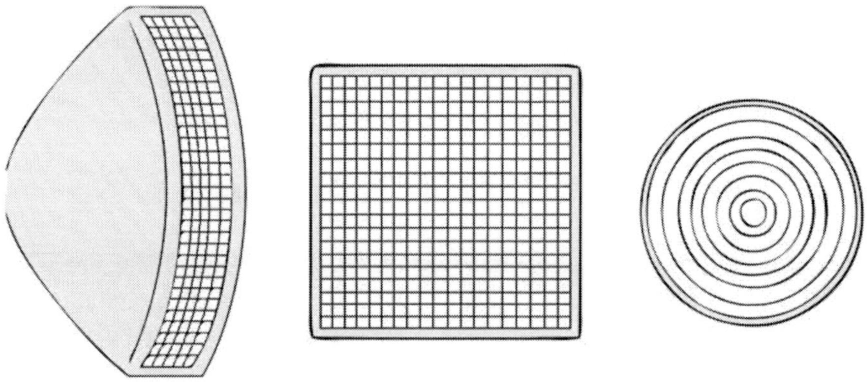

(a) (b) (c)

Figure 3. Some examples of arrays. (a) Concave linear array; (b) planar 2D array; (c) planar annular array. Each element could be excited individually.

4.2.3. Focused Transducers

Focalization is obtained by using spherically curved transducers that concentrate the energy in a point called focus (Figure 4a).

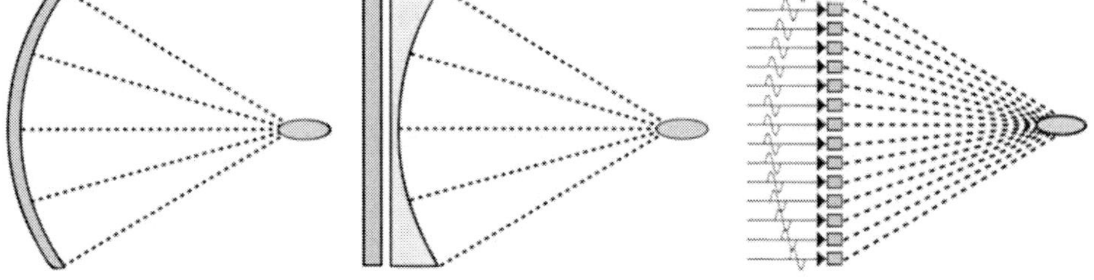

(a) (b) (c)

Figure 4. (a) Focused transducer. (b) Plane transducer focused using lens. (c) Electrical focalization using arrays. These focused transducers are used during HIFU therapies.

Also, focalization can be achieved by using planar transducers with lenses to concentrate the energy in a focus (as seen in Figure 4b); this approach permits the use of just one transducer to concentrate the energy at different distances by only changing the lens. However, efficiency can be decreased if the lens is not coupled correctly with the transducer or the material is not adequate. When focalization is obtained, the acoustic intensity at the focus usually is some hundred times the acoustic intensity at the transducer surface which is convenient when high temperatures in a very short time are required.

Another way to focalize the ultrasonic waves is by using electrical focalization. For this method, geometrically focused transducers are substituted by transducer phase arrays in which the excitation signals of every element are at different phase. Each element is excited by a specific signal at specific time in order to get the addition of that energy at the desired focus by constructive interference. In Figure 4c, a representation of the electrical excitation signal for each element is shown. This kind of focalization has been widely used because it is possible to get the focus at different distances with the same applicator by just varying the phase of excitation of each element. For continuous excitation, the delay time is translated to a signal phase.

5. Hyperthermia Parameters

The practice of ultrasonic hyperthermia requires regulation before being applied in clinic. The international organisms for standardization have created definitions for some hyperthermia applications that should be used in the design of commercial devices. There are several parameters to define the acoustic radiation produced by the transducers and others to define the power deposition in the tissue in terms of time and temperature. Next, the most used parameters in literature of ultrasonic hyperthermia are detailed.

5.1. Acoustic Intensity

The acoustic intensity initially appeared in texts as power density (W/m^2). It is calculated by the time-averaged instantaneous acoustic intensity that is the instantaneous rate per unit of area at which work is done by one element of matter to its adjacent element. It is equal to the product of the instantaneous acoustic pressured p and the particle velocity v, and then the acoustic intensity is given by

$$I = \langle pv \rangle, \tag{8}$$

where $\langle\ \rangle$ is the time average. When the acoustic wave is plane, the acoustic pressure is

$$p = \rho c v, \tag{9}$$

and the acoustic intensity can be expressed as

$$I = \frac{\langle p \rangle^2}{\rho c} = \frac{p_0{}^2}{2\rho c}, \tag{10}$$

here p_0 is the amplitude of the acoustic pressure, and ρ is the density of the media.

5.2 Acoustic Impedance

The characteristic acoustic impedance of a medium indicates how easy the acoustic waves propagate in that medium. The acoustic impedance relations are analogous to those of the electric impedance, but the relation of the energy flow is opposed, i.e. the electrical impedance indicates the media opposition of the electrons flow while acoustic impedance indicates how easy the ultrasonic wave travels in the medium. If it is considered the classical analogy between the mechanical force with the electric potential and the vibration velocity (particle velocity) with the electrical intensity, the characteristic acoustic impedance Z is given by

$$Z = \frac{p}{v}, \qquad (11)$$

where the variables are the same as before. If the medium is loss-free and the wave is plane, by using Eq. (9) the acoustic impedance could be represented by

$$Z = \rho c; \qquad (12)$$

for media with losses, the acoustic impedance is complex and should be calculated with Eq. (11).

5.3. Specific Absorption Ratio

The specific absorption ratio (*SAR*) is defined as the time rate energy absorption per unit of mass. It is strongly related to the media absorption coefficient and to the acoustic power, which means that this value is neither absolute for a transducer nor for a media. The *SAR* can be calculated with:

$$SAR = C\frac{\Delta T}{\Delta t}, \qquad (13)$$

where C is the heat capacity (J/kg·°C), ΔT is the change of the temperature (°C), Δt is the change of the time (s). If the process is isobaric and isochoric, the heat produced per unit of volume is $Q = \rho C\,\Delta T$; the *SAR* in Eq. (13) is then transformed to

$$SAR = \frac{2\alpha I}{\rho}, \qquad (14)$$

here α is the absorption coefficient (Np/m), I is the acoustic intensity (W/m^2), and ρ is the medium density (kg/m^3).

5.4. Thermal Dose

This parameter has been defined in relation to the temperature and time required to induce tissue damages. The establishment of this relation is not a simple task because the biological response mainly depends on time and temperature and not on the deposited energy as it occurs in other areas (Field and Bleehen, 1979). The thermal dose (*TD*) could be easily defined when one of the variables, temperature and time, were constant. However, this does not happen in real hyperthermia because the temperature is almost never constant and hence thermal dose should be defined using both variables. In treatments of high intensity focused ultrasound, the time is considered constant and the temperature is the parameter dependent on the acoustic intensity.

The most used empirical definition (Arora et al., 2005; Diederich and Hynynen, 1999) was developed by Sapareto and Dewey which establishes that the thermal dose should be determined with reference to a temperature of 43°C; the result, in minutes, is equivalent to the heating time at 43°C. They noticed, by multiple *in vitro* experiments, that a specific effect was produced at time t_1 when heating at temperature T_1, and the same effect was also produced at time t_2 when heating at temperature T_2. The ratio of the times was equivalent to (Church, 2007; Sapareto and Dewey, 1984)

$$t_2/t_1 = R^{T_1-T_2} \qquad \begin{array}{l} R=1 \text{ for } T>43°\text{C} \\ R\approx 4 \text{ for } T<43°\text{C}, \end{array} \tag{15}$$

where R is an empirical constant that depends on the tissue sensitivity, the animal species and more factors (Church, 2007). Another approach based on Sapareto and Dewey was used by Arora et al. to control the thermal dose during a hyperthermia treatment (Arora et al., 2005). They used equation (15) to find the cumulative equivalent minutes (CEM) of heating at 43°C as

$$TD(t) = \text{CEM at } 43°\text{C} = \int_0^t R^{[43-T(\tau)]} d\tau, \tag{16}$$

where R is given by

$$R = \begin{cases} 0 & \text{for } T < 39°C \\ 0.25 & \text{for } 39 < T < 43°C \\ 0.5 & \text{for } T > 43°C. \end{cases}$$

In equation (16), it is implicit the reference temperature of 43°C, while in equation (15), it is not. The thermal dose calculation by Arora is time-dependent and it can be used in real-time controllers using feedback of temperature sensors (Arora et al., 2005).

5.5. Standing Wave Ratio

Standing wave ratio (*SWR*) is a parameter that indicates the relationship between the energy dissipated in transmission line and that in the load. When the impedance matching between the transmission line and the load is optimum, the transferred energy is maximum. On the other hand, when matching is poor, less power arrives to the load and much energy is absorbed by the transmission line (Carr, 2001). If a transmission line is considered, the wave going to the load plus the wave going backward form together a total standing wave along the transmission line (Sengupta and Liepa, 2006). The amplitude of the maximum peak divided by the minimum peak denotes the *SWR* in the transmission line. The *SWR* gives the ratio between these opposite waves, incident and reflected waves (Balanis, 1989).

The *SWR* can be computed through measurements of incident and reflected current, voltage or power and through measurements of line and load impedances (Carr, 2001). When computing *SWR* from power measurements, the following expression could be used

$$SWR = \frac{1+\sqrt{P_r/P_i}}{1-\sqrt{P_r/P_i}}, \tag{17}$$

where P_r and P_i are the measured reflected and incident power. The *SWR* is expressed often in terms of voltage as *VSWR*. In a rigorous way, *VSWR* should be measured at each point of the transmission line to determine the overall maximum and minimum of the voltage signal, but practically, it is calculated by its relation with the reflection coefficient Γ

$$VSWR = \frac{V_{max}}{V_{min}} = \frac{1+|\Gamma|}{1-|\Gamma|}, \tag{18}$$

and

$$\Gamma = \frac{V_r}{V_i}, \tag{19}$$

where V_r is the "reflected" voltage signal and V_i is the "incident" voltage signal. The ideal *SWR* of a transmission line is 1:1, which indicates that all the energy is transmitted to the load; the superior limit of *SWR* is infinite when the reflection coefficient is equal to 1 in equation (18). A mismatch in the transmission line can be solved by increasing the dimensions of this in order to get a total length as multiples of signal wavelength; a mismatch between the line impedance and that of the load can be solved by the addition of inductances and capacitances to adjust load impedance to the characteristic impedance of the transmission line. The practical implications of *SWR* different from 1 are: loss of power in the line (heating in the transmission line), low power at the load, and heating in the amplifier or even damage.

5.6. Cavitation

During ultrasonic hyperthermia therapy, cavitation could be produced. Cavitation may be defined as the formation and activity of gas or vapour-filled cavities (bubbles) in a medium exposed to an ultrasonic field (Hill et al., 2004; Hodnett and Zeqiri, 2004). In literature, two types of cavitation are reported, commonly known as inertial and non-inertial cavitation. Inertial cavitation occurs when a gas-filled cavity in a liquid expands during a part of the acoustic cycle and then collapses rapidly to a small fraction of its initial value. Non-inertial cavitation describes the activities (translational motion, surface distortions, growth by rectified diffusion) of a range of gas-filled bodies in an ultrasonic field (Hill et al., 2004).

Cavitation may or may not be desirable during ultrasonic hyperthermia. In physiotherapy, "hot-spots" due to bubbles formation could be generated and cause damage to tissues or pain to patients. In high-temperature oncological hyperthermia, cavitation can lead to faster treatments and allow thermal ablation detection while monitoring, but also can complicate energy deposition and distort ablative lesion shapes (Salgaonkar et al., 2009).

Cavitation is directly related to ultrasound intensity and temperature in tissues. Hyperechoic regions, i.e. the apparition of echoes coming from a part without any physical reflector, can emerge when monitoring HIFU therapies; this effect has been related to the bubbles formation by cavitation and it can be useful in the detection of lesion formation during therapy (Rabkin et al., 2006).

5.7. Parameters for Non-Focused Transducers

Non-focused transducers for hyperthermia, commonly planar transducers, require some specific parameters to be characterized. Some of these parameters are defined by standards, and then those definitions could change.

5.7.1. Effective Radiating Area

There have been many definitions for the effective radiating area. One of these is given by the FDA, which defines the A_{ER} (but called *ERA*) as the area consisting of all points of the effective radiating surface (all points within 5 mm from the applicator face) at which the intensity is 5 percent or more of the maximum intensity at the effective radiating surface, expressed in square centimeters (FDA, 2008). The IEC standards also define the A_{ER} and the measuring method (IEC-61689, 1998). This technique consists in measuring and in registering the acoustic intensities in four planes parallel to the transducer face at four distances along the propagation edge. Four "partial A_{ER}'s" (called A_{BCS} by IEC standards, beam cross-sectional area) are determined and the transducer A_{ER} is gotten by extrapolation of the A_{BCS}'s at the transducer surface. For more detailed information, see (Gutierrez et al., 2010; Hekkenberg, 1998). The FDA definition of A_{ER} gives larger uncertainties (more than 20%) than the IEC definition (less than 10%).

5.7.2. Beam Non-uniformity Ratio

Called *BNR* in FDA standard, R_{BN} represents the relation between the square of the maximum acoustic pressure measured in the acoustic field and the spatial media of the square of the acoustic pressure for the entire measurements (Hekkenberg, 1998), where the spatial media is taken on the effective radiating area for FDA (FDA, 2008) or in the four measured planes for the IEC (IEC-61689, 1998).

5.7.3. Penetration Depth

The penetration depth (P_D) depends on the properties of the medium where the ultrasound is passing through. Its value is not defined by IEC standards, but it should be considered in planning in order to get appropriate heating at the desired depth. A definition for electromagnetic hyperthermia that could be applied to ultrasound was given by Hand et al. as the distance from the transducer where the *SAR* is 50% of its value at the A_{ER} (Hand et al., 1989). This definition could be interpreted as the depth where the acoustic intensity decreased to the half of its value at the transducer surface, or where the acoustic pressure amplitude decreased to $1/\sqrt{2}$ of that value at the transducer surface. Penetration depth depends on the ultrasound frequency because it is linearly related to the attenuation coefficient through the *SAR*. However, other factors affect the P_D in real tissues as: tissue composition, blood perfusion, transducer diffraction, acoustic intensity, etc. that make this parameter, at the time, unpredictable (Demmink et al., 2003).

5.8. Parameters for Focused Transducers

As well as planar transducer, focused transducers should require some parameters to be characterized. Even though focused ultrasound systems for thermal ablation therapy are being used in clinic, measurement standards for acoustic field characterization, acoustic power level, focused beam characteristics (length and width), intensity at focus, etc., must be established (Harris, 2009; Shaw and Ter Haar, 2006). Here, three parameters for spherically curved circular focused ultrasound transducer, in order to obtain a focused beam, are presented.

5.8.1. Radius of Aperture and Radius of Curvature

According to O'Neil's analysis (O'Neil, 1949) based on Figure 5 for focused fields, focused beam depends on the radius of aperture a, and the radius of curvature A. Some resulting parameters useful in acoustic field determination can be found, as: the shape parameter a/A, the relative size parameter a/λ (where λ is the wavelength) and the relative distance parameters z/A and z/a.

From Figure 5, some geometric identities can be found to express each parameter. The radius of aperture (a) can be given by

$$a = A\sin(\alpha); \tag{20}$$

the radius of curvature (A) can be expressed in terms of h and angle α by

$$A = \frac{h}{1-\cos(\alpha)}. \tag{21}$$

For the equations (20) and (21), it was considered that

$$h = \frac{b^2}{2A}, \tag{22}$$

where h is the depth of the concave surface and b is the chord from the center of the surface S to the transducer edge. The main propagation axis is on the z-axis and a point Q can be located using a vector from the origin O with components in terms the cylindrical coordinates r and z (O'Neil, 1949).

5.8.3. Focus Size

The main goal of focused ultrasound systems falls into the achievement of the concentration energy in the focus. Commonly focus shape is described as a "cigar" (Khokhlova et al., 2006) but dimensions vary in accordance to the device. A way to denote the focus size is by its length (along the beam axis) and width (transversal to the beam axis), called respectively as full length at half maximum (FLHM) intensity and full width at half maximum (FWHM) intensity (Diederich and Hynynen, 1999; Rivens et al., 1996; Wu et al., 2007).

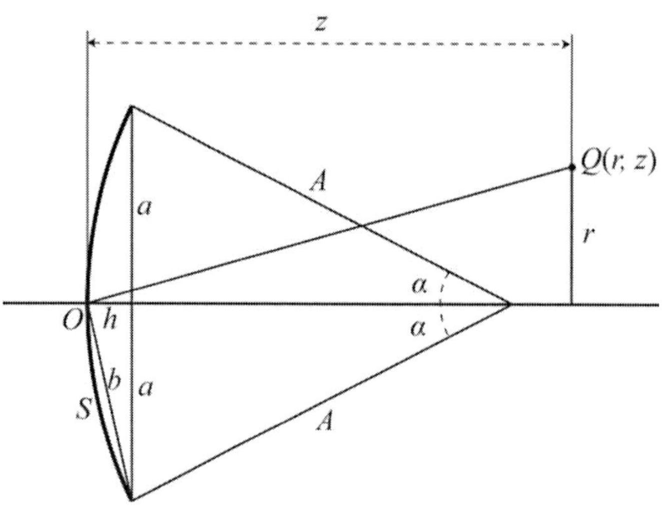

Figure 5. Geometry of a focused transducer.

6. Hyperthermia Systems

There are different ways to get temperature increases in tissues using ultrasound. The applicators can be manufactured in many shapes, and additionally, the driver signal modifies the acoustic field distribution. In order to easily analyze the hyperthermia systems, applications will be separated in two general parts in accordance to the kind of used transducer: planar transducers and focused transducers. The planar transducers considered next are those that do not have any kind of focalization (geometric, electrical, etc.) and their response is the result of an excitation "in phase" along the aperture; planar arrays are considered here when the driven signal is in phase. On the other hand, focused transducers are those that have a kind of focalization without considering the overall shape, i.e. focused transducers for interstitial hyperthermia are also included in this category.

6.1. Hyperthermia Using Planar Transducers

Monoelement planar transducers have been used since the initial reports in hyperthermia (Nyborg, 2001). This is the simplest way of producing ultrasound because it requires few electronic circuits for driving the device. Thermal patterns are related, among other variables, to the transducer acoustic field which is, by definition, non-uniform (see Figure 2). When the thermal effects are produced in near-field, the most irregular part, the therapy planning should consider this fact in order to compensate the effect. In practical transducers, it is required that the near-field irregularities are as low as possible in order to warrant a regular (and then predictable) thermal distribution (Hekkenberg, 1998); high non-uniformities in near-field could create "hot-spots" that can damage the tissue or at least cause pain in patient during therapy. Hyperthermia protocols should specify the acoustic intensities, the transducer operating frequency and the time of treatment in accordance to the application in order to accomplish the desired thermal effects. Also, protocols should include indications of using other specific accessories as water boluses, ice pads, water baths or simply acoustic coupling media. Physiotherapy, bone-healing and oncological superficial heating are the most common applications for hyperthermia with planar transducers (ter Haar, 1999).

6.1.1. Non-Oncological Applications

Hyperthermia has been used in treating muscular injuries and healing bones since the 50 decade (Schwan et al., 1954). The first application registered was carried out in Germany when a patient with sciatica was treated in 1938 (Nyborg, 2001). Since the 80s, many commercial ultrasound devices have been used in clinics although the research is not finished yet and even the therapeutic effects are not clearly known (Robertson and Baker, 2001; Speed, 2001). The devices for physiotherapeutic hyperthermia work mainly between 1 and 3 MHz at acoustic intensities up to 3 W/cm^2. The energy can be delivered continuously or by variable squared pulses determined with a duty cycle. The therapy session could last from 5 to 30 min as the treatment requires.

A number of reports about effects produced by ultrasound in tissues have been published along years (Baker et al., 2001; Speed, 2001; Watson, 2008). Initially, attention was put in determining the relation of temperature increase and muscular healing during hyperthermia (Watson, 2008). Theories about cells mechanisms involved in this process were planted, but no proofs of those theories had been clearly presented (Baker et al., 2001). At that time, it was concluded that hyperthermia (independently of heat source) had provoked increment of blood flow in the region and this had promoted a fast interchange of substances between cells and surrounding media (Robertson et al., 2005). This increment of blood flow also increased the quantity of blood cells (lymphocytes, macrophages, etc.) which have been known by their contribution during healing process (Watson, 2008). Recent investigations have been directed about the action of ultrasonic energy with no heat in muscular therapies. The effects have been separated in two general categories: thermal effects and non-thermal effects (Speed,

2001). Non-thermal effects have gain the attention recently because "apparently" they promote, in a stronger way, the tissue regeneration in comparison with the thermal effects (Baker et al., 2001; Speed, 2001; ter Haar, 1999). Main non-thermal effects are the cavitation and the microstreaming although it has been suggested that the latter is a consequence of the former (Watson, 2008). Because no-evidence of any effect has been presented in-vivo, their contribution in therapy is still unclear (Baker et al., 2001).

A number of non-thermal effects, with implications in tissue regeneration and some of them in its destruction, are reported. An undesired effect in physiotherapy is the cavitation, because it could destroy tissues by mechanical forces. In physiotherapy hyperthermia, the intensities are often considered cavitation free, but measurements in water tanks indicate that cavitation could occur at those intensities under specific conditions (Hekkenberg et al., 2001). An effect that could be beneficial in the tissue regeneration is the acoustic microstreaming, which refers to the production of flow currents around the cells that could promote better substance interchange. Acoustic microstreaming has been just observed in-vitro (Baker et al., 2001), but if this effect were present in-vivo, more significant contributions to therapy could be due to this (Speed, 2001; ter Haar, 1999). More theories that could explain the interaction of ultrasound with intracellular proteins and enzymes have been presented (Marchioni et al., 2009). Frequency resonance hypothesis, by Johns, talks about alteration of the enzymes function due to conformational modifications provoked by mechanical energy of ultrasound (Johns, 2002). How this change could be beneficial for therapy is unclear and more work should be done at this matter.

There are more practical conditions during an ultrasonic treatment that require better standardization rules. Ultrasonic energy can be transmitted efficiently from the transducer to tissue provided that a coupling media is used. Any air layer between the transducer and skin affects dramatically the transmission of energy and could provoke injuries to patient and damages to the transducer. The coupling media should be capable to displace the air by filling any space and should allow ultrasound transmission with no absorption and no distortion (Watson, 2008). Minerals in coupling media, according to those presented in tissues (calcium, potassium, etc.), are desirable to reduce osmosis.

It has been proved that the acoustic intensities during practical therapies are lower than those required to produce a therapeutic temperature increase (Watson, 2008); this is because the election of the maximum delivered intensity depends on patient temperature sensitivity. Because temperature receptors are mainly located in the skin, increase of temperature sensation does not imply the achievement of an acceptable temperature increase in deep tissues. Using of water boluses, as commonly done in oncological hyperthermia, could be a good idea to be capable of increasing acoustic intensities beyond the standards limits. However, this option is not applied in practice and standards even not take in account this possibility. Also, temperature measurement into the tissues could be required to correctly adjust the acoustic intensity and get the desired effects.

The temperature increase in deep tissues is affected also by penetration depth which is related to frequency. As seen before, *SAR* is related to absorption and penetration depth is related to *SAR*. The absorption coefficient depends on frequency. These relations are used to choose which frequency is more appropriated to heat at a given depth. The combination of acoustic intensity and frequency has been used in arrays of planar transducers to heat large regions at different depths (Diederich and Hynynen, 1999). At low frequencies (for example 1 MHz), the absorption is small and high intensities are required to get the desired temperature increase at deep tissues. On the other hand, at higher frequencies, the absorption is bigger and the required intensities to get the desired temperature are lower and penetration depth is smaller. In Figure 6, it is shown the effect of using two different frequencies under the same transducer size, and intensity; the figure is just to exemplify the effect of transducer frequency in penetration depth and no blood flow was included in the calculation. Other factors as tissue composition (Demmink et al., 2003), radiation beam patterns, and natural focalization of planar transducers (Johns et al., 2007) affect the penetration depths during physiotherapy.

As can be seen in Figure 2 and Figure 6, the acoustic field of planar transducers is non-homogeneous. Actually, acoustic fields from Figure 6 were calculated with a theoretical approach using ideal uniform distribution on the transducer face. Real radiators vibrate non-uniformly and in most cases in non-axisymmetric modes. Standardization organisms have created rules, as the R_{BN} magnitude defined before, to determine a safe range for acoustic field non-uniformities (IEC-61689, 1998). However, irregular radiation is unavoidable and the treatment protocols were adjusted to distribute the energy homogeneously and reduce the risk of "hot spots". During a therapy, dynamic application techniques should be preferred; the most used dynamic technique consist in continuously moving the transducer by forming circles as about twice the A_{ER} (Watson, 2008). The transducer size should be at least the half of the region to treat, and its incorrect election could result in absence of therapeutic effects (transducer smaller than required) or even injuries to patient (transducer bigger than required). A temperature feedback could be helpful in reducing the risk of undesired effects during this kind of therapies. Actual ultrasound machines do not provide media temperature feedback (Castel, 1995; Houghton and Obray, 1990).

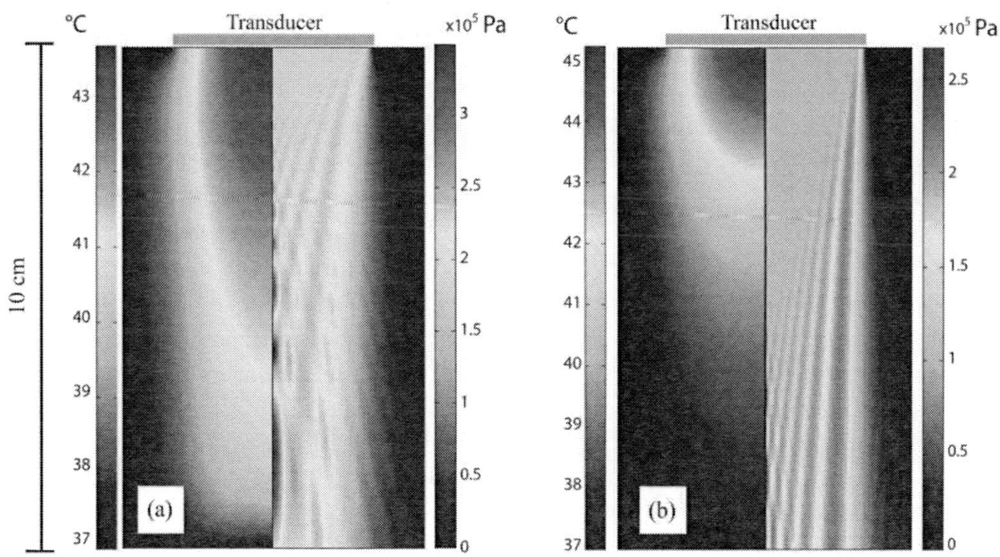

Figure 6. Finite element simulation of temperature increase vs. acoustic field at two frequencies. Transducer was simulated as a vibrating piston with radius of 2 cm. (a) 1 MHz transducer, and medium absorption coefficient of 0.4 dB/cm. (b) 3 MHz transducer and medium absorption coefficient of 1.2 dB/cm. The shown acoustic fields are not attenuated. The simulated acoustic intensity was of 1 W/cm^2 in both cases. Temperature distributions were gotten by Eq. (6) with no blood flow. Cooling at the transducer surface was not included.

6.1.2. Oncological Applications

Although the focused transducers and phase arrays dominate oncological hyperthermia, the planar transducers are used for the treatment of big size superficial tumors. The radiation produced by planar transducers is approximately collimated and the area depends on the transducer radius; this permits to treat large tumors efficiently. As in physiotherapy, the maximum penetration depth is limited by the transducer frequency and tissue absorption, which can be an advantage if planar transducer arrays of different frequency are used.

Temperature rise during planar transducer hyperthermia is commonly lower than 45°C. This means that planar ultrasound transducers can be applied to mild hyperthermia (40-45°C) although higher temperatures can be easily gotten. Temperature elevations depend directly on tissue absorption, (which depends on frequency), time, and intensity; the higher intensities, the higher temperature increments. The use of planar transducers for high temperature hyperthermia is limited by the difficulty of controlling the size of the radiated volume. The control of this size is made by combining multiple frequency transducers working at different intensities. Water boluses are required to reduce superficial heating and allow the use of high ultrasonic intensities with less patient pain.

Multielement planar arrays can be used to deliver energy at different size areas with higher homogeneity in comparison with monoelement applicators. Element size can be as required because it is not determined by the ultrasonic wavelength (as in electromagnetic applicators). It is possible to construct applicators without any near-field overlapping (as the produced by monoelement transducers in Figure 6) to generate homogeneously distributed radiation. Also, each element can be driven individually to generate the desired radiating pattern and control the radiated zone. The variation of acoustic intensities of each element in conjunction with the dynamic thermal properties of tissues (thermal conduction, blood perfusion, etc.) can be applied to control heating patterns. Diederich and Stauffer observed improvement in heating patterns when using multielement planar arrays in comparison with a similar geometry multielement 915 MHz microwave applicator (Diederich and Hynynen, 1999). Adaptations with other therapies, as radiotherapy, are possible.

The arrays of planar transducers have been used successfully in the treatment of large superficial tumors. The array can be composed of different sizes planar transducers working at different frequencies to control penetration depth. Multifrequency array can be gotten either by driving separate transducers at different frequencies or by multiplexing the frequency of less transducers among its fundamental and odd harmonic frequencies. The array is mechanically positioned to cover the entire tumor and the therapy planning is made off-line although imaging to monitor the therapy is desired. In Figure 7, it is shown a mechanically scanned array of planar transducer (Diederich and Hynynen, 1999). Transducers orientation permits to concentrate the different frequencies ultrasonic energy directly in the tumor in order to make use the different penetration depths and intensities to produce a desired heating pattern. The transducers are immersed in degassed temperature-controlled water to permit high acoustic intensities and reduced patient pain due to skin heating. Temperature and pain feedback algorithms to control the power deposition are desirable.

There are multiple desired effects of ultrasonic energy in tissues for oncological applications. The main objective is to eliminate the tumor, but the means to reach it are variable. It was demonstrated that mild hyperthermia does not significantly affect tumor cells but it does increases tumor cell sensitivity to radiotherapy (Field and Bleehen, 1979). Studies using low intensity ultrasound in combination with mild hyperthermia (not produced by ultrasound) indicate that ultrasonic non-thermal effects induce apoptosis and cell lysis *in vitro*. The combination of hyperthermia and ultrasound results in the enhancement of cell killing, in comparison with other hyperthermia techniques (Feril and Kondo, 2004). Using chemical agents, as echo contrast agents, increase ultrasound-induced cell killing due to microbubbles in their compositions. The bubbles resonant vibration provokes cell destruction, lysis and sonoporation in cell cultures with the most common echo constant agents used in clinics (Feril and Kondo, 2004). Increasing the immune response is another way to destroy the tumor. Publications about ultrasound-mediated immune response indicate the increasing of immune cells in the treated region during non-thermal ultrasound radiation (Paliwal and Mitragotri, 2008). Other effects not mentioned here that should be considered during therapy have been reported (Feril and Kondo, 2004; Marchioni et al., 2009; Nyborg, 2001; Paliwal and Mitragotri, 2008).

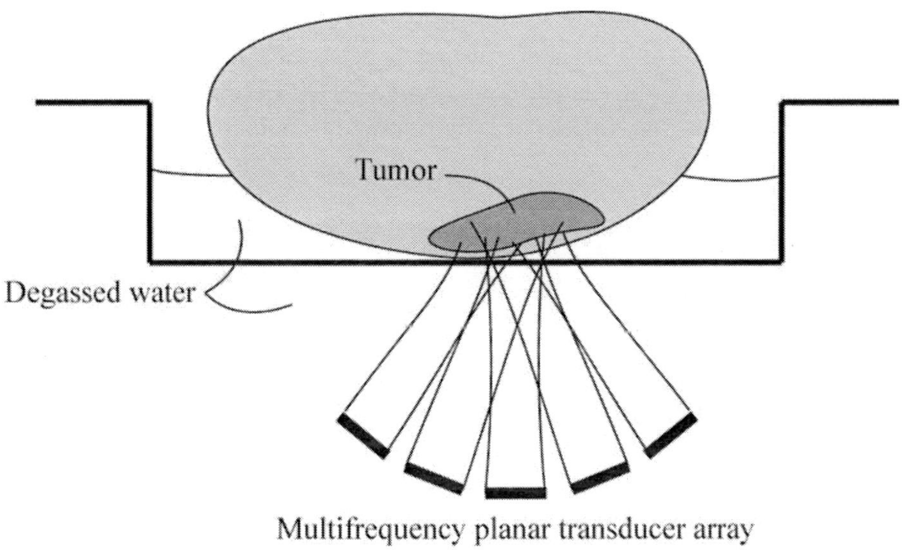

Figure 7. System setup for superficial tumors hyperthermia using a planar transducer arrays.

Although oncological hyperthermia using planar transducers has been investigated in the past years, most effort has been made in the use of focused transducer for oncological applications. The main advantage of planar transducers is that the heated region is as big as the transducer radiating area which can be useful in the treatment of big size superficial tumors. However, there are problems during heating of *in vivo* tissues that make the heating patterns irregular and difficult to determine. Some of these problems, that are going to be explained later in the next sections, are reduced or even avoided during high temperature hyperthermia using high intensity focused applicators. The use of planar transducers in oncology requires technologic advances that permit a better planning (radiation patterns, geometry with blood flow, large vessels, layers, etc.) in order to accomplish the desired heating patterns.

6.2. Hyperthermia Using Focused Transducers

Focused ultrasound effects were discovered in the 1940's by the Fry brothers, whom used a four quartz transducer array to focalize the ultrasonic beam (ter Haar and Coussios, 2007a). In 1942, Lynn et al., obtained paraffin models of focused ultrasound field due to voltage and current variation of a concave transducer (Lynn et al., 1942). Later, the Fry brothers and Lynn et al. could generate lesions in the brains of cats and monkeys (Kennedy et al., 2003; Lynn and Putnam, 1944). Ablation results on human brains were reported near 1960 (ter Haar and Coussios, 2007a, b), where the treated patients were suffering of neurological diseases. During 1970, first ophthalmological applications emerged for treating glaucoma and presented clinical results of 212 patients (Valtot et al., 1989). Despite of obtaining successful results of focused ultrasound therapy, this technique was displaced due to the introduction of pharmacological drugs to treat Parkinson and due to the introduction of laser for ophthalmological surgery (ter Haar and Coussios, 2007a, b). Focused ultrasound therapy was limited for many years due to the lack of real-time imaging guidance. In the last two decades, improved quality images of generated lesion, due to focused ultrasound, were obtained by magnetic resonance imaging (MRI) and diagnostic ultrasound imaging (ter Haar and Coussios, 2007b).

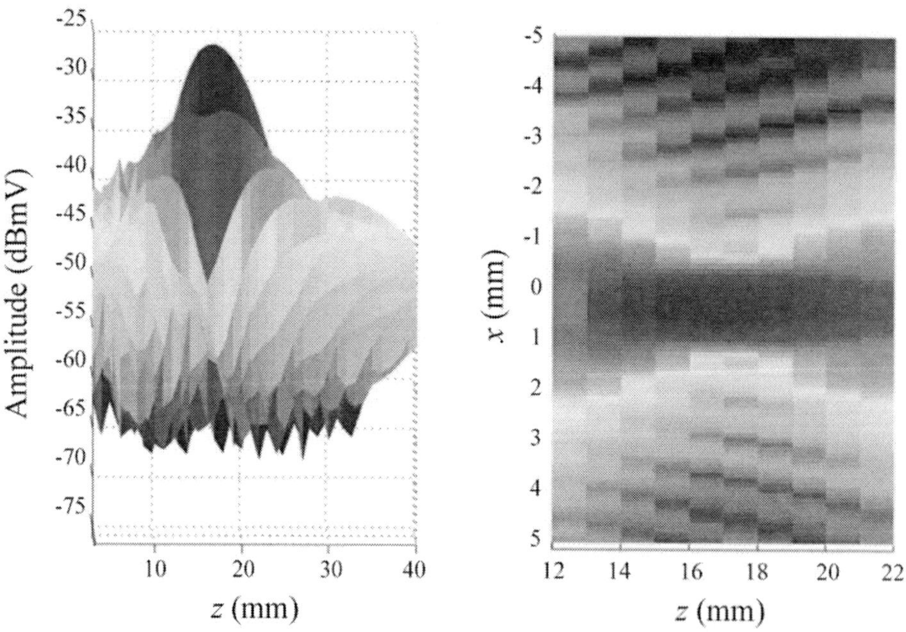

Figure 8. Acoustic field of a 2 MHz spherical focused transducer measured with a precise 3D positioning system. Step resolution for x-axis was 0.1016 mm and for z-axis was 1 mm. Focus distance from transducer face measured at 17 mm with a focus size of a FLHM intensity of 4 mm and a FWHM intensity of 0.6096 mm.

Heating systems have been developed for interstitial, intracavitary or external applications based on radiators features such as multiple transducer array, phased arrays, focused beams, mechanical positioning, dynamic frequency control and transducers of various shapes and sizes (Diederich and Hynynen, 1999). As seen in Figure 8, focused beams can be generated by means of spherically curved transducers, using acoustic lenses in conjunction with planar transducers (Driller and Lizzi, 1987), and with electronically phased arrays. Any of these focalizations allows concentrating the energy at a focus zone inside the organism without affecting outer tissues.

Focused beams can improve penetration depth from less than 1 cm to 10 cm or more by varying the transducer frequency and size. Phase arrays allow wider variety of shapes and permit to control the beam distribution. The resonant frequency of the transducer depends on material and shape (Driller and Lizzi, 1987). Focus size will depend on both beam focalization and transducer frequency and usually, focus geometry could show an elliptical shape (ter Haar and Coussios, 2007b). The lesion size will depend on focus, treatment time and tissue conductivity. For high temperature and short treatment time, lesion size will be close to focus size.

When using high-intensity focalization for high-temperature hyperthermia, energy concentration at focus provokes irreversible cell damage due to abrupt temperature increase at very short time exposure (ter Haar and Coussios, 2007a) which cause thermal coagulation of tissue (Diederich and Hynynen, 1999). Exposure time is usually short, less than 10 s, and focus intensity may vary from 14 W/cm^2 to 20,000 W/cm^2 in accordance with the application (ter Haar and Coussios, 2007b). Focused ultrasound for high-temperature hyperthermia was firstly thought for non-invasive brain surgery (Lynn and Putnam, 1944), but almost 90% of the ultrasonic beam is lost in the skull bone interface due to its high ultrasound attenuation and acoustic impedance (Wu and Nyborg, 2006). Nevertheless, a 500 phased array elements extracorporeal device was developed and tested in pig and monkey skulls to determine the maximum power possible for being used in humans to ablate deep tumors in the brain without removing the skull (Wu and Nyborg, 2006).

Focused ultrasound surgery or HIFU tissue ablation is a promising therapy for non-invasive oncological treatment by causing immediate cell death (ter Haar and Coussios, 2007b). The absorption of ultrasonic energy causes rapid temperature increase over 50°C and focalization minimizes surrounding tissue damage (Dubinsky et al., 2008). Tissue thermal coagulation is immediate and relatively independent of non-homogeneous media and tissue thermal parameters such as blood perfusion (Diederich and Hynynen, 1999; ter Haar and Coussios, 2007b). For the treatment of large volumes, interval in the order of 30 s to 60 s between each exposure may be required to avoid temperature buildup or accumulation of thermal dose in non intended regions (Diederich and Hynynen, 1999). Different configurations and devices (such as extracorporeal, interstitial and transrectal devices) are being used to treat breast cancer, uterine fibroids, liver cancer, pancreatic cancer, kidney cancer, benign prostatic hyperplasia (BPH) and prostatic cancer (Hwang and Crum, 2009; Illing and Chapman, 2007; Kennedy et al., 2003; ter Haar, 1999; ter Haar and Coussios, 2007a, b).

In oncological focused ultrasound, the most important effect to be considered is the thermal effect. However, Dubinsky et al. described the non-thermal effects that can be present during focused ultrasound therapy which can be divided in mechanical effects (such as cavitation, microstreaming and radiation forces), cell effects (apoptosis, lysis) and nonlinear propagation effects ("extra" heating) (Dubinsky et al., 2008). Cavitation occurs due to compression and expansion of tissue which causes bubble formation when gas is extracted from tissue during expansion. There are two types of cavitation: stable or non-inertial cavitation and inertial cavitation (as seen in section 5.6). When inertial cavitation occurs near a cell membrane, one may anticipate mechanical, rather than thermal, damage to cell. Microstreaming is a consequence of stable cavitation and can produce high shear forces close to the bubble that can disrupt cell membranes (ter Haar, 1999). Radiation forces are developed when a wave is either absorbed or reflected; these forces could be the cause of some mechanical effects in cells observed *in vitro* as membrane permeabilization (Hensley and Muthuswamy, 2002). Apoptosis is produced at low level of energy deposition, when nucleus of apoptotic cells self-destruct with rapid degradation of DNA by endonucleases (Dubinsky et al., 2008). Apoptosis and cell lysis were observed at low ultrasound intensities in combination with mild hyperthermia (produced by other source); this suggests that hyperthermia produced by ultrasound could be more effective than other hyperthermia modalities because of apoptosis and lysis induction (Feril and Kondo, 2004). If all the non-thermal effects are added, cell death from focused ultrasound exposure could cover a more extensive volume than original target volume due to this multiple combination effects (Feril and Kondo, 2004). Also, nonlinear propagation effects could be the responsible of other important changes in tissue. Nonlinear propagation produces the change of waveform (originally sinusoidal) due to the generation of higher harmonics that are rapidly absorbed by tissue and can vary the heating distribution (Wojcik, 1998). This means that higher temperature increases than the expected ones can be produced, which can be used to reduce the required acoustic intensities or the treatment duration.

6.2.1. Extracorporeal Techniques

Extracorporeal devices are used to treat targets within the abdomen, brain, breast and limbs (see Figure 9). These are commonly conformed by multi-element phased array arranged in such way to allow complex beams formation and penetration depth adjustment in order to produce well-localized thermal damage deep within body and to avoid affecting non-targeted surrounding tissue (Diederich and Hynynen, 1999). These arrays allow the formation of more complex beams in order to ensure power deposition delivery to the target. In addition to this, extracorporeal devices are electrically or mechanically positioned to spatially control the beam focus at higher scanning speeds with the aim of sequentially radiate the tumor at fixed equally spaced points; this ensures that the complete target has received the thermal dose (Hokland et al., 2006). The disadvantage of extracorporeal phase arrays is that these require more complex applicator systems using multi-channel amplifiers that allow phase and power control for the robust transducer arrays (Diederich and Hynynen, 1999).

Magnetic resonance (MR) or ultrasound (US) imaging are used with extracorporeal devices for treatment planning and heating monitoring (ter Haar and Coussios, 2007b). MRI guidance allows having anatomical views with good soft tissue contrast, and includes the ability to image temperature elevations and detect tissue changes after ultrasound radiation (Dubinsky et al., 2008; Wu and Nyborg, 2006). However, when using MRI-guide, the transducer arrays have to be MR compatible (transparent to magnetic fields), in order to eliminate possible magnetic field distortions like non-homogeneities and significant image artifacts (ter Haar and Coussios, 2007b; Wu and Nyborg, 2006) which increase the general system cost. MRI systems allow the calculation of thermal dose during a therapy to ensure cell irreversible damage cytotoxic levels on the anatomical region (ter Haar and Coussios, 2007b). This possibility has been used in experimentation for temperature mapping.

Ultrasound imaging is another technique used for planning the energy delivery and monitoring the treatment (Hwang and Crum, 2009; Vaezy et al., 2001). Ultrasound imaging can allow fast processing for real time monitoring and coagulated lesion identification (Anand and Kaczkowski, 2004; Pernot et al., 2004). Although lesion is better identified after coagulation, hyperechoes appear during HIFU exposures before coagulation which can be used for focus localization (Rabkin et al., 2006; ter Haar and Coussios, 2007b). *In vivo* experiments made by Rabkin et al. in pig and rabbits (Rabkin et al., 2006) showed that a bright hyperechoic region is observed in B-mode ultrasound during high-intensity focused ultrasound exposure due to a rapid increase of temperature (100°C) and presence of cavitation in the focal zone. This suggests that hyperecho detection can be used for guidance of focused ultrasound therapy when cavitation is a desired effect (Rabkin et al., 2006).

Due to the tissue heating induced by focused ultrasound, temperature control is needed to measure thermal dose delivered to the target in order to reach coagulative necrosis and avoid damage to surrounding tissue (ter Haar and Coussios, 2007a, b; Wu and Nyborg, 2006). Hence, it becomes essential to have a feedback regulation system, which would need, as a thermal reference, a procedure for measuring with accuracy the spatial distribution of temperature in the inner parts of tissue (Bazan et al., 2009).

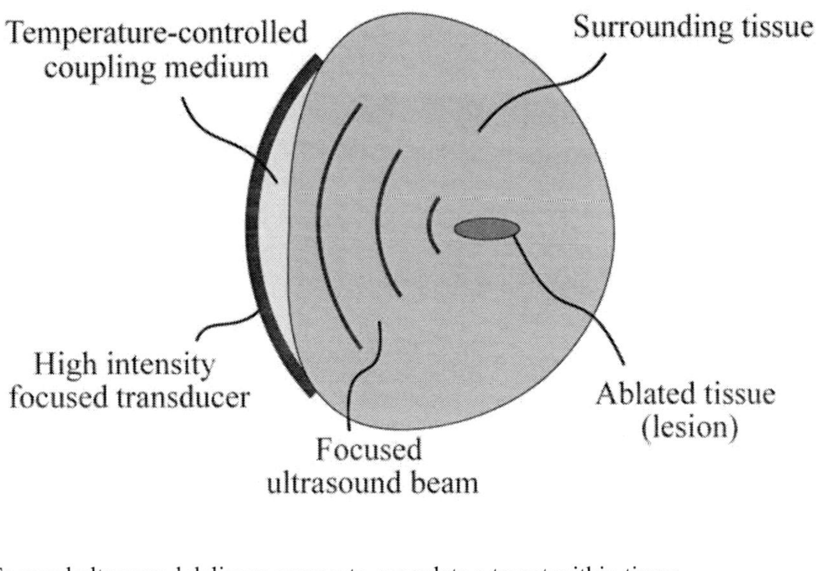

Figure 9. Focused ultrasound delivery energy to coagulate a target within tissue.

Current thermometry measurements are based on the temperature sensitivity of water proton resonance frequency (PRF) which provides good spatial resolution, linearity with temperature in the liquid phase, good approximation independent of tissue type and can be mapped rapidly using standard MR imaging sequences (Gorny et al., 2006; Hokland et al., 2006; Wu and Nyborg, 2006). However, patient respiration or accidental movement artifacts can limit the accuracy of MRI temperature maps (Hokland et al., 2006; Hwang and Crum, 2009). Gorny et al. mention, in a technical report, that in order to obtain adequate temperature levels during treatment, the physician may manually adjust therapy parameters such as ultrasonic beam focus size, acoustic output power, treatment duration and frequency of the ultrasonic beam (Gorny et al., 2006). Temperature estimation with ultrasound is being investigated based on the thermal dependence of some ultrasound characteristic parameters as: attenuation, acoustic velocity, etc. Bazan et al. presented a comparison of common approaches used to non-invasively estimate temperature with ultrasound in the hyperthermia range (Bazan et al., 2009).

During the therapy, applicator must be acoustically coupled to the patient skin in order to transmit efficiently the ultrasonic energy. This coupling can be made by using either ultrasound coupling gel, a patient drape, a pelvic gelatin based pad or a degassed water bolus (Gorny et al., 2006). The water bolus is the most used option because it permits water circulation to cool both the skin and transducer. This allows the use of higher acoustic intensities with less patient pain produced by high temperature at the skin. The cooled water circulating through the transducer should be degassed to avoid bubble formation that modify the acoustic intensity delivered to the patient (Hodnett and Zeqiri, 2004; Zeqiri et al., 2003).

6.2.2. Intracavitary and Interstitial Techniques

Intracavitary techniques are applied in zones where the target region is close to the walls of a natural body cavity. Carcinoma of the prostate is one example of a tumor accessible though a cavity: the rectus. This technique could also be applied in the treatment of vaginal tumors and others, but most of the developments have been made in prostatic tumor treatment. Ultrasound applicators are manufactured as transducer arrays in the shape of a cylinder with electrodes on its inner and outer surfaces (see Figure 10). Initially, these arrays were excited at same phase, but at different power to generate a desired radiating pattern. Last developments have used phase arrays applicators because of their advantage of controlling the focus position; with phase arrays, the energy can be distributed more uniformly and the penetration depth can be increased. As for other hyperthermia systems, a water bolus often couples the transducer and the cavity and controls the temperature inside at the cavity wall by circulating water. Transurethral applicators are commonly design in this way and during therapy they are inserted in a multilumen delivery system (of 5 cm to 6 cm outside diameter) reaching high temperatures (50°C to 80°C) in 10 min treatment (Diederich and Hynynen, 1999). It has been also proposed the use of intracavitary transducers to deliver focused ultrasound energy from the esophagus to the heart in order to treat cardiac arrhythmias (Wu and Nyborg, 2006) and diagnostic ultrasound imaging-guided extracorporeal systems, for the same purpose, (Diederich and Hynynen, 1999) through cardiac tissue ablation.

Among intracavitary techniques to deliver focused ultrasound energy, transrectal technique is being used in clinic for the treatment of benign prostatic hyperplasia (BPH) and prostate cancer (ter Haar and Coussios, 2007b). Transrectal applicators are commonly mechanically fixed focused transducer in conjunction with a B-mode transducer localized in cylindrical plastic probe. Diagnostic ultrasound imaging transducer may or may be not of the same frequency of the therapy transducer and are phase electronically and mechanically activated in order to apply focused ultrasound and monitor the process. Lesion formation occurs in few seconds and scanning and positioning probe control are required in order to ablate the entire target volume (Diederich and Hynynen, 1999; Lu et al., 2007; ter Haar and Coussios, 2007b).

Coupling between transducer and patient is made by some means. For transrectal devices, the use of a covering sheath like a condom is putted over the ultrasound probe. Then the probe is introduced into the rectum and the covering sheath is inflated and is filled with degassed water in order to couple between ultrasound probe and rectal wall (Lu et al., 2007).

Interstitial methods are used for treating tumors or sites that are difficult to reach by external or intracavitary methods. The heating source is implanted into the tumor to deliver the energy efficiently and avoid affect surrounding healthy tissues. The applicability of interstitial methods is wide, despite their invasive nature. There are two approaches for this method. The first one consists of using a wave-guide to deliver external ultrasound energy into the tumor from a planar transducer. The second approach, more used than the first, requires small cylindrical ultrasound sources as that shown in Figure 10.

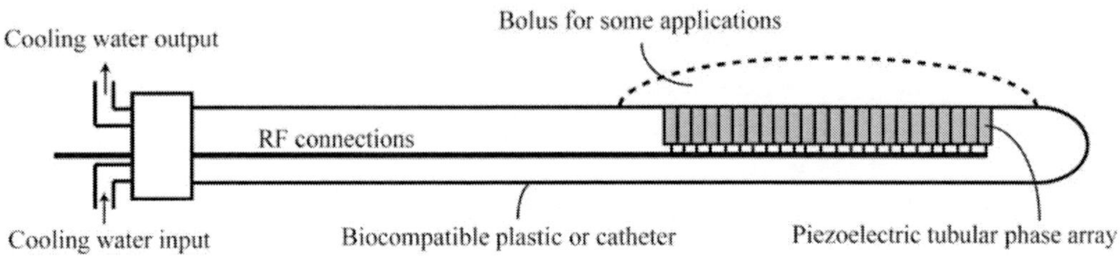

Figure 10. Piezoelectric tubular array for intracavitary and interstitial applications.

The configuration of interstitial devices is related to the application region. They could be water-cooled, air-cooled, no-cooled, and they could combine, in the same applicator, other therapies as electromagnetic hyperthermia or radiotherapy (Diederich, 1996). It was demonstrated that better performance can be gotten with cooled applicators than those no-cooled, because of the higher intensity possible on the former (Deardorff and Diederich, 1998). The Figure 10 shows a tubular arrays applicator mounted on plastic structure that facilitates support and placement when it is inserted (Vera et al., 2008); the array type and localization depends on the application. Although most of the applicators are composed by systems of water cooling, it is not general and some use air-cooling or even not cooling (Diederich and Hynynen, 1999). Thermocouples could be included in the interstitial applicator and combinations of therapies can be used, e.g. ultrasound hyperthermia with radiotherapy (Diederich, 1996).

6.3. Problems during Heating

There are many problems during ultrasonic hyperthermia that limit its applicability. Temperature increase produces, in tissues, nonlinear reactions to be considered in therapy planning. When the cells survive to heating, thermotolerance in those cells is induced; thermotolerance depends on the kind of tissue. This means that the cells can support posterior heating even for higher temperature than before. The magnitude of this effect is extremely large, e.g. the slope of a survival curve after thermotolerance may be decreased in a factor of 15 (Field and Bleehen, 1979). The same temperature for two cell cultures, one for the control and the second after a light heating without apparent effect, has produced same damage at 32 min (the control) and at 140 min (the second group). Thermotolerance could impede the posterior applications of hyperthermia for a determined tumor. Other effect that could be used in a beneficial way is the step down sensitization. When the temperature does not arrive to the threshold to produced thermotolerance, the cells often become more vulnerable to lower temperatures than before. Low temperature treatments could be more effective after a mild high temperature increase than those applied alone (Field and Bleehen, 1979).

Other purely thermodynamic difficulties are present during therapies (Diederich and Hynynen, 1999). Supposing that the acoustic field were homogenous, the heterogeneous blood flow would modify the thermal distribution considerably. This effect is highly appreciable during long time low temperature hyperthermia treatments, but it can be neglected in high temperature hyperthermia because of the shorter treatment times and the rapid temperature rises. When ultrasound energy is delivered in regions with large vessels, the fast thermal interchange around them reduces the temperature increments very fast (Arkin et al., 1994). The solution could be either to avoid those large vessels, to use higher acoustic intensities or even to physically occlude blood flow.

6.4. Excitation Systems for Hyperthermia Treatments

Ultrasonic hyperthermia treatments do not require huge equipments in order to generate fast temperature rises effectively. The general components of hyperthermia systems are shown in Figure 11 where the use of each part should depend on the design requirements for the therapy (Vera et al., 2008). For example, in portable systems, the control is made by a microcontroller by sacrificing the range of applicability; when the technique requires better control systems (3D positioner systems, multiple transducers, an anatomical 3D analysis, etc.), the control should be made using a computer. The signal in the block diagram in Figure 11 follows next direction:

1. The oscillation signal at the transducer resonance frequency is generated at high stability by the frequency synthesizer.
2. The low power oscillating signal is amplified by the RF amplifier that should be capable of not including, at the output, parasitic frequencies that reduce the overall efficiency.
3. If it is required by the application, the power measurement is carried out at the output of the amplifier.
4. Before the signal arrives to the transducer, some electrical matching networks are commonly included to match electrical impedances between amplifier and transducer and to transmit the electrical energy efficiently.
5. The temperature increase is measured by a thermometry system which is a feedback of the control system that participates in all the process described before, as can be seen in Figure 11.
6. The control system (or computer) receives feedbacks of power and temperature measurements and decides on what control action apply to each block in order to keep the efficiency as high as possible and maintain the temperature at the desired value. Some manufacturers of physiotherapy hyperthermia devices include in the control system algorithms to check the transducer resonance frequency at each specified time to better energy generation in long time treatments.

In physiotherapy hyperthermia, there are a wide variety of portable commercial devices with standard features that permit therapists to select the parameters manually and customize their own common therapies. Those devices generally do not have temperature feedback and thermal dose depends on patient pain threshold, therapist experience, among others.

Moreover, although the designs have most of the parts of Figure 11, the transducers are different (even for the same device model) and generate irregular acoustic fields that produce, by consequence, different therapeutic results (Gonzalez et al., 2002; Hand et al., 1989). Standardization organisms have made efforts to control those devices for medical applications (FDA, 2008; IEC-61689, 1998), but the studies should be deeper to increase the repeatability of results and reduce the variability between devices (Johns et al., 2007). Because of all this facts and some no mentioned here, this therapy is subjective, at this time, and then many works about its non-efficacy are in literature (Busse et al., 2002; Robertson and Baker, 2001; van der Windt et al., 1999).

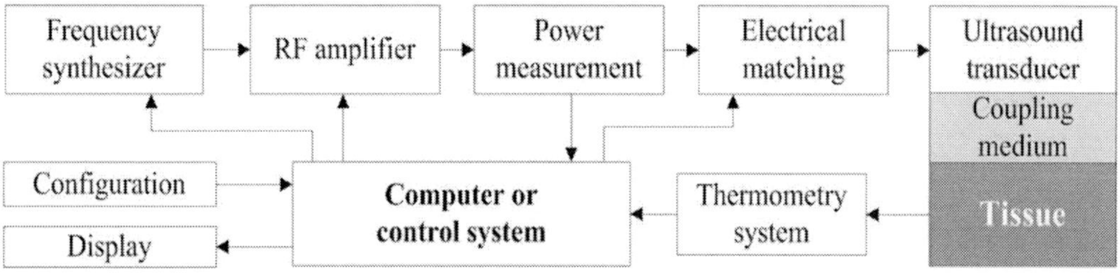

Figure 11. Block diagram of ultrasound hyperthermia systems. The general structure could change in accordace with a specific application (Vera et al., 2008).

The devices for excite ultrasound transducers for oncological hyperthermia differs from those for physiotherapy in the control system. Since the thermal dose should be effectively controlled to not producing tumor thermotolerance, the temperature feedback is greatly important (ter Haar, 1999). Also, the anatomical and tumor geometry should be considered in therapy planning and, if it is possible, during the therapy (Hand et al., 1989). Images from CT (Computer Tomography) and MRI (Magnetic Resonance Imaging) joint to 3D imaging design software have been used off-line for this purpose. Real time imaging and 3D real time thermography are desirable (or even required), to accomplish the therapy successfully. Actual systems of hyperthermia (for cancer of prostate) use off-line 3D therapy planning and mechanical 3D positioners to locate the transducer at each programmed position (Diederich and Hynynen, 1999). When the therapy is just mild hyperthermia, the temperature feedback is made using temperature sensitive MRI sequences in development (Diederich and Hynynen, 1999). During thermal surgery, hyperthermia systems do not have temperature feedback because the preferred approach is to ablate the tissue by increasing the temperature quickly (ter Haar, 1999); instead of that, the induced-lesion monitoring could be made by using ultrasonic imaging systems (Anand and Kaczkowski, 2004).

6.4.1. Equivalent Circuit of US Transducer

The ultrasound transducer transforms the electrical energy to mechanical energy and vice versa. Some models to represent this behavior have been created in order to predict the system response and find the configuration parameters for a better performance. Three main equivalent circuits for piezoelectric transducers are shown in Figure 12, in which the mechanical equations were transformed to electrical magnitudes by using the classical analogies: mechanical force to electrical potential and particle velocity to electrical current.

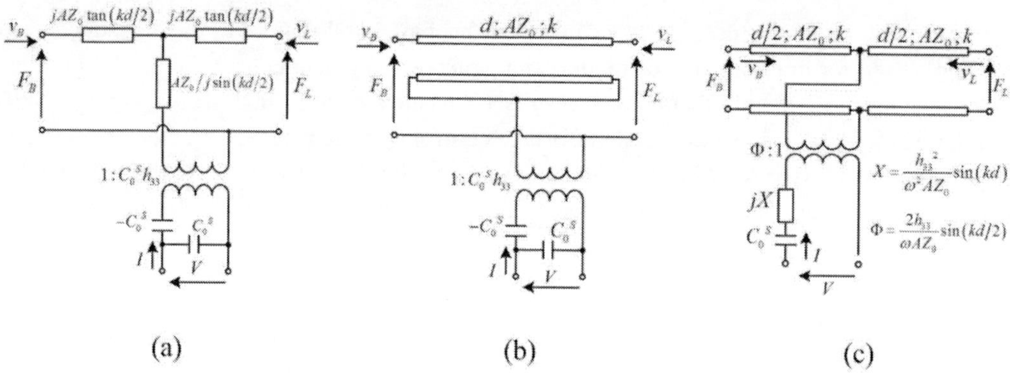

Figure 12. Equivalent circuits of thickness extensional piezoelectric transducers. (a) Mason circuit. (b) Redwood circuit. (c) KLM circuit.

In Figure 12, F_B and F_L are the forces at backing and load, respectively; v_B and v_L, the particle velocities at backing and load, respectively; A, the transducer surface; d, the transducer thickness; Z_0, the characteristic acoustic impedance of the piezoelectric material; h_{33}, the piezoelectric constant; k, the wavenumber or propagation constant of the piezoelectric material; C_0^S, the clamped (zero strain, high frequency) capacitance of the piezoelectric plate (San Emeterio and Ramos, 2008).

The first equivalent circuit was developed by W. P. Mason for a thickness extensional transducer Figure 12a. The electrical signal is connected to the mechanical part through a transformer with some capacitances at the primary (notice a negative capacitance). Redwood circuit in Figure 12b is similar to that of Mason but with the substitution of the impedance T-network of by a coaxial transmission line; the electrical signal is connected to the coaxial shield and not to the central wire. The KLM circuit (from Krimholtz, Leedom, and Matthaei) is derived from the others (see Figure 12c), where the transmission line represents the wave propagation in the piezoelectric material and the electrical part requires a frequency dependent transformer. The three circuits are equivalent, but their differences in the representation make them easier to be interpreted in determined situations: e.g. in Mason and Redwood circuits, the clamped capacitance provides a good representation of the electrical port of a transducer, while the KLM represents better the mechanical port because of the direct analogy of the transmission line.

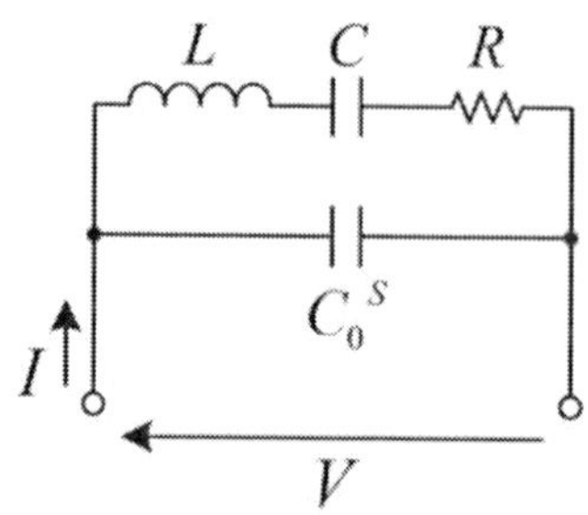

Figure 13. Simplified equivalent circuit for piezoelectric ultrasonic transducers for narrow band applications.

In some applications, when the transducer is excited exactly at the resonance frequency, its response could be approximated by the simplified equivalent circuit shown in Figure 13 which is derived from the Mason circuit. This circuit is only useful in thickness extensional mode at continuous excitation, and it does not represent the electrical response outside the resonance frequency. The mechanical port was simplified to just the values of the components *R*, *L* and *C*.

6.4.2. Frequency Synthesizer

The ultrasound piezoelectric transducers mainly are operated in a resonant frequency (thickness extensional vibration mode) clearly identifiable with impedance analyzers. This is the frequency in which the electrical energy is more efficiently converted to mechanical energy, and it is here where the ultrasound transducer should be operated. The frequency synthesizer is the part of the generator that permits to adjust the output frequency at the transducer resonant frequency. It is not just an oscillator, but a synthesizer, because it permits a range of adjustment with a desired resolution. In Figure 14, it is shown the general block diagram of a digital frequency synthesizer in which the frequency operation range and resolution depend on the PLL selection and the limits of the frequency dividers. The equation for the output frequency is given by

$$f_0 = f_{oscillator} \cdot \frac{M}{N}, \tag{23}$$

where $f_{oscillator}$ is a stable and constant frequency (in Figure 14 of 2 MHz), M and N should be determined in accordance with the system requirements. Practically, the step size for frequency adjustment is given by

$$\text{step size} = \frac{f_{oscillator}}{N}. \tag{24}$$

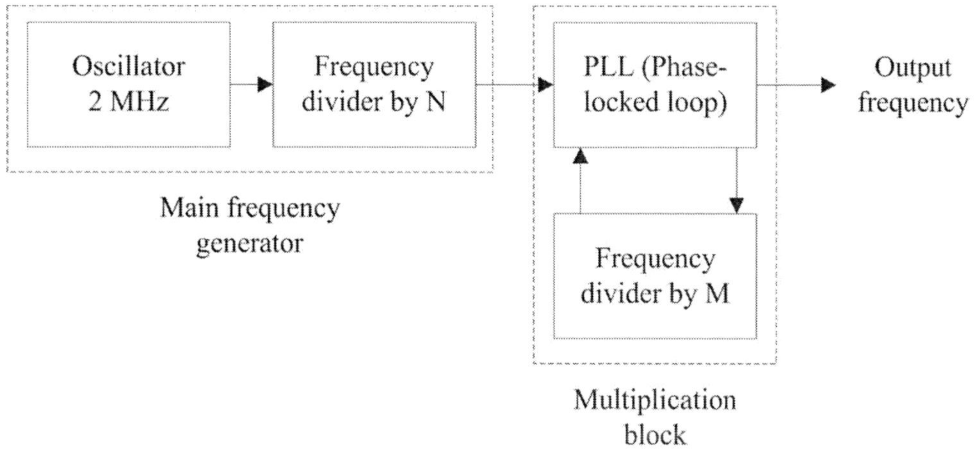

Figure 14. Block diagram of a classical frequency synthesizer.

According to the design specification and the final application, the block diagram shown above may require more stages, i.e. digital to analog converters, current to voltage converters (or vice versa), computer interfaces, etc. It could be necessary an impedance matching network at the frequency synthesizer output in order to guarantee maximum signal transference between the synthesizer and the radiofrequency amplifier.

6.4.3. RF Amplifier

After the signal is generated at the specified frequency, this must be amplified to supply the transducer. The designed requirements for the amplifier are simple, but hard to accomplish: high efficiency and low signal distortion. Many basic circuits (Albulet, 2001) can be used as linear amplifiers, but those could not provide good performance when they are used with ultrasound transducers. The theoretical efficiency of linear amplifiers cannot be bigger than 78.5% for class B amplifier (practically much lower) and the power losses are principally due to the heat produced at the main active device (transistors, FETs or MOSFETs). As seen before, transducer equivalent circuit is composed by frequency dependent parts that complicate the amplifier design and the matching networks. Common linear amplifiers (class A, B, AB according to (Albulet, 2001)) would not work appropriately because of the complex loading and their efficiency (already low for class A, B and AB) could be even lower. The reason for this is that linear amplifier output impedance is relatively high and it should be matched to the transducer impedance, which is much lower and variable (dependent on the medium impedance, temperature, frequency, reflected wave, etc); dynamic matching could be a solution for this problem, or using non-linear amplifier designs.

Non-linear amplifiers are those that the output signal amplitude is not a linear function of the input amplitude; moreover, in some types, output is a function of other parameters of the input signal, e.g. the triggered angle in class C amplifiers. Switched amplifiers, non-linear too, use the low power input signal to drive a switched element (transistor, MOSFET, etc.) into its up and down conduction limits (cut-off and saturation regions for transistors) in order to keep the power consumption in zero and reach an efficiency of 100%. Theoretically, when the element is cut-off (as an open switch), its current is zero and its voltage is that of the power supply, and when the element is saturated (as a closed switch) its current is maximum and its voltage is zero; the power consumption ($P = VI$) of the element would be zero if the transition times between cut-off and saturation conditions were zero. Because these times cannot be zero, the real efficiency of switched amplifiers is not 100% but it is close if fast switching elements are used. The output amplitude is that of the power supply but the output wave form is not a sinusoidal, but a square. Posterior filters are required in order to eliminate undesirable frequencies, but this filtering reduces the overall efficiency. Amplifiers class D, E and F work in switched mode (Albulet, 2001), but the class F amplifier owns a high efficiency even after filtering, and this is the most used in the design of commercial general purpose amplifiers (Albulet, 2001). Physiotherapy ultrasound commercial devices generally use simplified class D amplifiers with light filters at the output (Barsotti and Hance, 1988; Houghton and Obray, 1990).

6.4.4. Electrical Matching Amplifier-Transducer

This part is really important during the design of a system for power generation. A mismatch between amplifier and load could result in low efficiency, and even damages of the amplifier. The electrical signal coming from the amplifier should arrive entirely to the transducer and then be converted into mechanical energy. When the electrical impedances are not matched, a part of the signal goes back to the amplifier and increases the *SWR*. In some devices, there are protection systems to detect mismatch with the load and to protect the amplifier when the *SWR* is too large. However, this problem should be avoided (or reduced) by using electrical matching networks between amplifier and transducer in accordance with the electrical impedance of this latter. Conventionalisms establish values of "standard" impedances of transmission lines and commercial devices (amplifiers, generators, network analyzers, etc.) have already implemented those values at their inputs and outputs. Common coaxial wires are designed to have characteristic impedances of 50 Ω or 75 Ω, and when they are used, careful should be taken during amplifier design and adjustments of the output impedance should be made. Electrical signal coefficient reflection (Γ) is given by

$$\Gamma = \frac{Z_1 - Z_0}{Z_1 + Z_0}, \tag{25}$$

where Z_1 is the transducer impedance and Z_0 is the wire characteristic impedance. Matching techniques could be applied to get at the load same impedance of the wire and avoid signal reflections.

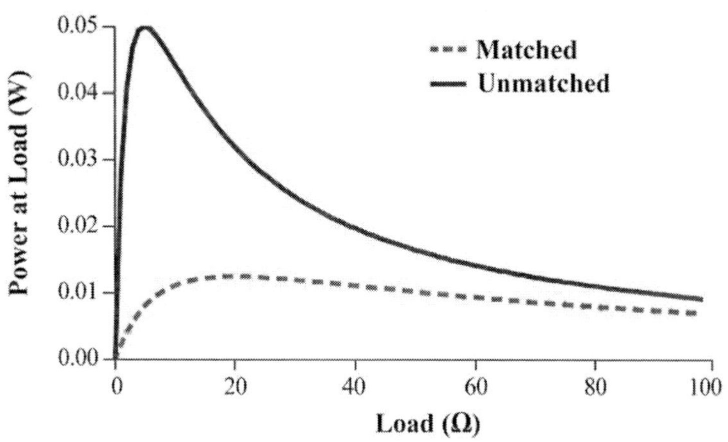

Figure 15. Comparative of power at load for two different amplifier output impedances. Dotted line, power at load with the amplifier output impedance matched with a load of 20 Ω; continue line, power at load with the amplifier output impedance unmatched but small (5 Ω). The power at load is better for the second case even at 20 Ω.

Matching techniques could be also applied to the amplifier output impedance which affects the energy delivered to the load. When the output impedance is equals to the load impedance, the amplifier is delivering the maximum power possible but with 50% of efficiency; on the other hand, when the load impedance is much larger than the output impedance, the efficiency is close to 100% (but it is not the maximum power). In Figure 15, it is shown the power at the load for two amplifier output impedances in which the maximum of both situations is clear; when the output impedance is small (in Figure 15 continue line, 5 Ω), the total power at the load increase more than the value at maximum power using the selective matching (in Figure 15 dotted line, the amplifier output impedance was matched with the load). The maximum with matching at 20 Ω is lower than the power for the same load with output resistance unmatched. However, matching output impedances is meaningful in complex impedance situations, and the election of matching the amplifier output should depend on system requirements. This should not be just a "rule".

6.4.5. Power Meter and Coupling Detector

In order to warrant a stable power emission, the system could require a power meter. In physiotherapy devices, power meter can be used to detect the contact between transducer and patient and to avoid the emission of ultrasonic energy to air. The electronic measurement of the emitted power (mechanical energy) is not really possible directly, but it could be inferred by using some assumptions and with an adequate calibration. Ideally, if the transducer is driven at its resonant frequency, all the electrical energy is converted to mechanical energy without losses. In practice, the emitted energy is a fraction of the electrical energy and the rest is converted into heat either in the transducer or in the amplifier due to the reflected acoustical and electrical wave, respectively. The efficiency of a transducer as high power emitted depends on the construction. Wideband transducers (commonly with backing) do not convert the electrical excitation efficiently to mechanical vibration. On the other hand, narrowband transducers (often not-backed) present efficiencies near to 90% in energy conversion.

Once the transducer emission is characterized, together with electrical signal excitation, the power measurement can be carried out. The electrical output power can be calculated by

$$P = v(t) \cdot i(t) = VI\cos(\phi), \tag{26}$$

where P is the electrical power, $v(t)$ is the instantaneous output voltage at the transducer, $i(t)$ is the instantaneous output current through the transducer, V and I are the amplitude of the output voltage and current, respectively, and ϕ is the phase angle in radians between the voltage and current signals. This equation permits the calculation of both instantaneous and stable power in just few steps and the approach is simple if a central processing unit (a microcontroller for example) is employed. The measurements of transducer voltage and current can be obtained with few loops transformers in parallel or serial as required; the quantity of loops depends on the transformer core and the signal frequency (Barsotti and Hance, 1988). One or two loops transformers could be enough to measure the output current at the transducer, without affecting significantly the waveform. Parallel resistors can be used to measure the output voltage (Sakurai and Gotanda, 1992), but self produced high frequency noise can affect the measurement. The phase angle can be calculated using an XOR operation between voltage and current saturated signals before an integrator circuit to transform the squared pulses in continuous voltage (Barsotti and Hance, 1988).

Coupling detections for physiotherapeutic applications are very important to protect the device, but mainly to protect the patient. Because of the different acoustic impedances of the transducer and air, this emission has almost zero efficiency. Under this condition, most acoustic radiation goes back to the transducer which produces heating that could injure the patient (Kollmann et al., 2005). In order to avoid emission to air, physiotherapy devices should (but not must) include a coupling detection system to reduce the emitted energy when the transducer is not coupled. This detection is carried out in some devices by using the emitted power, the phase between current and voltage, or even the spectral analysis of the measured output signal.

6.4.6. Cavitation Detectors

Cavitating bubbles behave as secondary sources of sound, emitting acoustic radiation whose spectral characteristics provide detailed information on the dynamics of the cavitation process (Zeqiri et al., 2003). Due to this characteristic, cavitation detectors have been designed in order to measure the acoustic signals to detect frequency components generated by cavitating bubbles which provide an indication of when the measured therapy power level exceeds a pre-defined threshold (Hodnett and Zeqiri, 2004).

Zeqiri et al. designed a sensor for monitoring acoustic cavitation (Zeqiri et al., 2003) which ideally should:

- be as non-perturbing as possible to acoustic field under test.
- exhibit a wide acoustic measurement bandwidth, in order to monitor bubble acoustic energy during motion and collapse.
- posses a degree of spatial resolution to determine the volume region associated to the bubble collapse.

The sensor was a cylinder of 38 mm diameter and 32 mm height with a hollow in the middle with a diameter of 30 mm. It was constructed using materials that allow shielding effect in the external surface (no adherence of bubbles) and acoustical impedance optimization by using absorber media. The cylindrical internal surface was made of a piezoelectric film that detects the acoustic signals generated from bubbles in a wide bandwidth centered at 10 MHz (Zeqiri et al., 2003).

6.4.7. Control System

This is a very important part of hyperthermia systems. Every algorithm to deliver efficiently the energy, to measure variables, as temperature and power, to position the transducer as required, and to process the data on-line and off-line should be managed for a central system of control. In accordance to the application, this part could be a microcontroller, e.g. physiotherapy hyperthermia, or a computer for most oncological applications. In therapies with mechanically positioned applicators, the control system manages the measurements, the intensities, some specific algorithms (as pain feedback for example), and even the scan-raster to cover a determined tumor volume. It is desirable to include in the control therapy guidance and temperature mapping using imaging with MRI and US, but this depends on the technological capabilities.

7. Characteristics of Commercial Devices

Hyperthermia has been used in clinic for cancer therapy. Reports about successful application of hyperthermia by itself or in combination with other therapies have been published (Diederich and Hynynen, 1999). However, before commercializing a device, it should pass the phases of the clinical trials (e.g. FDA clinical trials) in the respective country.

These trials require usually some years, and because of this, most of the hyperthermia possibilities are not commercial and widely applied yet. That is not the same for the physiotherapy hyperthermia devices, which have been in market for at least 50 years with few changes in their features (Barsotti and Hance, 1988; Güttner and Pichel, 1959; Houghton and Obray, 1990).

7.1. General Characteristics of Physiotherapy Ultrasound Devices

An extensive number of patents have been created since the first use of ultrasound in physiotherapy. In general, the equipment parts have been the same, but with just some modifications in the control systems (Barsotti and Hance, 1988; Güttner and Pichel, 1959). Transducers have been improved gradually and the efficiency in energy conversion (electric to mechanical) has been increased (Houghton and Obray, 1990). Actual designs include high performance active devices for high frequency signal generation, but the logical operation of this new circuit with previous used has not been noticeable changed (Güttner and Pichel, 1959; Hance and Barsotti, 1983). Also, novel devices include in their features algorithms of transducer resonance frequency determination for high efficiency transducer driving and load detection to avoid emission to the air (Houghton and Obray, 1990; Sakurai and Gotanda, 1992).

The general "desired" features of ultrasonic physiotherapy commercial devices are:

- Maximum acoustic intensity of 3 W/cm^2.
- Applicators frequency of 1 MHz or 3 MHz (same or different applicator).
- Selectable applicators with different *ERA* (effective radiating area) to treat different body zones.
- Beam non-uniformity ratio (*BNR*) of applicator of 1:6.
- Watertight applicator for underwater therapies.
- Two selectable therapy modes: continuous, to potentiate thermal effect, and pulsed with duty cycle selection (in %), to use only non-thermal effects.
- Selectable treatment time.
- Coupling detection between transducer and patient skin.
- Alarm system to indicate errors in coupling, end of therapy, etc.
- Transducer temperature detection (for non-efficient transducers).

7.2. General Characteristics of Oncological Hyperthermia Ultrasound Devices

Most oncological hyperthermia ultrasound devices are used together with radiotherapy and chemotherapy. Their focal intensity may reach 2,000 W/cm^2 (Diederich and Hynynen, 1999). Even when therapy is applied with different devices, it is very important to monitor the therapy for target detection, planning and guiding. Techniques usually are magnetic resonance (MR) and ultrasonic (US) imaging. Nowadays oncological hyperthermia devices used in clinic are classified in accordance to the transducer or array of transducers: extracorporeal and transrectal.

Extracorporeal commercial devices utilize composite arrays of 208 to 251 elements to be used in the treatment of uterine fibroids, liver tumors, renal tumors, bone metastasis, breast cancer and pancreatic cancer (Hwang and Crum, 2009). Transducer central frequency varies from 0.9 MHz to 1.3 MHz and the ultrasound penetration depth (focus localization) can be adjusted from 1 cm to 16.9 cm. The devices are MR compatible and some include an US imaging system for guidance. Degassed circulating-water acts like coupling media and regulates the temperature between the transducer array and the patient (Diederich and Hynynen, 1999). Transrectal commercial devices are US imaging guided and have both imaging and therapy transducer in the same probe. Imaging transducer central frequency is between 4 MHz and 7.5 MHz (Illing and Chapman, 2007) and therapy transducer central frequency is between 3 MHz and 4 MHz (Illing and Chapman, 2007). Devices can have manual power adjustment or predefined power levels. The ultrasonic penetration depth (focus localization) can be adjusted from 3 cm to 4 cm.

Every single device that is ready for clinical use has to be evaluated according to national or international standards before releasing it into market. Some of these regulation organisms are the European Community (CE), Food and Drug Administration (FDA), State Food and Drug Administration (SFDA), etc. Harris et al. made a review of the regulations required for high-intensity focused ultrasound devices (Harris, 2009). The apparatus are classified in three classes depending on the control necessary to assure the safety and the effectiveness of the device. In Class I, the low risk devices are evaluated; in Class II, the greater risk devices than Class I; and Class III, the high risk devices or the devices using new technology. According to the above classification, focused ultrasound surgery and high-intensity focused ultrasound devices are located in Class II or Class III in accordance to the penetration depth. Standardization is indispensable to guarantee that acoustic power level, beam/focusing characteristics, temperature distribution and low risk of damaging surrounding tissue are all adequate to clinical usage. However, methods to measure these parameters are not yet standardized (Shaw and Ter Haar, 2006).

Conclusion

Ultrasonic hyperthermia is a promising therapy that has demonstrated acceptable results in laboratory experiments. Although there are already some oncological commercial devices, most of the possible applications are yet in research. The technological requirements for ultrasound generation are not as expensive as other hyperthermia modalities, e.g. microwave hyperthermia, and the range of possibilities (multiple shape transducers, arrays, electronic focusing, etc.) are evidently bigger.

This fact together with *in vitro* evidence of non-thermal ultrasound effects on cell killing makes this therapy hopeful for tumor destruction in the next years. Other advantage of oncological ultrasonic hyperthermia is that it showed to be useful in cancer therapy in combination with other non-thermal therapies as radiotherapy and chemotherapy. Ultrasonic hyperthermia could be used then in selective tumor sensitization to radiotherapy, chemotherapy and even to electromagnetic radiation (Larkin et al., 2005)

In relation to the physiotherapeutic hyperthermia, the research in this area has continued at small steps in the last years. Although this modality is widely applied currently in clinics around the word, the therapeutic effects are not completely clear yet and the temperature increments are not adequately established. Improvements in the machine control, transducer radiation and acoustic field modeling are required to determine the security ranges and to produce appropriated standards. Work in creating therapeutic protocols in accordance with published results for each muscular suffering is yet needed in order to reduce subjectivities during therapy application.

REFERENCES

Albulet M (2001) *RF Power Amplifiers*, Noble Publishing Corporation, Atlanta, GA.

Anand A and Kaczkowski PJ (2004) Monitoring formation of high intensity focused ultrasound (HIFU) induced lesions using backscattered ultrasound. *Acoustics Research Letters Online*-Arlo 5:88-94.

Arkin H, Xu LX and Holmes KR (1994) Recent Developments in Modeling Heat-Transfer in Blood-Perfused Tissues. *IEEE Transactions on Biomedical Engineering* 41:97-107.

Arora D, Cooley D, Perry T, Skliar M and Roemer RB (2005) Direct thermal dose control of constrained focused ultrasound treatments: phantom and in vivo evaluation. *Physics in Medicine and Biology* 50:1919-1935.

Baker KG, Robertson VJ and Duck FA (2001) A review of therapeutic ultrasound: Biophysical effects. *Physical Therapy* 81:1351-1358.

Balanis CA (1989) *Advanced Engineering Electromagnetics*, John Wiley and Sons, Inc., USA.

Barsotti EJ and Hance RD, inventor; Dynawave Corporation, assignee. "Ultrasound therapy device," United States Patent, US 4791915, Dec 20 1988.

Bazán-Trujillo I, "Propuesta de un método para la caracterización de equipos de terapia ultrasónica mediante técnicas de medida de temperatura y de presión acústica," *Master Thesis in Ingeniería Eléctrica*, sección Bioelectrónica, Cinvestav, México, D.F. 2005.

Bazan I, Vazquez M, Ramos A, Vera A and Leija L (2009) A performance analysis of echographic ultrasonic techniques for non-invasive temperature estimation in hyperthermia range using phantoms with scatterers. *Ultrasonics* 49:358-376.

Busse JW, Bhandari M, Kulkarni AV and Tunks E (2002) The effect of low-intensity pulsed ultrasound therapy on time to fracture healing: a meta-analysis. *Canadian Medical Association Journal* 166:437-441.

Carr JJ (2001) *Practical Antenna Handbook*, McGraw-Hill.

Castel JC, inventor; PTI, Inc. (Topeka, KS), assignee. "Ultrasound therapy system with automatic dose control," *United States Patent*, US 5413550, May 9 1995.

Church CC (2007) A Proposal to Clarify the Relationship Between the Thermal Index and the Corresponding Risk to the Patient. *Ultrasound in Medicine and Biology* 33:1489-1494.

Deardorff DL and Diederich CJ (1998) Interstitial ultrasound applicators with internal cooling for controlled high temperature thermal therapy. In 1998 *IEEE Ultrasonics Symposium Proceedings IEEE*, Sendai, Japan.

Demmink JH, Helders PJM, Hobaek H and Enwemeka C (2003) The variation of heating depth with therapeutic ultrasound frequency in physiotherapy. *Ultrasound in Medicine and Biology* 29:113-118.

Diederich CJ (1996) Ultrasound applicators with integrated catheter-cooling for interstitial hyperthermia: Theory and preliminary experiments. *International Journal of Hyperthermia* 12:279-297.

Diederich CJ and Hynynen K (1999) Ultrasound technology for hyperthermia. *Ultrasound in Medicine and Biology* 25:871-887.

Draper DO, Castel JC and Castel D (1995a) Rate of temperature increase in human muscle during 1 MHz and 3 MHz continuous ultrasound. *J. Orthop Sports Phys. Ther* 22:142-150.

Draper DO, Schulthies S, Sorvisto P and Hautala AM (1995b) Temperature changes in deep muscles of humans during ice and ultrasound therapies: an in vivo study. *J. Orthop. Sports Phys. Ther.* 21:153-157.

Draper DO, Sunderland S, Kirkendall DT and Ricard M (1993) A comparison of temperature rise in human calf muscles following applications of underwater and topical gel ultrasound. *J. Orthop. Sports Phys. Ther.* 17:247-251.

Driller J and Lizzi FL (1987) Therapeutic Applications of Ultrasound - a Review. *IEEE Engineering in Medicine and Biology Magazine* 6:33-40.

Dubinsky TJ, Cuevas C, Dighe MK, Kolokythas O and Hwang JH (2008) High-intensity focused ultrasound: Current potential and oncological applications. *American Journal of Roentgenology* 190:191-199.

FDA (2008) Performance Standards for Sonic, Infrasonic, and Ultrasonic Radiation-Emitting Products International Electrotechnical Commission, United States of America.

Feril LB and Kondo T (2004) Biological effects of low intensity ultrasound: the mechanism involved, and its implications on therapy and on biosafety of ultrasound. *Journal of Radiation Research* 45:479-489.

Field SB and Bleehen NM (1979) Hyperthermia in the Treatment of Cancer. *Cancer Treatment Reviews* 6:63-94.

Gómez-Flores W, "Método para la caracterización de sistemas de terapia ultrasónica mediante una técnica de termografía basada en Cristales Líquidos Termocromáticos," Master Thesis in Ingeniería Eléctrica, sección Bioelectrónica, Cinvestav, México, D.F. 2006.

Gonzalez G, Azbaid A, Leija L, Ramos A, Rami X, San Emeterio JL and Moreno E (2002) Experimental Evaluation of Some Narrow-Band Ultrasonic Transducers as Therapy Applicators. In 3rd European Congress on Acoustics Forum, *Forum Acusticum* 2002, Sevilla, Spain.

Gorny KR, Hangiandreou NJ, Hesley GK, Gostout BS, McGee KP and Felmlee JP (2006) MR guided focused ultrasound: technical acceptance measures for a clinical system. *Physics in Medicine and Biology* 51:3155.

Gutierrez MI, Vera A and Leija L (2010) Methods for Characterization of Physiotherapy Ultrasonic Transducers. In *New Developments in Biomedical Engineering,* D Campolo, ed, pp 517-542, In-Teh, Croacia.

Güttner W and Pichel G, inventor; Siemens-Reiniger-Werke A.G., assignee. "Ultrasonic Treatment Apparatus," *United States Patent US* 2876764, March 10 1959.

Hance RD and Barsotti EJ, inventor; DynaWave Corporation, assignee. "Ultrasound therapy device," *United States Patent, US* 4368410, Jan 1 1983.

Hand JW, Lagendijk JJW, Andersen JB and Bolomey JC (1989) Quality Assurance Guidelines for ESHO Protocols. *International Journal of Hyperthermia* 5:421-428.

Harris GR (2009) FDA regulation of clinical high intensity focused ultrasound (HIFU) devices. *Conf. Proc. IEEE Eng. Med. Biol. Soc.* 1:145-148.

Hekkenberg RT (1998) Characterising ultrasonic physiotherapy systems by performance and safety now internationally agreed. *Ultrasonics* 36:713-720.

Hekkenberg RT, Beissner K, Zeqiri B, Bezemer RA and Hodnett M (2001) Validated ultrasonic power measurements up to 20 W. *Ultrasound in Medicine and Biology* 27:427-438.

Hekkenberg RT, Richards A, Beissner K, Zeqiri B, Prout G, Cantrall C, Bezemer RA, Koch C and Hodnett M (2004) Development of transfer standard devices for ensuring the accurate calibration of ultrasonic physical therapy machines in clinical use. *Journal of Physics: Conference Series* 2:90-95.

Hensley AB and Muthuswamy J (2002) "Ultrasound induced permeabilization of cell membranes as a therapy for cytotoxic neuronal edema," in Conference Proceedings Second Joint EMBS BMES Conference 2002 24th Annual International Conference of the Engineering in Medicine and Biology Society Annual Fall Meeting of the Biomedical Engineering Society Houston, TX, USA 23 26 Oct 2002.

Hill CR, Bamber JC and Haar GRt (2004) *Physical Principles of Medical Ultrasonics*, John Wiley and Sons, Ltd, Great Britain.

Hodnett M and Zeqiri B (2004) A detector for monitoring the onset of cavitation during therapy-level measurements of ultrasonic power. *Journal of Physics: Conference Series* 2:103-108.

Hokland SL, Pedersen M, Salomir R, Quesson B, Stodkilde-Jorgensen H and Moonen CTW (2006) MRI-guided focused ultrasound: methodology and applications. *Medical Imaging, IEEE Transactions on* 25:723-731.

Houghton RB and Obray DC, inventor; Mettler Electronics Corp., assignee. "Ultrasound power generating system with sampled-data frequency control," *United States Patent, US* 4966131, Oct 30 1990.

Hwang JH and Crum LA (2009) Current status of clinical high-intensity focused ultrasound. *Conf. Proc. IEEE Eng. Med. Biol. Soc.* 1:130-133.

IEC-61689 (1998) Ultrasonics - Physiotherapy systems - Performance requirements and methods of measurement in the frequency range 0,5 MHz to 5 MHz. International Electrotechnical Commission.

Illing R and Chapman A (2007) The clinical applications of high intensity focused ultrasound in the prostate. *International Journal of Hyperthermia* 23:183-191.

Johns LD (2002) Nonthermal effects of therapeutic ultrasound: The frequency resonance hypothesis. *Journal of Athletic Training* 37:293-299.

Johns LD, Demchak TJ, Straub SJ and Howard SM (2007) The role of quantitative schlieren assessment of physiotherapy ultrasound fields in describing variations between tissue heating rates of different transducers. *Ultrasound in Medicine and Biology* 33:1911-1917.

Kennedy JE, Ter Haar GR and Cranston D (2003) High intensity focused ultrasound: surgery of the future? *The British journal of radiology* 76:590-599.

Khokhlova VA, Bailey MR, Reed JA, Cunitz BW, Kaczkowski PJ and Crum LA (2006) Effects of nonlinear propagation, cavitation, and boiling in lesion formation by high intensity focused ultrasound in a gel phantom. *Journal of the Acoustical Society of America* 119:1834-1848.

Kollmann C, Vacariu G, Schuhfried O, Fialka-Moser V and Bergmann H (2005) Variations in the output power and surface heating effects of transducers in therapeutic ultrasound. *Archives of Physical Medicine and Rehabilitation* 86:1318-1324.

Larkin J, Soden D, Collins C, Tangney M, Preston JM, Russell LJ, McHale AP, Dunne C and O'Sullivan GC (2005) Combined electric field and ultrasound therapy as a novel anti-tumour treatment. *European Journal of Cancer* 41:1339-1348.

López-Haro SA, Leija L, Favari L and Vera A (2010) Measurement of Ultrasonic Properties into Biological Tissues in the Hyperthermia Temperature Range. *Physics Procedia* 3:551-558.

Lu J, Hu W and Wang W (2007) Sonablate-500 transrectal high-intensity focused ultrasound (HIFU) for benign prostatic hyperplasia patients. Journal of Huazhong University of Science and Technology Medical sciences = Hua zhong ke ji da xue xue bao Yi xue Ying De wen ban = Huazhong keji daxue xuebao 27:671-674.

Lynn JG and Putnam TJ (1944) Histology of Cerebral Lesions Produced by Focused Ultrasound. *American Journal of Pathology* 20:637-649.

Lynn JG, Zwemer RL, Chick AJ and Miller AE (1942) A New Method for the Generation and Use of Focused Ultrasound in Experimental Biology. *Journal of General Physiology* 26:179-193.

Marchioni C, Riccardi E, Spinelli S, Dell'Unto F, Grimaldi P, Bedini A, Giliberti C, Giuliani L, Palomba R and Castellano AC (2009) Structural changes induced in proteins by therapeutic ultrasounds. *Ultrasonics* 49:569-576.

Merrick MA, Bernard KD, Devor ST and Williams MJ (2003) Identical 3-MHz ultrasound treatments with different devices produce different intramuscular temperatures. *J. Orthop. Sports Phys. Ther.* 33:379-385.

Moros EG, Dutton AW, Roemer RB, Burton M and Hynynen K (1993) Experimental evaluation of two simple thermal models using hyperthermia in muscle in vivo. *International Journal of Hyperthermia* 9:581-598.

Nyborg WL (1981) Heat-Generation by Ultrasound in a Relaxing Medium. *Journal of the Acoustical Society of America* 70:310-312.

Nyborg WL (2001) Biological effects of ultrasound: Development of safety guidelines. Part II: General review. *Ultrasound in Medicine and Biology* 27:301-333.

O'Neil HT (1949) Theory of Focusing Radiators. *Journal of the Acoustical Society of America* 21:516-526.

Paliwal S and Mitragotri S (2008) Therapeutic opportunities in biological responses of ultrasound. *Ultrasonics* 48:271-278.

Pennes HH (1948) Analysis of Tissue and Arterial Blood Temperatures in the Resting Human Forearm. *Journal of Applied Physiology* 1:93-122.

Pernot M, Tanter M, Bercoff J, Waters KR and Fink M (2004) Temperature estimation using ultrasonic spatial compound imaging. *IEEE Transactions on Ultrasonics Ferroelectrics and Frequency Control* 51:606-615.

Pye SD and Milford C (1994) The performance of ultrasound physiotherapy machines in Lothian Region, Scotland, 1992. *Ultrasound in Medicine and Biology* 20:347-359.

Rabkin BA, Zderic V, Crum LA and Vaezy S (2006) Biological and physical mechanisms of HIFU-induced hyperecho in ultrasound images. *Ultrasound in Medicine and Biology* 32:1721-1729.

Rivens IH, Clarke RL and ter Haar GR (1996) Design of focused ultrasound surgery transducers. *Ultrasonics, Ferroelectrics and Frequency Control, IEEE Transactions on* 43:1023-1031.

Robertson VJ and Baker KG (2001) A review of therapeutic ultrasound: effectiveness studies. *Physical Therapy* 81:1339-1350.

Robertson VJ, Ward AR and Jung P (2005) The effect of heat on tissue extensibility: A comparison of deep and superficial heating. *Archives of Physical Medicine and Rehabilitation* 86:819-825.

Sakurai T and Gotanda M, inventor; Olympus Optical Co., Ltd., assignee. "Apparatus for generating ultrasonic oscillation," *United States Patent, US* 5151085, Sep 29 1992.

Salgaonkar VA, Datta S, Holland CK and Mast TD (2009) Passive cavitation imaging with ultrasound arrays. *Journal of the Acoustical Society of America* 126:3071-3083.

San Emeterio JL and Ramos A (2008) Models for Piezoelectric Transducers Used in Broadband Ultrasonic Applications. In *Piezoelectric Transducers and Applications,* AA Vives, ed, pp 97-116, Springer-Verlag Berlin Heidelberg, Berlin.

Sapareto SA and Dewey WC (1984) Thermal dose determination in cancer therapy. *International journal of radiation oncology, biology, physics* 10:787-800.

Schwan HP, Carstensen EL and Li-Kam (1954) Electric and ultrasonic deep-heating diathermy. *Electronics* 27:172-175.

Sengupta DL and Liepa VV (2006) *Applied Electromagnetics and Electromagnetic Compatibility,* John Wiley and Sons, Inc., New Jersey.

Shaw A and Ter Haar GR (2006) Requirements for measurement standards in High Intensity Focused (HIFU) *Ultrasound Fields*. National Physical Laboratory (NPL).

Speed CA (2001) Therapeutic ultrasound in soft tissue lesions. *Rheumatology* 40:1331-1336.

ter Haar G (1999) Therapeutic ultrasound. *European Journal of Ultrasound* 9:3-9.

ter Haar G (2007) Therapeutic applications of ultrasound. *Progress in Biophysics and Molecular Biology* 93:111-129.

ter Haar G and Coussios C (2007a) High intensity focused ultrasound: past, present and future. *International Journal of Hyperthermia* 23:85-87.

ter Haar G and Coussios C (2007b) High intensity focused ultrasound: physical principles and devices. *International Journal of Hyperthermia* 23:89-104.

Vaezy S, Andrew M, Kaczkowski P and Crum L (2001) IMAGE-GUIDED ACOUSTIC THERAPY. pp 375-390.

Valtot F, Kopel J and Haut J (1989) Treatment of glaucoma with high intensity focused ultrasound. *International Ophthalmology* 13:167-170.

van der Windt DAWM, van der Heijden GJMG, van den Berg SGM, ter Riet G, de Winter AF and Bouter LM (1999) Ultrasound therapy for musculoskeletal disorders: A systematic review. *Pain* 81:257-271.

Vázquez-Hernández M, "Sistema para la Caracterización Ultrasónica de Phantoms en Aplicaciones de Terapia," Doctorate Thesis in Ingeniería Eléctrica, sección Bioelectrónica, Cinvestav, México, D.F. 2006.

Vera A, Chong JE, Leija L, Hernández Y and Marchal C (2001) Hipertermia electromagnética, una alternativa para el tratamiento del cáncer: antecedentes, aspectos físicos y biológicos. *Revista Mexicana de Ingeniería Biomédica* 22:78-88.

Vera A, Leija L and Munoz R (2008) Ultrasonic Hyperthermia. In *Piezoelectric Transducers and Applications*, AA Vives, ed, pp 467-495, Springer-Verlag Berlin Heidelberg, Berlin.

Watson T (2008) Ultrasound in contemporary physiotherapy practice. Ultrasonics 48:321-329.

Wojcik J (1998) Conservation of energy and absorption in acoustic fields for linear and nonlinear propagation. *Journal of the Acoustical Society of America* 104:2654-2663.

Wojcik J, Filipczynski L and Kujawska T (1999) Temperature elevations computed for three-layer and four-layer obstetrical tissue models in nonlinear and linear ultrasonic propagation cases. *Ultrasound in Medicine and Biology* 25:259-267.

Wu J and Nyborg W (2006) *Emerging Therapeutic Ultrasound*, World Scientific Publishing Co. Pte. Ltd.

Wu X, Worthington AE, Gertner MR, Hunt JW and Sherar MD (2007) Characterization of changes in therapeutic ultrasound transducer performance over time using the angular spectrum method. *IEEE Transactions on Ultrasonics Ferroelectrics and Frequency Control* 54:1028-1035.

Zeqiri B, Gelat PN, Hodnett M and Lee ND (2003) A novel sensor for monitoring acoustic cavitation. Part I: Concept, theory, and prototype development. *IEEE Transactions on Ultrasonics Ferroelectrics and Frequency Control* 50:1342-1350.

In: Hyperthermia: Recognition, Prevention and Treatment ISBN 978-1-61942-275-9
Editor: Xing-Hua Gao and Hong-Duo Chen, 85-113

Chapter 3

INFLUENCE OF HYPERTHERMIA ON AEROBIC EXERCISE PERFORMANCE: MECHANISMS OF FATIGUE

Julien D. Périard
ASPETAR, Qatar Orthopaedic and Sports Medicine Hospital,
Research and Education Centre, Doha, Qatar

ABSTRACT

Human aerobic performance in hot climatic conditions is markedly influenced by adjustments in circulatory, thermoregulatory, metabolic and neuromuscular function. Characteristically, exercise performed in the heat increases core and skin temperature, cardiovascular strain and perceived exertion. As a result, exercise intensity and duration are compromised. Currently, three mechanisms are suggested to explain the potential pathway(s) by which performance may be impaired. These include i) an increase in cardiovascular strain limiting oxygen delivery and uptake in exercising muscles and possibly the brain, ii) the development of hyperthermia-induced central fatigue reducing neural drive to the contracting musculature, and iii) anticipatory regulation of muscle recruitment ensuring the maintenance of whole-body homeostasis. These hypotheses have led to a diversity of experimental models investigating systemic hemodynamic responses during exercise in the heat, to evaluating force production capacity and voluntary muscle activation during both passive and exercise-induced hyperthermia. Research has shown that voluntary activation in previously active muscles is reduced and force production decreased when core temperature increases to ~40°C. Maximum heart rates have also been reported at these high temperatures on reaching exhaustion, indicating that the cardiovascular system may have attained functional limitations. Hence, the etiology of the impairment to aerobic exercise performance in hot climatic conditions remains contentious. This chapter explores the current state of knowledge regarding the aforementioned mechanisms, and contextualizes fatigue during constant rate and self-paced exercise, as well as identifies avenues of enquiry that remain unresolved.

Introduction

Thermal homeostasis in humans is modulated by the hypothalamus via thermolytic (vasoconstriction and shivering) and thermogenetic (vasodilation and sweating) pathways to maintain core temperature within a narrow functional range. In thermoneutral conditions, sensible (conduction, convection and radiation) and insensible (evaporation) heat loss mechanisms dissipate heat to the environment due to a negative thermal gradient between the skin and prevailing ambient conditions. However, when temperature and humidity are elevated, heat dissipation, especially via sweat evaporation, becomes less effective as the thermal gradient narrows. During exercise, heat accumulation increases as most of the energy derived from metabolism is released as heat (70–80%), rather than converted into movement (Gisolfi and Mora, 2000). The thermal compensability of the environment and the production of metabolic heat, which is proportional to exercise intensity (i.e. the percentage of maximal oxygen uptake utilized: % $\dot{V}O_2max$) (Nielsen, 1938; Saltin and Hermansen, 1966; Davies, 1979), therefore mediate the development of hyperthermia. When prolonged exercise is performed beyond the prescriptive zone (Lind, 1963), core and skin temperature rise (i.e. thermal strain increases) (Johnson and Park, 1979; González-Alonso et al., 1999b; Périard et al., 2011c), cardiovascular strain develops (Rowell, 1974; González-Alonso and Calbet, 2003; Périard et al., 2011b) and ratings of perceived exertion (RPE) increase (Nybo and Nielsen, 2001b; Armada-da-Silva et al., 2004). Consequently, work rate declines and performance decreases (Martin and Buoncristiani, 1999; Tatterson et al., 2000; Tucker et al., 2004; Ely et al., 2007), or is terminated altogether (Rowell, 1974; González-Alonso et al., 1999b). Although hydration strategies (González-Alonso et al., 1997; Cheuvront et al., 2010; Maughan and Shirreffs, 2010), pre-cooling methods (Arngrimsson et al., 2004; Wingo and Cureton, 2006b) and heat acclimation (Senay et al., 1976; Sawka et al., 1985; Nielsen, 1998) improve exercise-heat tolerance, the development of hyperthermia ultimately impairs endurance performance. This impairment (i.e. fatigue or exhaustion) can be defined as a failure to maintain the required power output/running speed (Edwards, 1983).

Currently, the precise physiological mechanism(s) mediating the decrement in performance remains contentious. Several postulates have identified potential sources of fatigue associated with impaired neuromuscular and/or cardiovascular function. From a neuromuscular perspective, it is suggested that central nervous system (CNS) drive to exercising muscles is reduced when a critical core temperature of ~40°C is reached (i.e. hyperthermia-induced central fatigue) (Nybo and Nielsen, 2001a; Nielsen and Nybo, 2003; Nybo, 2008, 2010). This inability to maintain force production capacity due to a decline in voluntary activation during a maximal voluntary isometric contraction (MVC) has also been observed during passive heating (Morrison et al., 2004; Todd et al., 2005; Thomas et al., 2006; Racinais et al., 2008). As a result, the attainment of a 'critical core temperature' that is intrinsic to the individual has received a great deal of focus as the principal explanation for impaired aerobic performance in hot conditions (González-Alonso et al., 1999b; Nielsen et al., 2001; Nybo and Nielsen, 2001a; Nielsen and Nybo, 2003; Nybo, 2007, 2008, 2010). A similar proposed neural mechanism suggests that an anticipatory central governor determines the extent of skeletal muscle recruitment based on prior experience, feedback received from peripheral receptors, and performance expectations to ensure whole-body homeostasis is maintained and thermal catastrophe is avoided (Noakes et al., 2004; St Clair Gibson and Noakes, 2004; Noakes, 2011).

Conversely, exercise in the heat may be impaired by the attainment of cardiovascular limitations associated with the development of hyperthermia. An increase in thermal strain creates competition for blood flow between metabolic and thermoregulatory processes (Rowell, 1977, 1986). It is a fundamental tenet that sustained exercise requires the maintenance of blood perfusion to active muscles. However, a thermoregulatory-mediated rise in skin blood flow causes a peripheral redistribution of blood volume, which can compromise venous return and central blood volume (Rowell, 1974, 1986). The consequent progressive increase in heart rate and decreases in stroke volume, cardiac output and mean arterial pressure (Asmussen, 1940; Rowell et al., 1966; González-Alonso et al., 1999b; González-Alonso and Calbet, 2003), can significantly reduce $\dot{V}O_2$max (Nybo et al., 2001; Arngrimsson et al., 2003; Wingo et al., 2005) and attenuate performance. These cardiovascular adjustments are further influenced by several factors, including dehydration (González-Alonso et al., 1995, 1997; González-Alonso et al., 2000; Cheuvront et al., 2007; Maughan and Shirreffs, 2010) and exercise modality (i.e. self-paced and constant rate exercise) (Abbiss and Laursen, 2008; Schlader et al., 2010; Périard et al., 2011b; Schlader et al., 2011). In consideration of these conflicting etiologies, this chapter will review historical and contemporary evidence supporting each potential pathway of fatigue. Both neuromuscular and cardiovascular mechanisms of fatigue will be discussed in the context of self-paced and constant rate exercise. Apparent gaps in the research literature will be identified as directions for future research.

Hyperthermia and Neuromuscular Function

Exercise-Induced Hyperthermia and Central Fatigue

It is well established that a moderate increase in core, and particularly muscle temperature, enhances explosive short duration skeletal muscle performance (e.g. sprinting and jumping) by improving metabolic and contractile function, nerve conduction and conformational changes associated with muscle contraction (Asmussen and Boje, 1945; Edwards et al., 1972; Bergh and Ekblom, 1979; Fitts, 1994; Allen et al., 2008). Conversely, prolonged aerobic exercise in warm to hot conditions is markedly impaired by the progressive development of hyperthermia (Galloway and Maughan, 1997; González-Alonso et al., 1999b; Nybo and Nielsen, 2001a). Several human and animal studies have suggested that CNS dysfunction is the primary source of impairment limiting aerobic exercise in hot climatic conditions when core temperature is deemed to have reached a 'critical point' (Fuller et al., 1998; Walters et al., 2000; Nybo and Nielsen, 2001a; Nielsen and Nybo, 2003; Nybo, 2007). This hyperthermia-induced central fatigue is manifested in a decreased force production due to reductions in motor drive, failing to maintain muscle activation. In hamsters, it has been shown that direct warming of the anterior hypothalamus impairs running performance (Gordon, 1993), whereas goats fatigue when brain temperature reaches 42°C (Caputa et al., 1986). Similarly, dogs, cheetahs and rats cease exercising when brain and/or abdominal temperature attain a critical level, irrespective of the environmental heat load or initial core temperature (Young et al., 1959; Taylor and Rowntree, 1973; Fuller et al., 1998; Walters et al., 2000). Although circumstantial evidence from these comparative studies supports the hypothesis that exercise is impaired at the attainment of an intrinsic critical core temperature, it remains to be determined whether this phenomenon is related to an elevated core or hypothalamic temperature (Walters et al., 2000). Moreover, these studies are limited by the absence of other measures that may equally be associated with the cessation of exercise (e.g. cardiovascular strain).

In humans, Brück and Olschewski (1987) were among the first to postulate that heat stress might affect brain function and cause fatigue. They identified three factors of discomfort that might counteract motivation and progressively reduce the drive to exercise during hyperthermia. They suggested that the interaction of circulatory, thermal and muscular discomfort determined endurance time and work rate. Subsequently, Nielsen et al. (1990; 1993) investigated locomotor muscle blood flow and metabolism during exercise to exhaustion in the heat to determine if a reduction in perfusion limited aerobic exercise. Results indicated that cardiac output and exercising muscle blood flow did not decline, substrate utilization and availability remained unchanged, and blood lactate did not increase. It was concluded that hyperthermia per se, rather than a circulatory limitation, was the critical factor causing exhaustion during exercise under heat stress. It was further proposed that the attainment of a high core temperature (39.2-39.7°C) might influence the CNS by reducing mental drive (motivation) for motor performance (Nielsen et al., 1990; Nielsen et al., 1993). Interestingly, the capacity to generate force during a brief (3-5 s) MVC of the knee extensor and elbow flexor muscle groups was unaltered (Nielsen et al., 1993). This capacity to maintain force production has also been shown during 40 consecutive brief MVCs of the exercised knee extensors following exhaustive cycling in the heat (Nybo and Nielsen, 2001a). However, force production was progressively impaired during a sustained (120 s) MVC when compared with measures taken after exercise in cool conditions. The impairment was associated with a hyperthermia-induced reduction in voluntary activation, as total force (MVC + electrical stimulation) was similar between conditions. A sustained MVC performed by the non-exercised forearm flexors followed the same pattern of response. Thus, it was suggested that hyperthermia (~40°C) impairs the ability to sustain a prolonged MVC, but that the capacity to produce brief maximal force remains unaltered (Nybo and Nielsen, 2001a).

Conflicting research findings challenge the conceptual view that fatigue may either be of central or peripheral origin (i.e. arising proximal or distal to the neuromuscular junction). Following cycling exercise to exhaustion during which core temperature was moderately elevated (38.8°C), Saboisky et al. (2003) reported that the capacity to produce force in the non-exercised forearm flexors was maintained during a 5 s MVC, whereas it was attenuated in the exercised quadriceps. They suggested that afferent feedback from core and muscle thermoreceptors inhibited voluntary activation to specific skeletal muscles in order to maintain cellular integrity. More recently, it was shown that following self-paced exercise (40 km time trial) in hot and cool conditions, voluntary activation was similarly attenuated during a 20 s MVC (Périard et al., 2011c). It was also reported that the reduction in voluntary activation was maintained throughout the contraction (i.e. the extent and degree of decline was sustained) (Figure 1A). Consequently, the reduction in force production was similar between conditions (Figure 1B). The authors calculated that the post-exercise decline in voluntary activation accounted for ~20% of the decrease in total force production. Thus, the difference in core temperature (0.8°C) at time trial completion did not exacerbate the loss of force production in the heat, which was mediated by fatigue of both central and peripheral origin (Périard et al., 2011c). This provides evidence that fatigue is partly a consequence of the prolonged contractile activity of previously active muscles, and offers insight into the mechanism(s) of fatigue during self-paced exercise. Furthermore, this was the first study to demonstrate in a number of subjects, the capacity to maintain complete CNS activation of previously active muscles during a sustained MVC, despite core temperatures ≤39.9°C (Périard et al., 2011c).

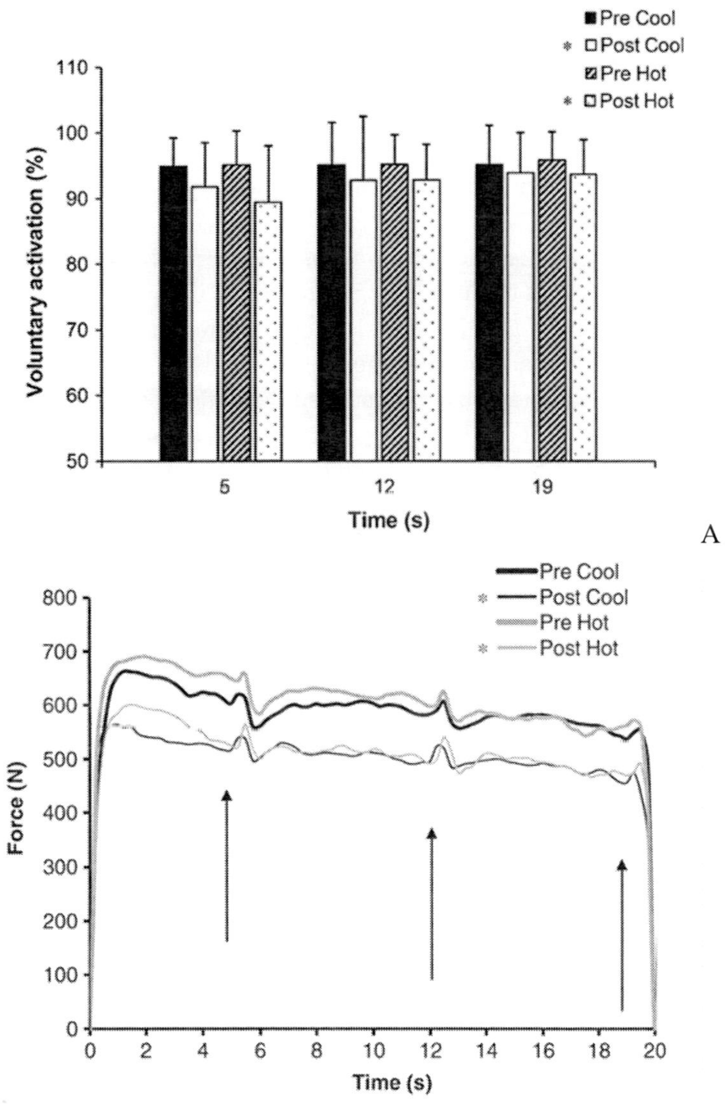

Figure 1. Voluntary activation percent (A) and force production (B) during a 20 s maximal voluntary isometric contraction of the knee extensors with superimposed electrical stimulation at 5, 12 and 19 s. The MVC was performed prior to and following self-paced exercise in hot and cool conditions. Values are means ± SD for 9 subjects. Force figure is simplified by omitting error bars (SD ranged from 64.2 to 83.4 N during all MVCs).*Indicates a significantly lower post-exercise mean voluntary activation/force production compared with pre-exercise ($P < 0.05$). Reproduced with permission from Périard et al. (2011c).

This ability had only been demonstrated in normothermic subjects without the influence of prior exercise (Merton, 1954; Bigland-Ritchie et al., 1978; Bigland-Ritchie et al., 1983b). It was speculated that providing feedback, verbal encouragement and a visual display of force enhanced or at least maintained subject motivation. In a follow up study, Périard et al. (2011a) further demonstrated the greater fatigability of previously active muscles when investigating neuromuscular function after active (39.8°C) and passive hyperthermia (39.5°C). During a 45 s MVC of the quadriceps, force production capacity was shown to decline at a faster rate following cycling to exhaustion in the heat, compared with passive heating via water immersion. As with the aformentionned study, brief maximal force production in the non-exercised hand grip flexors was unaffected by exercise or hyperthermia (Périard et al., 2011a; Périard et al., 2011c). Voluntary activation was also similar following active and passive hyperthermia. The noted decline was maintained throughout the sustained MVCs. The loss of force therefore originated from central and peripheral fatigue factors, with the combination of heat stress and prior contractile activity exacerbating the rate of decline (Périard et al., 2011a).

Hence, while it is clear that sustained maximal force production is depressed following prolonged aerobic exercise in the heat (e.g. Figure 1B), it is equally evident that the analysis of neuromuscular function is greatly dependent on numerous factors, including the duration of contraction, the muscle group utilized and the method by which hyperthermia is induced. In addition, several recent studies have questioned the notion of a 'critical core temperature' (Ely et al., 2009; Cheuvront et al., 2010; Lee et al., 2010). Historically, temperatures in excess of 40°C and up to 41.9°C have been recorded in road runners, marathoners, as well as in soccer and football players with no signs or symptoms of heat-related illness (Robinson, 1963; Pugh et al., 1967; Costill et al., 1970; Maron et al., 1977; Richards et al., 1979a; Richards et al., 1979b; Sutton, 1984; Roberts, 1989; Bangsbo, 1993; Thompson et al., 1997; Cheuvront and Haymes, 2001; Byrne et al., 2006; Yeargin et al., 2006). While the difference in core temperature between field studies/sporting events and laboratory experiments may be attributed to differing levels of fitness, exercise modality and motivation, the values attained outside the laboratory far exceed the range (39.2 - 40.0°C) associated with the 'critical core temperature' hypothesis (Nielsen et al., 1990; Nielsen et al., 1993; Nybo and Nielsen, 2001a).

Passive Hyperthermia and Central Fatigue

In order to isolate the precise influence of heat stress on neuromuscular function, passive heating techniques (e.g. water immersion, exposure to hot environments and liquid conditioned suits) have been utilized to induce hyperthermia (Morrison et al., 2006; Thomas et al., 2006; Périard et al., 2011a). By manipulating skin and core temperature, the confounding effect of exercise on skeletal muscle metabolism and contractile function can be circumvented. Employing a liquid conditioned suit, Morrison et al. (2004) evaluated knee extensor exercise during a 10 s MVC at 0.5°C temperature intervals from a core temperature range of 37.5°C to 39.5°C. The progressive increase in core temperature was paralleled by a decrease in voluntary activation (11%) and force production (13%). When water temperature circulating in the suit was decreased to lower core temperature, voluntary activation and force production remained depressed. Only when core temperature returned to baseline, was neuromuscular function restored (Morrison et al., 2004). Utilizing a similar protocol, body core and peripheral muscle temperature were isolated by maintaining one leg in a thermoneutral environment while allowing the body and other leg to warm-up and cool-down (Thomas et al., 2006). As with the previous study, similar reductions in voluntary activation and torque were noted between legs, and sudden cooling did not immediately restore CNS drive and torque, despite a significant decrease in skin temperature. Thus, it was proposed that neuromuscular fatigue is not influenced by afferent input from skin and peripheral muscle thermoreceptors, or cardiovascular and psychological strain (Morrison et al., 2004; Thomas et al., 2006). Moreover, the cardiovascular response associated with passive heating does not appear to influence cerebrovascular regulation. Although force production and voluntary activation are decreased when core temperature is passively elevated, prefrontal cortex oxygenation is maintained (Morrison et al., 2009). Therefore, it appears that central fatigue develops concurrently with the progressive increase in core temperature, irrespective of aerobic fitness/training status (Morrison et al., 2006), or heat acclimation (Brazaitis and Skurvydas, 2010).

These findings however do not preclude fatigue originating distal to the neuromuscular juntion in response to temperature-induced alterations in contractile function (Bigland-Ritchie et al., 1982; Bigland-Ritchie et al., 1983a; Bigland-Ritchie et al., 1983b). Using transcranial magnetic stimulation (TMS), Todd et al. (2005) demonstrated that moderate passive hyperthermia (38.5°C) is associated with decrements in brief (2.4%) and sustained (12%) maximal torque. Throughout the sustained MVC, it was calculated that the rate of peak muscle relaxation was 20% faster under hyperthermia than in control, which is consistent with passive increases in muscle temperature (~3°C) decreasing time to peak force and half-relaxation time in brief MVCs (Davies and Young, 1983). However, such adjustments in contractile function require greater motor unit firing rates to maintain the fusion of force. While such rates can be transiently achieved (e.g. a brief MVC), they cannot be maintained for prolonged durations (Todd et al., 2005), as the modified contractile state may exceed the centrally mediated rate of activation (Thompson, 2006). Moreover, it is well established that isometric exercise involves partial to complete occlusion of blood flow to active muscles (Barcroft and Millen, 1939; Edwards et al., 1972b), further increasing muscle temperature and stimulating chemoreflexes and mechanoreflexes (Barnes, 1980; Sejersted et al., 1984). The afferent limb of these reflexes (i.e. group III and IV polymodal fibers) responds to chemical, mechanical and thermal stimulation (Kalia et al., 1981; Kniffki et al., 1981; Mitchell and Schmidt, 1983), which increases muscle sympathetic nervous activity (MSNA) (Seals and Victor, 1991). Depending on the intensity of MVC, the rise in MSNA can alter motor unit excitability, modifying the relationship between central neural drive, motor unit recruitment and firing rate coding (Bigland-Ritchie et al., 1986; Woods et al., 1987). As such, impairment of neuromuscular function may relate to failure in the peripheral transmission of neural drive at any level from cortical activity to sarcolemma depolarization (Racinais et al., 2008).

Cerebral Perturbations and Fatigue

Research into potential links between the development of hyperthermia-induced central fatigue and cerebral electrical activity has indicated that RPE is highly correlated with electroencephalogram (EEG) activity over the prefrontal cortex ($r2 = 0.98$) and the increase in core temperature ($r2 = 0.98$) (Nybo and Nielsen, 2001b). The ratio of α/β frequencies, an index of suppressed arousal when elevated (Kandel and Schwartz, 1985), also increases during exercise in the heat (Nielsen et al., 2001), suggesting that alterations in CNS activity under heat stress may be associated with the development of fatigue. A lack of oxygen delivery and heat dissipation are pathways by which cerebral activity might be altered (Nybo and Rasmussen, 2007). Cerebral blood flow reductions of 20% during exercise have been shown to maintain brain temperature ~0.2°C higher than body core temperature (Nybo et al., 2002b). However, these reductions in cerebral blood flow were observed prior to a decline in performance. Exhaustion during maximal exercise in thermoneutral and hyperthermic environments is also related to a decline in cerebral perfusion, manifested in a decrease in middle-cerebral artery mean blood velocity and frontal cortex tissue oxygenation (González-Alonso et al., 2004). During a 6 min maximal rowing trial, performance was shown to increase when a greater fraction of oxygen was inspired (Nielsen et al., 1999). It was suggested that the prevention of cerebral desaturation by oxygen supplementation and the maintenance of arterial oxygen saturation enhanced performance. However, these improvements may be linked to an enhanced oxygen delivery to the exercising musculature preventing fatigue (Amann et al., 2006).

More recently, cerebral deoxygenation and declines in cerebral mitochondrial oxygen tension during maximal exercise in hyperthermic and thermoneutral conditions have been suggested to resemble those of hypoxic exercise (Rasmussen et al., 2010a; Rasmussen et al., 2010b). As such, strenuous exercise may be dependent on the maintenance of cerebral oxygenation to avoid the development of central fatigue. However, a reduction in cerebral perfusion at exhaustion is associated with an enhanced cerebral metabolic rate for oxygen, as well as glucose and lactate uptake (González-Alonso et al., 2004). Furthermore, cerebral glucose utilization increases by 7% during hyperthermia (Nybo et al., 2002a). A significant oxygen reserve is thus required by the brain on reaching exhaustion, which is suggested to protect it against a decline in oxygen delivery (González-Alonso et al., 2004; Crandall and González-Alonso, 2010). Moreover, prefrontal cortical activity is associated with higher-order cognitive behavior, which influences motivation and the drive to perform goal-oriented activities (Miller and Cohen, 2001; Ramnani and Owen, 2004). A reduction in oxygenation and EEG activity in the prefrontal region might therefore signal an unwillingness to continue exercising, rather than an inability to voluntarily activate sufficient motor units to produce a required force. Furthermore, increases in RPE correlate strongly with the rise in heart rate ($r2 = 0.97$) during exercise-heat stress (Nybo and Nielsen, 2001b), which at exhaustion can exceed 96% of maximum (González-Alonso et al., 1999b; Nielsen et al., 2001; Périard et al., 2011a). Hence, the precise influence of cerebral activity and circulation on fatigue remains contentious.

The interaction of numerous factors – cardiovascular, biochemical, mechanical, metabolic and psychological – could alter the development of fatigue in the heat (Rowell, 1986; Febbraio et al., 1994b; Davis and Bailey, 1997; Gandevia, 1998; González-Alonso, 1998; Cabanac, 2006). While the role of neurotransmitters (i.e. serotonin, dopamine and noradrenaline) has previously been implicated in the control of thermoregulation and the potential development of central fatigue (Newsholme, 1987; Bridge et al., 2003), recent evidence suggests that it is unlikely that a particular neurotransmitter system mediates the onset or development of fatigue (Meeusen and Roelands, 2010).

Peripheral Perturbations and Fatigue

From a peripheral perspective, fatigue may relate to the influence of high temperatures on muscle and cellular metabolism (Edstom and Kugelberg, 1968; Burke and Levine, 1971; Richter, 1984; Jansson et al., 1986; Larsson et al., 1991; Febbraio et al., 1994a; Febbraio et al., 1996). Muscle lactate content in dogs is positively correlated to muscle temperature and the development of hyperthermia is known to enhance muscle glycogen depletion (Kozlowski et al., 1985). In humans, exercise in the heat causes a greater post-exercise accumulation of ammonia as well as muscle and blood lactate (Dolny and Lemon, 1988; Febbraio et al., 1994a; Febbraio et al., 1994b). During maximal exercise, muscle lactate accumulation can exceed that of the blood by 10 mM (Diamant et al., 1968; Karlsson, 1971; Jacobs et al., 1983). This high muscle lactate production is linked with muscle fatigue, and the decline in force observed during work at high glycolytic rates (Fitts, 1994). It is also highly correlated to the release of force-depressing free hydrogen ions (H^+) (Sahlin et al., 1978; Metzger et al., 1989). Temperature-induced impairments in sarcoplasmic reticulum function or structural damage compromising sarcoplasmic reticulum calcium ion (Ca^{+2}) regulatory capacity may also influence skeletal muscle force production (Fitts, 1994; Schertzer et al., 2002). In addition, the occlusion of blood flow during a prolonged MVC and the accumulation of metabolites associated with muscular contractions may influence motivation. Considerable effort is required to sustain an MVC along with a willingness to withstand discomfort and pain. Mild sensations of discomfort are generally sensed at the onset of a contraction, which eventually develop into severe pain that alters the perception of sensations in the contracting musculature (Bigland-Ritchie et al., 1978). As a result, mental fatigue, which involves tiredness, limited attention span and an aversion or decreased commitment to continuing a task or activity (Holding, 1983; Hockey, 1997), may contribute to decrements in voluntary activation. Conscious signals originating from both central and peripheral afferent pathways could mediate behavior and reduce motivation in order to minimize discomfort (Cabanac, 2006) and cause subjects to abandon a task in which the energetic demands (i.e. effort) outweigh the perceived benefits of continued performance (Boksem and Tops, 2008). Essentially, a lack of motivation causes CNS drive and motoneurone firing to decline, leading to a loss of force production (Enoka and Stuart, 1992). Currently, the techniques used to assess central fatigue (i.e. the central activation ratio and interpolated twitch technique) fail to distinguish whether the reduction in voluntary activation is of physiological or psychological origin. In essence, they do not differentiate whether subjects cannot or will not fully activate contracting muscles during a sustained MVC (Thompson, 2006). Alternatively, when TMS evokes an increase in force production it indicates a supraspinal component to central fatigue (Gandevia, 2001). This signifies that additional voluntary drive remains to be utilized. However, TMS cannot evoke a maximal control twitch in the relaxed muscle, nor can it determine the fraction of relevant corticospinal axons excited, or establish whether voluntary corticospinal output has increased or decreased during an MVC (Gandevia, 2001). Therefore, central fatigue appears to represent a psychophysiological phenomenon whereby CNS drive to exercising muscles is reduced under the influence of alterations in neuromuscular function and motivation. Furthermore, it remains to be determined if fatigue during aerobic exercise is caused by a high internal temperature per se leading to central fatigue, or simply coincides with the factors that precipitate fatigue under heat stress.

Importantly, caution is required when interpreting and inferring changes in a form of exercise where all available motor units are recruited and maximally activated (i.e. maximal isometric contraction), to one where motor unit recruitment is submaximal and intermittent (i.e. submaximal dynamic exercise). High intramuscular tension during an MVC also causes the mechanical constriction of the blood vessels, which impinges on group III and IV nociceptors and impedes blood flow and the clearance of metabolic by-products (e.g. lactate and H^+). As such, disproportionately large increases in heart rate and blood pressure relative to oxygen uptake occur. In contrast, cardiovascular, hemodynamic and metabolic activity is matched to intensity during dynamic exercise.

HYPERTHERMIA AND CARDIOVASCULAR FUNCTION

Thermal Strain and Blood Flow

The circulatory requirements associated with aerobic exercise under heat stress include an increase in skin blood flow and the maintenance of cerebral and exercising muscle perfusion. It is well established that a high skin temperature, near that of the core, is related to a reflex increase in skin blood flow (Johnson, 1977; Johnson and Park, 1979; Johnson, 1986). For recent, comprehensive and non-exhaustive reviews on the reflex control of human skin blood flow, the reader is referred to the following manuscripts (Charkoudian, 2010; Johnson and Kellogg, 2010). Although a vasoconstrictor-mediated reduction in skin blood flow occurs at the onset of exercise, cutaneous blood vessels eventually vasodilate to allow the dissipation of accumulating metabolic heat. The core temperature threshold at which skin blood flow begins to rise is directly related to exercise intensity, and a delay in active vasodilation shifts this threshold to the right relative to rest (Kellogg et al., 1991; Smolander et al., 1991; Johnson, 1992; Kenney and Johnson, 1992). During prolonged exercise in the heat, the rate of rise in skin blood flow markedly decreases after 20-30 min, reaching a virtual plateau (~50% of maximum flow capacity) when core temperature nears 38°C (Johnson et al., 1974b; Brengelmann et al., 1977). At this point, the perfusion requirements of exercising muscles take precedence over thermoregulatory control. Despite this attenuation and the potential for splanchnic and renal vasoconstriction to redistribute 600-800 $ml \cdot min^{-1}$ of blood to the periphery, the large displacement of blood (6-8 $l \cdot min^{-1}$) to cutaneous vascular beds mediates a progressive decline in arterial pressure, central venous pressure and stroke volume, that is accompanied by an increase in heart rate (Rowell, 1974; Johnson and Rowell, 1975). Interestingly, the relationship between skin temperature and skin blood flow is minimally influenced by core temperature (Brengelmann et al., 1977; Johnson and Park, 1979). However, a strong link exists between the core-to-skin temperature gradient and skin blood flow. A 10°C increment in ambient temperature is associated with a 4.5°C decrease in the core-to-skin temperature gradient (Kenefick et al., 2010). This narrowing of the gradient is suggested to reflect a rise in skin blood flow to enhance evaporative heat loss with the environment. Historically, skin temperature (Iampietro and Goldman, 1965; Iampietro, 1971) and the convergence of skin and core temperature (Pandolf and Goldman, 1978) have been useful criteria in predicting tolerance time for work performed under heat stress. However, in highly motivated individuals these predictive tolerance models may depend more on the precise temperature of convergence (Nunneley et al., 1992).

In exercising muscles, blood flow requirements are associated with relative exercise intensity. Essentially, an increase in oxygen demand is matched by an increase in systemic and muscle oxygen delivery, as well as perfusion pressure (Holmgren, 1956; Astrand et al., 1964; Ekelund and Holmgren, 1967). In absolute terms, muscle blood flow can increase from ~0.3 $l \cdot min^{-1}$ at rest, to 10 $l \cdot min^{-1}$ during maximal exercise in less than 10 s (Saltin et al., 1998). If work rate is stable, a steady state is reached within 30-90 s and blood flow rises only slightly when exercise intensity increases (Radegran and Saltin, 1998). As such, muscle blood flow and oxygen uptake are quite stable during prolonged exercise at fixed intensities in thermoneutral conditions (Savard et al., 1987). Under conditions of heat stress however, a rise in thermal strain exacerbates the cardiovascular response as metabolic and thermoregulatory processes compete for cardiac output (Rowell, 1974; Montain and Coyle, 1992a; Crandall and González-Alonso, 2010). The classic hypothesis of cardiovascular limitations proposed by Rowell (1974, 1983, 1986) lies with this redistribution and pooling of blood to the periphery, which reduces central blood volume and concomitantly ventricular filling pressure, end-diastolic volume and stroke volume. A more contemporary hypothesis suggests that the reduction in stroke volume is primarily due to an increase in intrinsic heart rate (Fritzsche et al., 1999; Coyle and Gonzalez-Alonso, 2001; Trinity et al., 2010), mediated by the direct effects of temperature on the sinoatrial node, and/or baroreflex modulation of sympathetic and parasympathetic activity (Wilson and Crandall, 2011). In effect, the decline in stroke volume during moderate prolonged exercise in the heat likely originates from both a reduction in central blood volume reducing cardiac filling pressure, and a shorter diastolic filling time reducing ventricular end-diastolic volume. Regardless, when exercise is sustained and motivation is high, the attainment of maximum or near maximum heart rate is a well documented response. Concurrently, maximum cardiac output is decreased (Bassett and Howley, 2000; diPrampero, 2003) and the cardiovascular system is forced towards a functional limit at "submaximal" workloads and oxygen uptake (Sawka et al., 1985; Arngrimsson et al., 2003; Wingo et al., 2005). It is suggested that this reduction in cardiovascular reserve is the primary factor limiting constant rate aerobic exercise in the heat, and is manifested as an increase in relative exercise intensity (% $\dot{V}O_2max$) and perceived exertion (Robertson, 1982; Bassett and Howley, 2000; diPrampero, 2003; Cabanac, 2006).

Muscle Perfusion during Hyperthermia

Recently, Cheuvront et al. (2010) noted that the core temperature tolerated at exhaustion is inversely related to whole body skin blood flow requirements, suggesting that fatigue may correspond with adjustments in muscle and cerebral perfusion. However, numerous previous studies have demonstrated that muscle blood flow increases or is maintained during prolonged moderate exercise (50-80% $\dot{V}O_2max$) to exhaustion in the heat (Kirwan et al., 1987; Savard et al., 1988; Nielsen et al., 1990; Nielsen et al., 1993; Nielsen et al., 1997). Notably, Savard et al. (1988) manipulated skin temperature by circulating water at different temperatures through a liquid conditioned suit during three consecutive 25 min bouts of exercise. When heat stress was imposed, blood flow to exercising muscle was maintained, despite heart rate and core temperature increasing significantly. Others have since demonstrated that even at volitional fatigue, cardiac output and exercising muscle blood flow are not reduced during moderate exercise in the heat (Nielsen et al., 1990; Nielsen et al., 1993; Nielsen et al., 1997). Consequently, fatigue was attributed to the attainment of a critical core temperature, irrespective of the rate of heat storage, pre-exercise core temperature, or heat acclimation (Nielsen et al., 1993; González-Alonso et al., 1999b). However, recent evidence suggests that only when heart rate increases above 95% of maximum is muscle perfusion limited (González-Alonso and Calbet, 2003; Mortensen et al., 2005; Mortensen et al., 2008). Gonzalez-Alonso and Calbet (2003) noted that intense cycling (80% of peak power output) to exhaustion with or without heat stress is associated with a reduction in systemic and exercising muscle blood flow, oxygen delivery and uptake. These reductions are exacerbated by heat stress, which accelerates the decline in mean arterial pressure and cardiac output, reducing $\dot{V}O_2max$ (González-Alonso and Calbet, 2003). It has also been shown that during maximal incremental and constant rate exercise, systemic oxygen delivery is blunted by a plateau or decrease in cardiac output at intensities below $\dot{V}O_2max$ (Mortensen et al., 2005). From 50–90% $\dot{V}O_2max$, systemic and exercising muscle blood flow, along with oxygen delivery, match the rise in oxygen uptake. However, beyond 90% $\dot{V}O_2max$ a levelling off occurs that attenuates the rate of rise in oxygen uptake, despite maximum increases in arteriovenous oxygen difference and heart rate. This impairment indicates an inability of the cardiovascular system to sustain a linear increase in oxygen delivery to exercising muscles. Aerobic power and capacity are therefore impaired by reductions in cardiac output and oxygen delivery to exercising muscles, partly due to an enhanced MSNA which attenuates leg blood flow (Knight et al., 1992; Rosenmeier et al., 2004; Mortensen et al., 2005). These findings have been corroborated during maximal and supramaximal intensity cycling (Mortensen et al., 2008). Although metabolic energy requirements are greater in supramaximal exercise, a plateau in cardiac output and exercising muscle vascular conductance was noted at similar levels of oxygen delivery at both maximal and supramaximal intensities. Together, these observations challenge the postulate that oxygen delivery increases linearly with oxygen uptake from rest to $\dot{V}O_2max$ (Astrand et al., 1964; Saltin and Stensberg, 1964; Ekblom, 1968; Ekblom and Hermansen, 1968). Presently, it is speculated that a similar response compromises oxygen delivery during prolonged submaximal exercise in the heat when maximum heart rate is attained in motivated subjects.

It is well established that cardiovascular adjustments are influenced by hydration status. Gradual dehydration beyond 2% body mass during moderate intensity exercise in the heat leads to further increases in core and skin temperature, progressive declines in systemic, muscle and skin blood flow, a greater reliance on muscle glycogen and anaerobic metabolism, as well as a reduction in exercising muscle $\dot{V}O_2$ at exhaustion (Sawka et al., 1979; Montain and Coyle, 1992a; Febbraio et al., 1994b; Hargreaves et al., 1996; González-Alonso et al., 1998, 1999a). González-Alonso et al. (1998, 1999a) demonstrated that muscle blood flow was reduced during prolonged exercise in the heat when dehydrated. The reduction, accompanied by decreases in cardiac output and systemic vascular conductance, resulted in exhaustion at a core temperature of 39.7°C. Conversely, euhydrated trials terminated in the same time frame resulted in a final core temperature of 38.3°C and a heart rate 26 beats·min^{-1} lower. The 39% reduction in blood flow was attributed to a decline in perfusion pressure and systemic blood flow. This resulted in a much higher muscle glycogen utilization and blood lactate concentration, consistent with previous observations (Fink et al., 1975; Febbraio et al., 1994a; Febbraio et al., 1994b; Febbraio et al., 1996; Hargreaves et al., 1996; Parkin et al., 1999). Although these observations provide circumstantial evidence of a reduction in oxygen delivery and uptake, they do not infer a causal relationship between a decline in muscle perfusion and fatigue in the heat, nor do they indicate that the increase in muscle glycolysis is an attenuating factor to performance. Additional research is therefore required to directly measure hemodynamic responses in vivo. For a greater description of the influence of dehydration on performance in the heat, the reader is referred to the following reviews (Sawka et al., 1992; Sawka et al., 1998; Cheuvront et al., 2007; Maughan and Shirreffs, 2010).

Exercise Modality, Anticipatory Regulation and Fatigue

The suggestion that performance in the heat is impaired by cardiovascular limitations is supported by both constant rate and self-paced exercise studies (Ely et al., 2009; Ely et al., 2010; Kenefick et al., 2010; Périard et al., 2011a; Périard et al., 2011b). Several investigations have reported near maximum heart rates at exhaustion during prolonged constant rate exercise (Cheung and McLellan, 1998; Selkirk and McLellan, 2001; Périard et al., 2011a) with significant declines in central blood volume, stroke volume and cardiac output (Asmussen, 1940; Rowell et al., 1966; Périard et al., 2011a). Following either pre-cooling, pre-heating or normothermic control conditions, it was shown that subjects cycling to exhaustion at a fixed intensity fatigued at a heart rate between 98-99% of maximum, and that the rate of rise in cardiovascular strain was graded in proportion with the magnitude of hyperthermia (Table 1) (González-Alonso et al., 1999b).

Alternatively, fatigue during exercise in the heat has been proposed to develop because of an anticipatory mechanism (Tucker et al., 2004; Tucker et al., 2006; Tucker, 2009). Derived from the central governor model of exercise (Noakes et al., 2004; St Clair Gibson and Noakes, 2004), it is suggested that anticipatory regulation of muscle recruitment ensures the prevention of thermal injury. It is hypothesised that the moment exercise is initiated, afferent feedback regarding exercise intensity, climatic conditions and prior experience is integrated to forecast the safe duration of exercise. This initial prescription of duration, along with the rate of increase in RPE, are continuously modified via input received from various physiological systems (Tucker, 2009; Noakes, 2011). Therefore, the maximum tolerable RPE occurs before harmful disturbances can disrupt homeostasis and exercise is terminated (e.g. constant rate exercise) or power output decreased (e.g. self-paced exercise). However, the calculations performed to develop this premise and demonstrate anticipatory regulation based on rate of heat storage (Tucker et al., 2006) have been shown to be erroneous, thus questioning the evidence for the proposition of a feedforward mechanism associated with the rate of heat storage (Jay and Kenny, 2009). Moreover, the anticipatory mechanism is clearly not a “fail-safe” protection strategy as athletes occasionally succumb to exertional heat stroke during training and competition (Armstrong et al., 2007).

Table 1. Heart rate, stroke volume, and cardiac output at 10 min of exercise and exhaustion during cycling in 40°C heat at different initial core temperatures (Cool = 35.9 ± 0.2°C, Control = 37.4 ± 0.1°C, and Hot = 38.2 ± 0.1°C). 1999b)

	Heart rate (beats·min^{-1})			Stroke volume (ml·beat^{-1})			Cardiac output (l·min^{-1})		
	Cool	Control	Hot	Cool	Control	Hot	Cool	Control	Hot
10 min*	140 ± 5	166 ± 5	182 4	152 ± 7	126 ± 6	109 ± 4	22.5 ± 1.1	21.8 ± 0.8	20.3 ± 0.5
Exhaustion†	198 ± 4	197 ± 4	196 ± 4	106 ± 5	104 ± 4	103 ± 4	20.8 ± 0.7	20.2 ± 0.6	19.9 ±0.4
% Difference	↑ 29	↑ 16	↑ 7	↓ 30	↓ 17	↓ 6	↓ 8	↓ 7	↓ 2

*Indicates a significant difference between Cool, Control and Hot trials within each measured parameter ($P < 0.05$).

† Indicates no difference between Cool, Control and Hot trials within each measured parameter. Mean (SE).

Oxygen uptake was similar (3.05 to 3.11 l·min^{-1}) across conditions at 10 min despite a 2.4°C difference in core temperature between cool and hot conditions. Heart rate and stroke volume responded in proportion to the level of hyperthermia elicited (up to ~40.1°C). Calculated from (González-Alonso et al.,

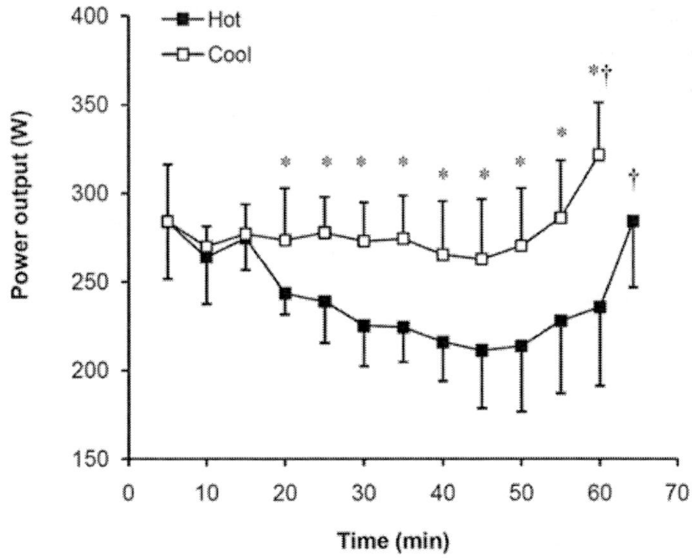

Figure 2. Mean power output at 5 min intervals and during the final kilometre at maximal effort during a 40 km cycling time trial in hot and thermoneutral (cool) conditions. Values are means ± SD for 8 subjects. *Significantly higher than hot ($P < 0.05$). †Significantly higher than previous ($P < 0.01$). Reproduced with permission from Périard et al. (2011b).

In addition, anticipatory reductions in power output attributed to large differences in skin temperature (7-8°C) during exercise in the heat may be misleading, especially while occurring at similar core temperatures, heart rates and perceived exertions to those in cool climatic conditions (Tucker et al., 2004; Tucker et al., 2006). Such elevated skin temperatures and narrow core-to-skin temperature gradients are associated with a large redistribution of blood to cutaneous vascular beds during the first 20 min of exercise (Rowell et al., 1969; Rowell, 1974; Johnson et al., 1974b; Johnson and Rowell, 1975; Johnson, 1992; Kenney and Johnson, 1992; Nose et al., 1994; Fritzsche et al., 1999), which exacerbates the cardiovascular response (Périard et al., 2011b). Field based evidence has shown that self-paced exercise is not impaired by anticipatory regulation, as no correlation was established between the rate of heat storage and running speed (Ely et al., 2009). Runners were also capable of accelerating despite high rates of heat storage in hot and cool conditions, and a core temperature of 40°C. In laboratory settings, even brief (15 min cycling time trial) aerobic performance is suggested to deteriorate because of hyperthermia-induced alterations in cardiovascular function (Ely et al., 2010). The impairment seems to stem from physiological adjustments associated with a reduction in $\dot{V}O_2$peak, despite moderate hyperthermia (38.2°C) (Ely et al., 2010). Accordingly, during the final maximal effort in a simulated 40 km time trial in hot and cool conditions, it was shown that $\dot{V}O_2$peak declined by 23% and 5%, respectively (Périard et al., 2011b). Throughout exercise in the heat, the decline in power output (Figure 2) was paralleled by decreases in stroke volume, mean arterial pressure and cardiac output (Figure 3). However, RPE remained elevated and heart rate was higher than in the cool condition. It was therefore suggested that the development of cardiovascular strain mediated the utilization of a progressively greater % $\dot{V}O_2$peak. Consequently, sustainable work rate and peak power output were reduced (Périard et al., 2011b). Although power output was depressed upon reaching $\dot{V}O_2$peak in the heat, the authors suggested there was a similar level of motor outflow/neural drive to the contracting musculature during the final maximal effort during both conditions (Périard et al., 2011b). Accordingly, the magnitude of central command, which is associated with the parallel activation of motor control and cardiovascular control centres, is largely dictated by RPE and independent of power output or force production (Williamson, 2010).

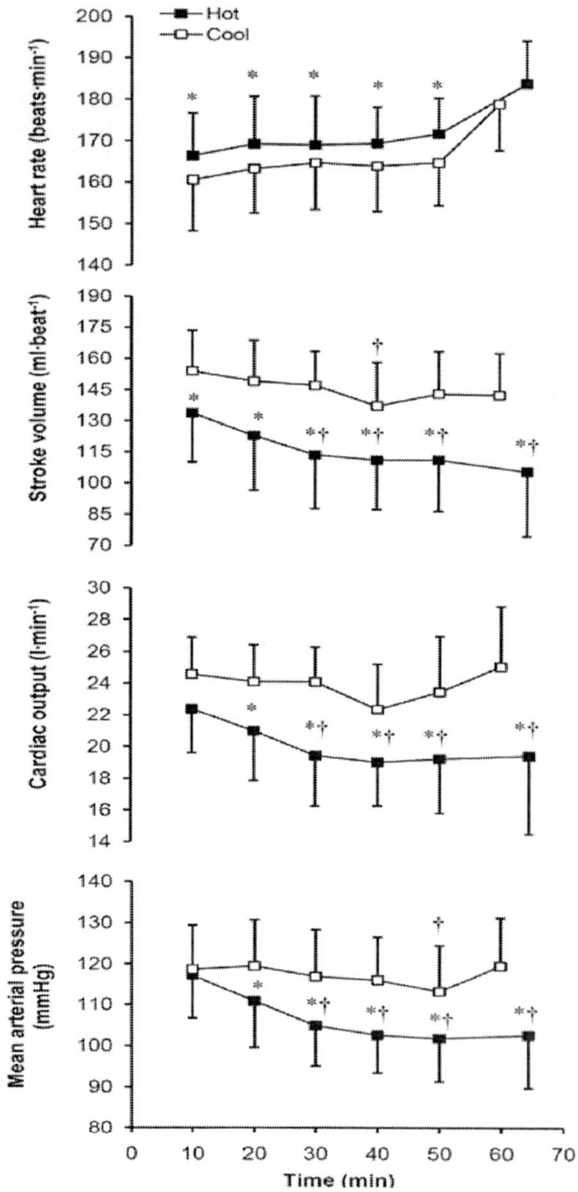

Figure 3. Heart rate, stroke volume, cardiac output, and mean arterial pressure at 10 min intervals and during the final kilometre at maximal effort during a 40 km cycling time trial in hot and thermoneutral (cool) conditions. Values are means ± SD for 8 subjects. *Significantly different from thermoneutral ($P < 0.05$). †Significantly lower than 10 min ($P < 0.05$). Reproduced with permission from Périard et al. (2011b).

It is also well established that RPE, or the sense of effort, can be manipulated to modify the cardiovascular response independent of alterations in muscle afferent input or motor activity. For example, an attempted handgrip contraction in humans under transient paralysis has been shown to result in a similar increase in heart rate and mean arterial pressure to that of an actual contraction, as well as a marked sense of effort (Gandevia et al., 1993). Therefore, the degree of central command (i.e. motor outflow) representing similar heart rates and RPE at exhaustion in hot and cool conditions may not necessarily result in identical muscle recruitment patterns.

Summary

Clearly, the regulation of aerobic exercise performance in the heat is a complex phenomenon mediated by the interplay of several physiological systems and influenced by psychological parameters (e.g. motivation and thermal comfort). A potential pathway of impairment to prolonged dynamic exercise may lie with a reduction in central neural drive to contracting muscles, occurring in conjunction with the attainment of an intrinsically elevated core temperature (Nielsen et al., 2001; Nybo and Nielsen, 2001a; Nybo, 2008). Although passive heating and exercise studies have associated the decrease in force production with a hyperthermia-induced decline in voluntary activation (Saboisky et al., 2003; Morrison et al., 2004; Thomas et al., 2006), there is controversy as to whether this impairment is systemic and extends to both brief and sustained MVCs. Moreover, fatigue can occur on a continuum of core temperatures (Sawka et al., 2001) and is known to be task-dependent. Recently, it was demonstrated that voluntary activation can be maintained during a prolonged contraction in highly motivated subjects, despite significant hyperthermia (Périard et al., 2011c). As such, there is only indirect or circumstantial evidence at present for a centrally-mediated muscle fatigue response that develops during prolonged exercise in the heat. Further investigation of the specific role of central fatigue during dynamic exercise is thus warranted to determine the extent to which performance is impaired by a reduction in neural drive to contracting muscles, and whether this reduction is of physiological or psychological origin. The influence of temperature-induced adjustments on skeletal muscle contractile function (e.g. enhanced rate of muscle relaxation) also requires clarification to determine whether the centrally mediated rate of activation is sufficient to maintain voluntary drive during sustained contractions.

Although the proposition of a central governor model provides an interesting hypothesis for describing the impairment of exercise in the heat, there is limited published research supporting the concept of a feedforward mechanism regulating performance as an anticipatory strategy to avert catastrophe. Current testing methods are inadequate in directly corroborating the existence of this complex intelligent system and separating the influence of the system from a model based on physiological limits (Cheung, 2009; Shephard, 2009). Direct experimental support is therefore required to substantiate the circumstantial evidence utilized to describe this model. In contrast, there is strong evidence to suggest that prolonged aerobic exercise is impaired by an increase in cardiovascular strain associated with the development of hyperthermia. A wealth of research has strongly established the cardiovascular limitations to strenuous exercise in the heat (Rowell et al., 1966; Rowell et al.,

1969; Pirnay et al., 1970; Brengelmann, 1983; Johnson and Proppe, 1996; Nybo et al., 2001; González-Alonso and Calbet, 2003). During maximal exercise, decreases in stroke volume and cardiac output have been noted at the attainment of maximum heart rate (González-Alonso and Calbet, 2003; Mortensen et al., 2005). These decreases were accompanied by reductions in blood flow to exercising muscles. Similar increases in heart rate and declines in stroke volume, cardiac output and mean arterial pressure have been indirectly observed during submaximal constant rate and self-paced exercise in motivated individuals (Périard et al., 2011a; Périard et al., 2011b). In addition, the convergence of core and skin temperature during heat stress and the redistribution of blood to the periphery compromises $\dot{V}O_2$max and limits exercise performance. Thus, it appears that in motivated individuals, impaired prolonged aerobic exercise performance in the heat is associated with the attainment of limits to cardiovascular function, preventing adequate systemic and exercising muscle oxygen delivery. Future research endeavours should strive to provide further evidence of this attenuation via direct measurements of active muscle blood flow, especially during self-paced exercise and on reaching exhaustion during submaximal constant rate exercise.

REFERENCES

Abbiss CR and Laursen PB. (2008). Describing and Understanding Pacing Strategies during Athletic Competition. *Sports Med.* 38, 239-252.

Allen DG, Lamb GD and Westerblad H. (2008). Skeletal muscle fatigue: cellular mechanisms. *Physiol. Rev.* 88, 287-332.

Amann M, Eldridge MW, Lovering AT, Stickland MK, Pegelow DF and Dempsey JA. (2006). Arterial oxygenation influences central motor output and exercise performance via effects on peripheral locomotor muscle fatigue in humans. *J. Physiol.* 575, 937-952.

Armada-da-Silva PA, Woods J and Jones DA. (2004). The effect of passive heating and face cooling on perceived exertion during exercise in the heat. *Eur. J. Appl. Physiol.* 91, 563-571.

Armstrong LE, Casa DJ, Millard-Stanfford M, Moran DS, Pyne SW and Roberts WO. (2007). Exertional heat illness during training and competition. *Med. Sci. Sports Exerc.* 39, 556-572.

Arngrimsson SA, Petitt DS, Stueck MG, Jorgensen DK and Cureton KJ. (2004). Cooling vest worn during active warm-up improves 5-km run performance in the heat. *J. Appl. Physiol.* 96, 1867-1874.

Arngrimsson SA, Stewart DJ, Borrani F, Skinner KA and Cureton KJ. (2003). Relation of heart rate to percent VO2peak during submaximal exercise in the heat. *J. Appl. Physiol.* 94, 1162-1168.

Asmussen E. (1940). The cardiac output in rest and work in humid heat. *Am. J. Physiol.* 131, 54-59.

Asmussen E and Boje O. (1945). Body temperature and capacity for work. *Acta Physiol. Scand* 10, 1-22.

Astrand PO, Cuddy TE, Saltin B and Stenberg J. (1964). Cardiac output during submaximal and maximal work. *J. Appl. Physiol.* 19, 268-274.

Bangsbo J. (1993). *The Physiology of Soccer-With Special reference* to *Intense Intermittent Exercsie*. August Krogh Institute, University of Copenhagen, Copenhagen.

Bassett DRJ and Howley ET. (2000). Limiting factors for maximum oxygen uptake and determinants of endurance performance. *Med. Sci. Sports Exerc.* 32, 70-84.

Bergh U and Ekblom B. (1979). Physical performance and peak aerobic power at different body temperatures. *J. Appl. Physiol.* 46, 885-889.

Bigland-Ritchie B, Dawson NJ, Johansson RS and Lippold OC. (1986). Reflex origin for the slowing of motoneurone firing rates in fatigue of human voluntary contractions. *J. Physiol.* 379, 451-459.

Bigland-Ritchie B, Johansson RS, Lippold OC, Smith S and Woods JJ. (1983a). Changes in motoneurone firing rates during sustained maximal voluntary contractions. *J. Physiol.* 340, 335-346.

Bigland-Ritchie B, Johansson RS, Lippold OC and Woods JJ. (1983b). Contractile speed and EMG changes during fatigue of sustained maximal voluntary contractions. *J. Neurophysiol.* 50, 313-324.

Bigland-Ritchie B, Jones DA, Hosking GP and Edwards RHT. (1978). Central and peripheral fatigue in sustained maximum voluntary contractions of human quadriceps muscle. *Clin. Sci. Mol. Med.* 54, 609-614.

Bigland-Ritchie B, Kukulka CG, Lippold OC and Woods JJ. (1982). The absence of neuromuscular transmission failure in sustained maximal voluntary contractions. *J. Physiol.* 330, 265-278.

Boksem MAS and Tops M. (2008). Mental Fatigue: Costs and Benefits. *Brain Res. Rev.* I59, 125-139.

Brazaitis M and Skurvydas A. (2010). Heat acclimation does not reduce the impact of hyperthermia on central fatigue *Eur. J. Appl. Physiol.* 109, 771-778.

Brengelmann GL. (1983). Circulatory adjustments to exercise and heat stress. *Annu Rev. Physiol.* 45, 191-212.

Brengelmann GL, Johnson JM, Hermansen L and Rowell LB. (1977). Altered control of skin blood flow during exercise at high internal temperatures. *J. Appl. Physiol.* 43, 790-794.

Bridge MW, Weller AS, Rayson M and Jones DA. (2003). Responses to exercise in the heat related to measures of hypothalamic serotonergic and dopaminergic function. *Eur. J. Appl. Physiol.* 89, 451-459.

Brück K and Olschewski H. (1987). Body temperature related factors diminishing the drive to exercise. *Can. J. Physiol. Pharmacol.* 65, 1247-1280.

Burke RE and Levine DN. (1971). Mammalian motor units: physiological-histochemical correlation in three types in cat gastrocnemius. *Science Wash DC* 174, 709-712.

Byrne C, Lee JKW, Chew SAN, Lim CL and Tan EYM. (2006). Continuous thermoregulatory responses to mass-participation distance running in heat. *Med. Sci. Sports Exerc.* 38, 803-810.

Cabanac M. (2006). Sensory pleasure optimizes muscular work. *Clin. Invest. Med.* 29, 110-116.

Caputa M, Feistkorn G and Jessen C. (1986). Effects of brain and trunk temperatures on exercise performance in goats. *Pflugers Arch – Eur. J. Physiol.* 406, 186-189.

Charkoudian N. (2010). Mechanisms and modifiers of reflex induced cutaneous vasodilation and vasoconstriction in humans. *J. Appl. Physiol.* 109, 1221-1228.

Cheung SS. (2009). The mind-body connection remains key to understanding fatigue. *J. Appl. Physiol.* 107, 633-634.

Cheung SS and McLellan TM. (1998). Heat acclimation, aerobic fitness, and hydration effects on tolerance during uncompensable heat stress. *J. Appl. Physiol.* 84, 1731-1739.

Cheuvront SN and Haymes EM. (2001). Thermoregulation and marathon running: biological and environmental influences. *Sports Med.* 31, 743-762.

Cheuvront SN, Kenefick RW, Montain SJ and Sawka MN. (2010). Mechanisms of aerobic performance impairment with heat stress and dehydration. *J. Appl. Physiol.* 109, 1989-1995.

Cheuvront SN, Montain SJ and Sawka MN. (2007). Fluid replacement and performance during the marathon. *Sports Med* 37, 353-357.

Costill DL, Kammer WF and Fisher A. (1970). Fluid ingestion during distance running. *Arch. Environ. Health* 46, 795-800.

Coyle EF and Gonzalez-Alonso J. (2001). Cardiovascular drift during prolonged exercise: new perspectives. *Exerc. Sport Sci. Rev.* 29, 88-92.

Crandall CG and González-Alonso J. (2010). Cardiovascular function in the heat stressed human. *Acta Physiol.* 199, 407-423.

Davies CT and Young K. (1983). Effect of temperature on the contractile properties and muscle power of triceps surae in humans. *J. Appl. Physiol.* 55, 191-195.

Davies CTM. (1979). Influence of skin temperature on sweating and aerobic performance during severe work. *J. Appl. Physiol.* 47, 770-777.

Davis JM and Bailey SP. (1997). Possible mechanisms of central nervous system fatigue during exercise. *Med. Sci. Sports Exerc.* 29, 45-57.

Diamant B, Karlsson J and Saltin B. (1968). Muscle tissue lactate after maximal exercise in man. *Acta Physiol. Scand* 72, 383-384.

diPrampero PE. (2003). Factors limiting maximal performance in humans. *Eur. J. Appl. Physiol.* 90, 420-429.

Dolny DG and Lemon PWR. (1988). Effect of ambient temperature on protein breakdown during prolonged exercise. *J. Appl. Physiol.* 64, 550-555.

Edstom L and Kugelberg E. (1968). Histochemical composition, distribution of fibers and fatiguability of single motor units. *J. Neurol. Neurosurg Psychiatry* 31, 424-433.

Edwards RH, Harris RC, Hultman E, Kaijser L, Koh D and Nordesjö LO. (1972). Effect of temperature on muscle energy metabolism and endurance during successive isometric contractions, sustained to fatigue, of the quadriceps muscle in man. *J. Physiol.* 220, 335-352.

Edwards RHT. (1983). Biochemical bases of fatigue in exercise performance: catastrophe theory of muscular fatigue. In *Biochemistry of Exercise,* ed. Knuttgen HG, pp. 3-28. *Human Kinetics*, Champaign, IL.

Ekblom B. (1968). Effect of physical training on oxygen transport system in man. *Acta Physiol. Scand Suppl.* 328, 1-45.

Ekblom B and Hermansen L. (1968). Cardiac output in athletes. *J. Appl. Physiol.* 25, 619-625.

Ekelund LG and Holmgren A. (1967). Central hemodynamics during exercise. *Circ. Res.* 21, I33-I43.

Ely BR, Cheuvront SN, Kenefick RW and Sawka MN. (2010). Aerobic performance is degraded, despite modest hyperthermia, in hot environments. *Med. Sci. Sports Exerc.* 42, 135-141.

Ely BR, Ely MR, Cheuvront SN, Kenefick RW, DeGroot DW and Montain SJ. (2009). Evidence against a 40°C core temperature threshold for fatigue in humans. *J. Appl. Physiol.* 107, 1519-1525.

Ely MR, Cheuvront SN, Roberts WO and Montain SJ. (2007). Impact of weather on marathon-running performance. *Med. Sci. Sports Exerc.* 39, 487-493.

Enoka RM and Stuart DG. (1992). Neurobiology of muscle fatigue. *J. Appl. Physiol.* 72, 1631-1648.

Febbraio MA, Carey MF, Snow RJ, Stathis CG, Harrison MH and Hargreaves M. (1996). Influence of elevated muscle temperature on metabolism during intense, dynamic exercise. *Am. J. Physiol.* 271, R1251-1255.

Febbraio MA, Snow RJ, Hargreaves M, Stathis CG, Martin IK and Carey MF. (1994a). Muscle metabolism during exercie and heat stress in trained men: effect of acclimation. *J. Appl. Physiol.* 76, 589-597.

Febbraio MA, Snow RJ, Stathis CG, Hargreaves M and Carey MF. (1994b). Effect of heat stress on muscle energy metabolism during exercise. *J. Appl. Physiol.* 77, 2827-2831.

Fink WJ, Costill DL and Van Handel PJ. (1975). Leg muscle metabolism during exercise in the heat and cold. *Eur. J. Appl. Physiol. Occup. Physiol.* 15, 183-190.

Fitts RI. (1994). Cellular Mechanisms of Muscle Fatigue. *Physiol. Rev.* 74, 49-94.

Fritzsche RG, Switzer TW, Hodgkinson BJ and Coyle EF. (1999). Stroke volume decline during prolonged exercise is influenced by the increase in heart rate. *J. Appl. Physiol.* 86, 799-805.

Fuller A, Carter RN and Mitchell D. (1998). Brain and abdominal temperature at fatigue in rats exercising in the heat. *J. Appl. Physiol.* 84, 877-883.

Galloway SD and Maughan R, J. (1997). Effects of ambient temperature on the capacity to perform prolonged cycle exercise in man. *Med. Sci. Sports Exerc* 29, 1240-1249.

Gandevia SC. (1998). Neural control in human muscle fatigue: changes in muscle afferents, moto neurones and moto cortical drive. *Acta Physiol. Scand.* 162, 275-283.

Gandevia SC. (2001). Spinal and Supraspinal Factors in Human Muscle Fatigue. *Physiol. Rev.* 81, 1725-1789.

Gandevia SC, Killian K, McKenzie DK, Crawford M, Allen GM, Gorman RB and Hales JP. (1993). Respiratory sensations, cardiovascular control, kinaesthesia and transcranial stimulation during paralysis in humans. *J. Physiol.* 410, 85-107.

Gisolfi CV and Mora F. (2000). *The Hot Brain: Survival, Temperature, and the human body. The MIT Press.*,Cambridge, Massachusetts.

González-Alonso J. (1998). Separate and combined influences of dehydrationn and hyperthermia on cardiovascular responses to exercise. *Int. J. Sports Med.* 19, S111-S114.

González-Alonso J and Calbet AL. (2003). Reductions in systemic and skeletal muscle blood flow and oxygen delivery limit maximal aerobic capacity in humans. *Circulation* 107, 824-830.

González-Alonso J, Calbet AL and Nielsen B. (1998). Muscle blood flow is reduced with dehydration during prolonged exercise in humans. *J. Physiol.* 513, 895-905.

González-Alonso J, Calbet AL and Nielsen B. (1999a). Metabolic and thermodynamic responses to dehydraion-induced reductions in muscle blood flow in exercising humans. *J. Physiol.* 520, 577-589.

González-Alonso J, Dalsgaard MK, Osada T, Volianitis S, Dawson EA, Yoshiga CC and Secher NH. (2004). Brain and central haemodynamics and oxygenation during maximal exercise in humans. *J. Physiol.* 557, 331-342.

González-Alonso J, Mora-Rodriguez R, Below PR and Coyle EF. (1995). Dehydration reduces cardiac output and increases systemic and cutaneous vascular resistance during exercise. *J. Appl. Physiol.* 79, 1487-1496.

González-Alonso J, Mora-Rodriguez R, Below PR and Coyle EF. (1997). Dehydration markedly impairs cardiovascular function in hyperthermic endurance athletes during exercise. *J. Appl. Physiol.* 82, 1229-1236.

González-Alonso J, Mora-Rodriguez R and Coyle EF. (2000). Stroke volume during exercise: interaction of environment and hydration. *Am. J. Physiol. Heart Circ. Physiol.* 278, H321-330.

González-Alonso J, Teller C, Andersen SL, Jensen FB, Hyldic T and Nielsen B. (1999b). Influence of body temperature on the development of fatigue during prolonged exercise in the heat. *J. Appl. Physiol.* 86, 1032-1039.

Gordon CJ. (1993). Temperature Regulation in Laboratory Rodents. Cambridge Univ Press, Cambridge, UK.

Hargreaves M, Dillo P, Angus D and Febbraio M. (1996). Effect of fluid ingestion on muscle metabolism during prolonged exercise. *J. Appl. Physiol.* 80, 363-366.

Hockey GRJ. (1997). Compensatory control in the regulation of human performance under stress and high workload: A cognitive-energetical framework. *Biol. Psychol.* 45, 73-93.

Holding D. (1983). Fatigue. In *Stress and fatigue in human performance,* ed. Hockey GRJ, pp. 154-164. John Woley and Sons, Durnham.

Holmgren A. (1956). Circulatory changes during muscular work in man: with special reference to arterial and central venous pressure in the systemic circulation. *Scand. J. Clin. Lab. Invest.* 8, 1-97.

Iampietro PF. (1971). Use of skin temperature to predict tolerance to thermal environments. *Aerosp. Med.* 42, 396-399.

Iampietro PF and Goldman RF. (1965). Tolerance of men working in hot, humid environments. *J. Appl. Physiol.* 20, 73-76.

Jacobs I, Tesch PA, Bar-Or O, Karlsson J and Dotan R. (1983). Lactate in human skeletal muscle after 10 and 30 s of supramaximal exercise. *J. Appl. Physiol.* 55, 365-367.

Jansson EP, Hjemdahl P and Kaijser L. (1986). Epinephrine-induced changes in muscle carbohydrate metabolism during exercise in male subjects. *J. Appl. Physiol.* 60, 1466-1470.

Jay O and Kenny GP. (2009). Current evidence does not support an anticipatory regulation of exercise intensity mediated by rate of body heat storage. *J. Appl. Physiol.* 107, 630-631.

Johnson JM. (1977). Regulation of skin circulation during prolonged exercise. *Ann. NY Acad. Sci.* 301, 195-212.

Johnson JM. (1986). Nonthermoregulatory control of human skin blood flow. *J. Appl. Physiol.* 61, 1613-1622.

Johnson JM. (1992). Exercise and the cutaneous circulation. *Exerc. Sport Sci. Rev.* 20, 59-97.

Johnson JM and Kellogg DL. (2010). Local thermal control of the human cutaneous circulation. *J. Appl. Physiol.* 109, 1229-1238.

Johnson JM and Park MK. (1979). Reflex control of skin blood flow by skin temperature: role of core temperature. *J. Appl. Physiol.* 47, 1188-1193.

Johnson JM and Proppe DW. (1996). Cardiovascular adjustments to heat stress. In Handbook of Physiology: Environmental Physiology, ed. Fregly MJ and Blatteis CM, pp. 215-243. *Am. Physiol. Soc.*, Bethesda, MD.

Johnson JM and Rowell LB. (1975). Forearm skin and muscle vascular responses to prolonged leg exercise in man. *J. Appl.Physiol.* 39, 920-924.

Johnson JM, Rowell LB and Brengelmann GL. (1974b). Modification of the skin blood flow-body temperature relationship by upright exercise. *J. Appl. Physiol.* 37, 880-886.

Kandel ER and Schwartz JH. (1985). *Principals of neural science*. Elsevier, Amsterdam.

Karlsson J. (1971). Lactate and phosphagen concentrations in working muscle of man. *Acta Physiol. Scand.* 81, 1-72.

Kellogg DLJ, Johnson JM and Kosiba WA. (1991). Control of internal temperature threshold for active cutaneous vasodilation by dynamic exercise. *J. Appl. Physiol.* 71, 2476-2482.

Kenefick RW, Cheuvront SN, Palombo LJ, Ely BR and Sawka MN. (2010). Skin temperature modifes the impact of hypohydration on aerobic performance. *J. Appl. Physiol.* 109, 79-86.

Kenney WL and Johnson JM. (1992). Control of skin blood flow during exercise. *Med. Sci. Sports Exerc* 24, 303-312.

Kirwan JP, Costill DL, Kuipers H, Burrell MJ, Fink WJ, Kovaleski JE and Fielding RA. (1987). Substrate utilization in leg muscle of men after heat acclimation. *J. Appl. Physiol.* 63, 31-35.

Knight DR, Poole DC, Schaffartzik W, Gay HJ, Prediletto R, Hogan MC and Wagner P. (1992). Relationship between body and legVO2 during maximal cycle ergometry. *J. Appl. Physiol.* 73, 1114-1121.

Kozlowski S, Brzezinska Z, Kruk B, Kacibu-Uscillo H, Greenleaf JF and Nazar K. (1985). Exercise hyperthermia as a factor limiting physical performance: temperature effect on muscle metabolism. *J. Appl. Physiol.* 59, 766-773.

Larsson L, Edstom L, Lindegren B and Gorza L. (1991). MHC composition and enzyme-histochemical and physiological properties of a novel fast-twitch motor unit type. *Am. J. Physiol. Cell Physiol.* 261, C93-101.

Lee J, Nio A, Lim C, Teo E and Byrne C. (2010). Thermoregulation, pacing and fluid balance during mass participation distance running in a warm and humid environment. *Eur. J. Appl. Physiol.* 109, 887-898.

Lind AR. (1963). A physiological criterion for setting thermal environmental limits for everyday work. *J. Appl. Physiol.* 18, 51-56.

Maron MB, Wagner JA and Horvath SM. (1977). Thermoregulatory responses during competitive marathon running. *J. Appl. Physiol.* 42, 909-914.

Martin DE and Buoncristiani J. (1999). The effects of temperature on marathon runners' performance. *Chance* 12, 20-24.

Maughan RJ and Shirreffs SM. (2010). Dehydration and rehydration in competative sport. *Scand. J. Med. Sci. Sports* 20, 40-47.

Meeusen R and Roelands B. (2010). Central fatigue and neurotransmitters, can thermoregulation be manipulated? *Scand. J. Med. Sci. Sports* 20, 19-28.

Merton PA. (1954). Voluntary strength and fatigue. *J. Physiol.* 123, 553-564.

Metzger JM, Greaser ML and Moss RL. (1989). Variations in cross-bridge attachment rate and tension with phosphorylation of myosin in mammalian skinned skeletal muscle fibers. *J. Gen. Physiol.* 93, 855-883.

Miller EK and Cohen JD. (2001). An integrative theory of prefrontal cortex function. *Annu. Rev. Neurosci.* 24, 167-202.

Montain S and Coyle EF. (1992a). Influence of graded dehydration on hyperthermia and cardiovascular drift during exercise. *J. Appl. Physiol* 73, 1340-1350.

Morrison S, Sleivert GG and Cheung SS. (2004). Passive hyperthermia reduces voluntary and isometric force production. *Eur. J. Appl. Physiol* 91, 729-736.

Morrison S, Sleivert GG and Cheung SS. (2006). Aerobic influence on neuromuscular function and tolerance during passive hyperthermia. *Med. Sci. Sports Exerc.* 38, 1754-1761.

Morrison SA, Sleivert GG, Neary JP and Cheung SS. (2009). Prefrontal cortex oxygenation is preserved and does not contribute to impaired neuromuscular activation during passive hyperthermia. *Appl. Physiol. Nutr. Metab.* 34, 66-74.

Mortensen SP, Damsgaard R, Dawson EA, Secher NH and González-Alonso J. (2008). Restrictions in systemic and locomotor skeletal muscle perfusion, oxygen supply and VO2 during high-intensity whole-body exercise in humans. *J. Physiol.* 586, 2621-2635.

Mortensen SP, Dawson EA, Yoshiga CC, Dalsgaard MK, Damsgaard R, Secher NH and Gonzalez-Alonso J. (2005). Limitations to systemic and locomotor limb muscle oxygen delivery and uptake during maximal exercise in humans. *J. Physiol.* 566, 273-285.

Newsholme E. (1987). Amino acids, brain neurotransmitters and a function link between muscle and brain that is important in sustained exercise. In *Advances in myochemistry*, ed. Benzi G, pp. 127-133. John Libbey Eurotext, London.

Nielsen B. (1998). Heat acclimation - mechanisms of adaptation to exercise in the heat. *Int. J. Sports Med.* 19, S154-S156.

Nielsen B, Hales JRS, Strange S, Christensen NJ, Warberg J and Saltin B. (1993). Human circulatory and thermoregulatory adaptations with heat acclimation and exercise in a hot, dry environment. *J. Physiol.* 460, 467-485.

Nielsen B, Hyldig T, Bidstrup F, González-Alonso J and Christoffersen GR. (2001). Brain activity and fatigue during prolonged exercise in the heat. Pflugers Arch – *Eur. J. Physiol.* 442, 41-48.

Nielsen B and Nybo L. (2003). Cerebral changes during exercise in the heat. *Sports Med.* 33, 1-11.

Nielsen B, Savard G, Richter EA, Hargreaves M and Saltin B. (1990). Muscle blood flow and musle metabolism during exercise and heat stress. *J. Appl. Physiol.* 69, 1040-1046.

Nielsen B, Strange S, Christensen NJ, Warberg J and Saltin B. (1997). Acute and adaptive responses in humans to exercise in a warm, humid environment. *Pflugers Arch – Eur. J. Physiol.* 434, 49-56.

Nielsen HB, Boushel R, Madsen P and Secher NH. (1999). Cerebral desaturation during exercise reversed by O2 supplementation. *Am. J. Physiol.* 377, H1045-1052.

Nielsen M. (1938). Die Regulation der Körper temperatur bei Muskelarbeit. *Skand Arch. Physiol.* 79, 193-230.

Noakes TD. (2011). Time to move beyond a brainless exercise physiology: the evidence for complex regulation of human exercise performance. *Appl. Physiol. Nutr. Metab.* 36, 23-35.

Noakes TD, St Clair Gibson A and Lambert EV. (2004). From catastrophe to complexity: a novel model of integrative central neural regulation of effort and fatigue during exercise in humans. *Br. J. Sports* Med 38, 511-514.

Nose H, Takamata A, Mack GW, Oda Y, Kawabata T, Hashimoto S, Hirose M, Chihara E and Marimoto T. (1994). Right atrial pressure and forearm blood flow during prolonged exercise in hot environment. Pflugers Arch – *Eur. J. Physiol.* 426, 177-182.

Nunneley SA, Antuñano MJ and Bomalaski SH. (1992). Thermal convergence fails to predict heat tolerance limits. *Aviat Space Environ Med.* 63, 886-890.

Nybo L. (2007). Exercise and heat stress: cerebral challenges and consequences. *Prog. Brain Res.* 162, 29-43.

Nybo L. (2008). Hyperthermia and Fatigue. *J. Appl. Physiol.* 140, 871-878.

Nybo L. (2010). Cycling in the heat: performance perspectives and cerebral challenges. *Scand. J. Med. Sci. Sports* 20, 71-79.

Nybo L, Jensen T, Nielsen B and Gonzalez-Alonso J. (2001). Effects of marked hyperthermia with and without dehydration on VO2 kinetics during intense exercise. *J.Appl. Physiol.* 90, 1057-1064.

Nybo L, Moller K, Volianitis S, Nielsen B and Secher NH. (2002a). Effects of hyperthemia on cerebral blood flow and metabolsim during prolonged exercise in humans. *J. Appl. Physiol.* 93, 58-64.

Nybo L and Nielsen B. (2001a). Hyperthermia and central fatigue during prolonged exercise in humans. *J. Appl. Physiol.* 91, 1055-1060.

Nybo L and Nielsen B. (2001b). Perceived exertion is associated with an altered brain activity during exercise with progressive hyperthermia. *J. Appl. Physiol.* 91, 2017-2023.

Nybo L and Rasmussen P. (2007). Inadequate cerebral oxygen delivery and central fatigue during strenuous exercise. *Exerc. Sport Sci. Rev.* 35, 110-118.

Nybo L, Secher NH and Nielsen B. (2002b). Inadequate heat release from the human brain during prolonged exercise with hyperthermia. *J. Physiol.*575, 697-704.

Pandolf KB and Goldman RF. (1978). Convergence of skin and rectal temperatures as a criterion for heat tolerance. *Aviat Space Environ. Med.* 49, 1095-1101.

Parkin JM, Carey MF and Febbraio MA. (1999). Effect of ambient temperature on human skeletal muscle metabolism during fatiguing submaximal exercise. J Appl Physiol 86, 902-908.

Périard JD, Caillaud C and Thompson MW. (2011a). Central and peripheral fatigue during passive and exercise-induced hyperthermia. *Med. Sci. Sports Exerc.* 43, 1657-1665.

Périard JD, Cramer MN, Chapman PG, Caillaud C and Thompson MW. (2011c). Neuromuscular function following prolonged intense self-paced exercise in hot climatic conditions. *Eur. J. Appl. Physiol.* 111, 1561-156.

Périard JD, Cramer MN, Chapman PG, Caillaud C and Thompson MW. (2011b). Cardiovascular strain impairs prolonged self-paced exercise in the heat. *Exp. Physiol.* 96, 134-144.

Pirnay F, Deroanne R and Petit JM. (1970). Maximal oxygen consumption in a hot environment. *J. Appl. Physiol.* 28, 642-645.

Pugh LGCE, Corbett JL and Johnson RH. (1967). Rectal temperatures, weight losses, and sweat rates in marathon running. *J. Appl. Physiol.* 23, 347-352.

Racinais S, Gaoua N and Grantham J. (2008). Hyperthermia impairs short-term memory and peripheral motor drive transmission. *J. Physiol.* 586, 4751-4762.

Radegran G and Saltin B. (1998). Muscle blood flow at onset of dynamic exercise in humans. *Am. J. Physiol. Heart Circ Physiol.* 274, H314-322.

Ramnani N and Owen AM. (2004). Anterior prefrontal cortex: insights into function from anatomy and neuroimaging. *Nat. Rev. Neurosci.* 5, 184-194.

Rasmussen P, Nielsen J, Overgaard M, Krogh-Madsen R, Gjedde A, Secher NH and Petersen NC. (2010a). Reduced muscle activation during exercise related to brain oxygenation and metabolism in humans. *J. Physiol.* 588, 1985-1995.

Rasmussen P, Nybo L, Volianitis S, Møller K, Secher NH and Gjedde A. (2010b). Cerebral oxygenation is reduced during hyperthermic exercise in humans. *Acta Physiol.* 199, 63-70.

Richards D, Richards R, Schofield PJ, Ross V and Sutton JR. (1979a). Management of heat exhaustion in Sydney's The Sun City-to-Surf fun runners. *Med. J. Aust* 2, 457-461.

Richards R, Richards D, Schofield PJ, Ross V and Sutton JR. (1979b). Organization of The City-to-Surf fun run, Sydney, 1979. *Med. J. Aust* 2, 470-474.

Richter EA. (1984). Influence of sympahto-adrenal system on some metabolic and hormonal responses to exercise in the rat. *Acta Physiol. Scand Suppl.* 528, 5-42.

Roberts WO. (1989). Exercise-associated collapse in endurance events: a classification system. *Phys. Sportsmed.* 17, 49-55.

Robertson RJ. (1982). Central signals of perceived exertion during dynamic exercise. *Med. Sci. Sports Exerc.* 14, 390-396.

Robinson S. (1963). Temperature regulation in exercise. *Pediatrics* 32, 691-702.

Rosenmeier JB, Hansen J and Gonzalez-Alonso J. (2004). Circulating ATP-induced vasodilation overrides sympathetic vasoconstrictor activity in human skeletal muscle. *J. Physiol.* 558, 351-365.

Rowell LB. (1974). Human cardiovascular adjustments to exercise and thermal stress. *Physiol. Rev.* 54, 75-159.

Rowell LB. (1977). Competition between skin and muscle for blood flow during exercise. In *Problems with temperature regulation during exercise,* ed. Nadel ER, pp. 49-76. Academic Press, New York.

Rowell LB. (1983). Cardiovascular adjustments to heat stress. In *Handbook of Physiology The Cardiovascular System: Peripheral Circulation and Organ Blood Flow*, sect 2, ed. Shepherd JT, Abboud FM and Geiger SR, pp. 967-1023. American Physiological Society, Bethesda, MD.

Rowell LB. (1986). *Human Circulation: Regulation during Physical Stress*. Oxford University Press New York.

Rowell LB, Marx HJ, Bruce RA, Conn RD and Kusumi F. (1966). Reductions in cardiac output, central blood volume, and stroke volume with thermal stress in normal men during exercise. *J. Clin. Invest* 45, 1801-1816.

Rowell LB, Murray JA, Brengelmann GL and Kranning KK. (1969). Human cardiovascular adjustments to rapid changes in skin temperature during exercise. *Circ. Res.* 24, 711-724.

Saboisky J, Marino FE, Kay D and Cannon J. (2003). Exercise heat stress does not reduce central activation to non-exercised human skeletal muscle. *Exp. Physiol.* 88, 783-790.

Sahlin K, Alverstand A, Brandt R and Hultman E. (1978). Intracellular pH and bicarbonate concentration in human muscle during recovery from exercise. *J. Appl. Physiol.* 45, 474-480.

Saltin B and Hermansen L. (1966). Esophageal, rectal, and muscle temperature during exercise. *J. Appl. Physiol.* 21, 1757-1762.

Saltin B, Radegran G, Koskolou MD and Roach RC. (1998). Skeletal muscle blood flow in humans and its regulation during exercise. *Acta Physiol Scand* 162, 421-436.

Saltin B and Stensberg J. (1964). Circulatory response to prolonged severe exercise. *J. Appl. Physiol.* 19, 833-838.

Savard GK, Kiens B and Saltin B. (1987). Limb blood flow in prolonged exercise; magnitude and implication for cardiovascular control during muscular work in man. *Can. J. Appl. Sports Sci.* 12, S89-S101.

Savard GK, Nielsen B, Laszczynska J, Larsen BE and Saltin B. (1988). Muscle blood flow is not reduced in humans during moderate exercise and heat stress. *J. Appl. Physiol.* 64, 649-657.

Sawka MN, Knowlton RG and Critz JB. (1979). Thermal and circulatory responses to repeated bouts of prolonged running. *Med. Sci. Sports* 11, 177-180.

Sawka MN, Latzka WA, Matott RP and Montain SJ. (1998). Hydration effects on temperature regulation. Int J Sports Med 19, S108-S110.

Sawka MN, Latzka WA, Montain S, Cadarette BS, Kolka MA, Kraning II KK and Gonzalez RR. (2001). Physiologic tolerance to uncompensable heat: intermittent exercise, field vs laboratory. *Med. Sci. Sports Exerc.* 33, 422-430.

Sawka MN, Young AJ, Latzka WA, Neufer PD, Quigley MD and Pandolf KB. (1992). Human tolerance to heat strain during exercise: influence of hydration. *J. Appl. Physiol.* 73, 368-375.

Sawka MN, Young K, Cadarette BS, Levine L and Pandolf KB. (1985). Influence of heat stress and acclimation on maximal aerobic power. *Eur. J. Appl. Physiol. Occup. Physiol.* 53, 294-298.

Schertzer JD, Green HJ and Tupling AR. (2002). Thermal instability of rat muscle sarcoplasmic reticulum Ca2+-ATPase function. *Am. J. Physiol. Endocrinol. Metab.* 283, E722-728.

Schlader ZJ, Raman A, Morton R, Stannard SR and Mündel T. (2010). Exercise modality modulates body temperature regulation during exercise in uncompensable heat stress. *Eur. J. Appl. Physiol.* DOI: 10.1007/s00421-010-1692-3, 1-10.

Schlader ZJ, Stannard SR and Mündel T. (2011). Exercise and heat stress: performance, fatigue and exhaustion—a hot topic. *Br. J. Sports Med.* 45, 3-5.

Seals DR and Victor RG. (1991). Regulation of muscle sympathetic nerve activity during exercise in humans. *Exerc. Sport Sci. Rev.* 19, 313-349.

Selkirk GA and McLellan TM. (2001). Influence of aerobic fitness and body fatness on tolerance to uncompensable heat stress. *J. Appl. Physiol.* 91, 2055-2063.

Senay LC, Mitchell D and Wyndham CH. (1976). Acclimatization in a hot, humid environment: body fluid adjustments. *J. Appl. Physiol.* 40, 786-796.

Shephard RJ. (2009). Is it Time to Retire the 'Central Governor'? *Sports Med.* 39, 709-721.

Smolander J, Saalo J and Korhonen O. (1991). Effect of work load on cutaneous vascular response to exercise. *J. Appl. Physiol.* 71, 1614-1619.

St Clair Gibson A and Noakes TD. (2004). Evidence for complex system integration and dynamic neural regulation of skeletal muscle recruitment during exercise in humans. *Br. J. Sports Med.* 38, 797-806.

Sutton JR. (1984). Heat illness. In *Sports Medicine*, ed. Strauss RH, pp. 307-322. W.B. Saunders, Philadelphia.

Tatterson AJ, Hahn AG, Martin DT and Febbraio MA. (2000). Effects of heat stress on physiological responses and exercise performance in elite cyclists. *J. Sci. Med. Sport* 3, 186-193.

Taylor CR and Rowntree VJ. (1973). Temperature regulation and heat balance in running cheetahs: a strategy for sprinters? *Am. J. Physiol.* 224, 848-851.

Thomas MM, Cheung SS, Elder GC and Sleivert GG. (2006). Voluntary muscle activation is impaired by core temperature rather than local muscle temperature. *J. Appl. Physiol.* 100, 1361-1369.

Thompson MW. (2006). Cardiovascular drift and critical core temperature: factors limiting endurance performance in the heat? *J. Exerc. Sci. Fit* 4, 15-24.

Thompson MW, Ward JJ, Atkins K, Brotherhood J and Sutton JR. (1997). Maximising sport performance in hot, humid environmental conditions. In *The Atlanta Heat Study*. Scientific Report, Australian Sports Commission.

Todd G, Butler JE, Taylor JL and Gandevia SC. (2005). Hyperthermia: a failure of the motor cortex and the muscle. *J. Physiol.* 563, 621-631.

Trinity JD, Pahnke MD, Lee JF and Coyle EF. (2010). Interaction of Hyperthermia and Heart Rate on Stroke Volume during Prolonged Exercise. *J. Appl. Physiol.* 109, 745-751.

Tucker R. (2009). The anticipatory regulation of performance: The physiological basis for pacing strategies and the development of a perception-based model for exercise performance. *Br. J. Sports Med.* 43, 392-400.

Tucker R, Marle T, Lambert EV and Noakes TD. (2006). The rate of heat storage mediates an anticipatory reduction in exercise intensity during cycling at a fixed rating of perceived exertion. *J. Physiol.* 574, 905-915.

Tucker R, Rauch L, Harley YXR and Noakes TD. (2004). Impaired exercise performance in the heat is associated with an anticipatory reduction in skeletal muscle recruitment. Pflugers Arch – *Eur. J. Physiol.* 448, 422-430.

Walters TJ, Ryan KL, Tate LM and Mason PA. (2000). Exercise in the heat is limited by a critical internal temperature. *J. Appl. Physiol.* 89, 799-806.

Williamson JW. (2010). The relevance of central command for the neural cardiovascular control of exercise. *Exp. Physiol.* 95, 1043-1048.

Wilson TE and Crandall CG. (2011). Effect of thermal stress on cardiac function. *Exerc. Sport Sci. Rev.* 39, 12-17.

Wingo JE and Cureton KJ. (2006b). Body cooling attenuates the decrease in maximal oxygen uptake associated with cardiovascular drift during heat stress. *Eur. J. Appl. Physiol.* 98, 97-104.

Wingo JE, Lafrenz AJ, Ganio MS, Edwards GL and Cureton KJ. (2005). Cardiovascular drift is related to reduced maximal oxygen uptake during heat stress. *Med. Sci. Sports Exerc.* 37, 248-255.

Woods JJ, Furbush F and Bigland-Ritchie B. (1987). Evidence for a fatigue-induced reflex inhibition of motoneuron firing rates. *J. Neurophysiol.* 58, 125-137.

Yeargin BE, Casa DJ, Armstrong LE, Watson G, Judelson DA, Psathas E and Sparrow S. (2006). Heat acclimatization and hydration status of American football players during innital summer workouts. *J. Strength Cond. Res.* 20, 463-470.

Young DR, Mosher R, Erve P and Spector H. (1959). Body temperature and heat exchange during treadmill running in dogs. *J. Appl. Physiol.* 14, 839-843.

This chapter was reviewed by:

Associate Professor Martin W. Thompson
Discipline of Exercise and Sport Science,
The University of Sydney, Sydney, Australia

Dr. Sébastien Racinais
Research Scientist, Aspetar,
Qatar Orthopaedic and Sports Medicine Hospital,
Research and Education Centre, Doha, Qatar

In: Hyperthermia: Recognition, Prevention and Treatment ISBN 978-1-61942-275-9
Editor: Xing-Hua Gao and Hong-Duo Chen, 115-164

Chapter 4

Liver Responsiveness and Flexibility in the Vertebrata Comparative Row in Normal Condition and after Hyperthermia

Elena Ivanovna Antonova
Doctor of Biology, Professor
Major – histology, cytology, cellular biology
Omsk State Pedagogical University, Russia
Department of Cytology and Genetics
Tukhachevski Embankment, Omskaya oblast, Russia

Abstract

Rules of the evolutionary formation of responsiveness and flexibility of functionally similar tissues from the parallelism theory standpoint, levels of biological systems' organization (molecular, subcellular, cellular, inter-tissular, organ), metabolism strategies (hypercatabolic - resistant; tolerant – hypometabolic, tolerant hypobiosis), differences in DNA concentration, difference in the organisms' standings on the evolution ladder were studied on in situ model in the normal state and after hyperthermia action. Studying this problem in the light of the global warming is becoming especially topical. Hereby the state of organs directly involved in the organism homeostasis maintenance acquires a special importance. Liver is one of such organs.

In particular it was found that liver acinus histotopograohy in the comparative row of intact animals changes from tubular type (fishes, amphibias, reptiles) to tubular-trabecular (birds) and trabecular (mammals) type. Number density of liver acini, volumetric density of mitochondrias, hepatocyte apoptosis/necrosis ratios increase from ectothermal animals to endothermal ones. In the hepatocyte ultrastructure species specificity of the nucleolus structure, volumetric, surface and number indicators of mitochondrias, granular endoplasmic reticula, lysosomes, glycogen and lipids is identified.

In the evolution polyplodization due to acytolkinetic mitosis bocomes the leading form of hepatocytes' physiological regeneration. And, if in mammals, reptiles and fishes this hepatocyte population dominanation zones are not present, in birds this is portal tract zone and in amphibias – central veins' zone. Hereby the number of PCNA-positive hepatocytes in birds in all acinus zones is by an order of magnitude higher. In other animal groups their number is smaller with the predominant localization in the periporal zone.

The number of desmin-positive Ito cells grows from ectothermal animals to endothermal ones; whereas the maximum number of organ-specific macrophages is found in fishes and mammals.

In intact fishes and amphibias NAD- and NADP-dependent mitochondrial oxidation paths are expressed to an equal degree; in reptiles and birds a higher activity of NADP-dependent dehydrogenases and in mammals – NAD-dependent dehydrogenases is found. The highest activity of hepatocyte apoptosis biochemical markers is found in birds and the lowest activity – in amphibias and fishes.

2. Single hyperthermia detects two animal groups “amphibias/birds” and “fishes/reptiles/mammals” causes species-specific changes of the liver acinus bloodstream morphometrical indicators, increased number of stromal liver cells with the change of the topography thereof within the acinus; change hepatocytes’ physiological regeneration and death pathways. Hyperthermia determines the change of hepatocyte cell cycle structure: in “amphibias/birds” group the number of hepatocytes in the cell cycle G0-G1-stage increases and in the “fishes/reptiles/ mammals” group the increase is observed in S- and G2-M-stages. The number of hypodiploid hepatocytes in all the animal groups increases, the main death pathway is necrosis/autophagy.

Hepatocytes’ ultrastructural changes are of mosaic and oppositely-directed nature. Volumetric, number and surface density of the mitochondrias, nucleolus system, granular endoplasmic reticula, lysosomes as well as lipids/glycogen volumetric ratios change. Activation of the liver stromal cells manifested as the increase of their number, transformation of Ito cells into myofibroblasts and increase of autophagolysosomes and cell debris fragments in the Kuppfer cells is marked.

Depending on the development of hypo- or hypermetabolic responses these changes reflect the increase of the number of binucleate or PCNA-positive hepatocytes or both the hepatocyte populations.

In reptiles hyperthermia causes increased activity of NAD- and NADP-containing mitochondric oxidoreductases; in mammals – increased activity of NADP-containing mitochondric oxidoreductases and in fishes - NAD-containing mitochondric oxidoreductases which reflects hypercatabolism development. In birds and amphibias the activity of NAD- and NADP-dependent mitochondric oxidoreductases is inhibited which reflects hypometabolism development. Species differences demonstrate cytosol oxidoreductases’ activity, ions’ concentration, activities of biochemical markers of hepatocytes’ proliferation and death pathways.

Consequently, hypercatabolic metabolism strategy (resistant) after overheating is identified in reptiles and mammals and, to a smaller degree, in fishes. Phylogenetically more ancient hypometabolic strategy (tolerant hypobiosis, tolerant) – in amphibias and birds. Metabolic depression in intact ectothermal animals has natural basis when hypoxic state is changed by hyperoxic state due to reoxygenation. Nevertheless, in birds hypometabolism develops which is manifested at the level of both mitochondrial and cytosol oxidoreductases as well as in volumetric “lipids/glycogen” ratios, cell cycle indicators. It reflects the provision that in the evolution new possibilities/ functions arise on the basis of the already attained level, due to the activation of the evolutionally older hepatocyte functioning and energy-supply mechanisms. In its turn, the identified activation of oxidoreductases of tricarboxylic acids’ cycle and cytosol in reptiles is due to the fact that being descendants of the ancient amphibias, reptiles, particularly, tortoises gave birth two major branches of the present-day vertebrates – birds and mammals. Tortoise order descends directly from cotylosaurs and, consequently, their position is the closest to the reptile-like ancestors of mammals. Consequently, it is possible to suggest that in extreme conditions metabolic adaptation strategies’ switching is observed:

endothermal animals may activate poikilothermal thermal regulation mechanisms and vice versa.

INTRODUCTION

Study of the laws of the evolution stabilization of functionally similar tissues responsiveness and flexibility from the parallelism theory standpoint is a fundamental problem in biology nowadays (J.B. Pritchard, 2002; H. Scholz, 2002).

In the evolution temperature is one of the factors ensuring development, growth, proliferation, functioning and death of the cells in the holistic organism in ecto- and endothermal animals with different DNA index (A.E. Vinogradov, 2005; J. Frampton, 2006). Studies of this problem in the light of the global warming is becoming especially topical. In its turn, the studying of hyperthermia effect on biological systems is adequately topical both physiologically (climatic and geographic migration, high-temperature technology conditions) and clinically (S.C. Chen, 2001; H. Talbot, 2001; M. Muraca, 2002). Hereby the state of organs directly involved in the organism homeostasis maintenance acquired a special importance. Liver is one of such organs. Its structure complexity, multi-functionality, rate of involvement into destructive and reparative processes are the factors determining the unfading interest scientist experience to liver regeneration problems: sources, scales, mechanisms. The high metabolism intensity in endothermal animals, development of new stress stability mechanisms in these animals promoted morphology improvement and complexity development of the liver, which, in effect, is "peripheral metabolism integrator". Simultaneously, numerous responsiveness and flexibility aspects of its endo- and ectogenic de-stressing have not been studied sufficiently, even in the mammals. Data of the experimental studies of hyperthermia impact on the organisms of other ecto- and endothermic vertebrata are fragmentary (P.P. Warren, 2000; M.M. Kalašnikova, 2003; A.N. Rice, 2006). It does not enable a full-scale analysis of the evolution trend laws in the development of the urgently implemented general and particular re-organizations, liver enzyme modifications in the course of development of the reactive and plastic responses. In the framework of this fundamental problem detailed analysis of destructive and reparative consequences as well as methods of the liver structural and functional recovery after an extreme thermal load enjoy a special interest. It seemed appropriate to conduct a comparative morpho-functional research of this organ in two classes of homeothermal animals — mammals and birds in which thermoregulation developed in the course of evolution independently. In order to identify and single out liver structures and specific mechanisms of "opposition" to the thermal stress it is necessary to study the effects of the general overheating in the row of ectothermic vertebrates – fishes, amphibias and reptiles – as well. Direct comparison of structural and functional events identified in the liver of animals from different taxons having different thermoregulation, detoxication systems and different DNA indices enables a significant extension of the current understanding of the liver responsiveness and flexibility in the vertebrates' comparative row.

The Purpose of the Work is to identify histological and enzymatic laws of the liver responsiveness and flexibility manifestation after a short-term hyperthermia in the vertebrates' comparative row.

The Objectives of the Study. To study functional morphology of the liver acinus, stromal/ parenchymal relationships of the liver cytotypes, identify hepatocytes biological indices (ultrastructure, cell cycle, death ways), study the sources of the hepatocytes' physiological and reparative regeneration, study the biochemical foundations of reactive and plastic responses in the normal state and after an acute hyperthermia in the vertebrata comparative row at different levels of the biological systems' hierarchy in the normal state and after acute hyperthermia in the vertebrates' comparative row.

Research Objects and Methods

The research was conducted on five groups of vertebrates:

1. Fishes – crucian carps of the species Carassius auratus gibilio Bloch, 1782, gold carps (two- year-olds – 2+).
2. Amphibias – frogs of the species Rana terrestris Andrzejewski, 1832 (Rana arvalis Nilsson, 1842), male moor frogs (three-year-olds).
3. Reptiles – tortoises of the species Pscudemia (Chrysemys scripta var. Elegans 1889, Trachemys scripta elegans) viripotent male pond terrapins (five years of development).
4. Birds – doves of the species Columba livia (forma domestica) Gmelin, 1789, viripotent male rock pigeons (6 months of development).
5. Mammals – rats of the species Rattus norvegicus (var. alba) Berkcnhout, 1769, viripotent male outbred male rats (2 months of postnatal development).

The experiment was conducted on 180 animals (including 75 control ones). The acute overheating of the endothermal animals was achieved by means of their keeping in an air-ventillated chamber for 30 minutes at 42°C. For ectothermal animals the overheating was conducted in the water medium in free floating conditions with additional aeration for 30 minutes at 32°C for fishes and amphibias and at 42°C for reptiles. In these modes the temperature impact determined the development of medium-severity thermal shock for these species of animals (G.N. Somero, 2002; A.E. McKechnie, 2004; Y.-M. Chang, 2005).

Research Methods

The liver samples were fixed in Carnoy's fluid, 10% - neutral buffered formaldehyde, poured into paraffin, the slices were stained on Van Gieson method.

Liver cells' immunophenotyping. Stromal/parenchymal ratios of the liver cytotypes, hepatocytes' proliferative activity and polyploidization was determined by means of detection of the antibodies to these cells' protein markers (A.P. Kijasov, 2000). Intensity and topography of the proliferation and polyploidization processes in the liver acinus zones were determined by counting binuclear forms and PCNA-positive hepatocytes per 1,000 hepatocytes. The paraffin sections were stained with antibodies to PCNA - Proliferating Cell Nuclear Antigen (Ig – mouse MAb; Clone PC 10; dilution 1:100; DAKO) hepatocytes using streptavidin-biotin method. Ito cells were stained with antibodies to desmin also using streptavidin-biotin method (Ig – monoclonal antibodies; clone – desmin D 33; dilution 1:30; DAKO, DENMARK). The antigens were identified after preliminary retrieval on HIAR (Heat-Induced Antigen Retrieval) method. After the incubation with primary antibodies they were incubated with biotin-modified second antibodies (Link, DAKO LSAB+Kit Peroxidase), with streptavidin conjugated with horseradish peroxidase (Streptavidin, DAKO LSAB+Kit Peroxidase). AEC (aminoethyl carbazole) and hydrogen peroxide solution was used as peroxidase reaction substrate. Histochemical reaction of endogenic peroxidase activity determination on paraffin sections wasx conducted to identify Kuppfer cells using MacFie method. Cytorype and proliferating hepatocytes count was performed at the area of 0.04 mm2.

Hepatocyte DNA Cytophotometry. Using DNA opto-structural analysis method the share of hepatocytes in the cell cycle phases was determined- G0, S, G2+M, hypodiploids. Liver smears were Feulgen stained (G. Olson modification) in Schiff's reagent. On the preparations the state of both chromatin fractions were determined: integral optical density in standard units (DI – decondensed chromatin; FI - condensed chromatin), total integral optical density (TI), distribution area in square microns (DP –decondensed chromatin; FP – condensed chromatin), total nucleus area (TP), average optical density in standard units (DM – decondensed chromatin; FM –condensed chromatin, T_SR - both chromatin fractions). DNA was measured in 200 nuclei. The preparations were photographed and processed on automatic morphometric Axioplan system (manufactured by Karl Zeiss, Germany).

Fluorescence Microscopy. Live and dead assay (C.C. Liu et al., 2000) was conducted using hepatocytes vital staining with nuclear stains Hoechst 33342 (Sigma, USA; solution in 20 μg/ml solution in M3 medium for 15 mintues), iodic propidium (Sigma, USA; solution in 20 μg/ml solution in M3 medium for 25 mintues) to classify the hepatocytes into groups, based on the following criteria: cell plasmalemma and nucleus status followed by the detection of apoptonecrotic ratio of hepatocytes' programmed cells' death. Axioscope fluorescence microscope (manufactured by Karl Zeiss, Germany) was used with "blue" filter unit, wavelength λ=500-550 and 600-700.

Electronic Microscopy. Animal liver samples were fixed in 4% paraform solution on 0.1M phosphate buffer (pH 7.4) adding sucrose (5%). The material was placed into paraffin, the slices were stained in 0.1% toluidine blue. Then the liver samples were finally fixed in 1% osmium tetroxide solution and then epon-araldite was poured into the mixture. The ultrafine slices were contrasted using uranyl acetate, lead citrate. The ultrathin slices were viewed and photographed using Hitachi-600H electronic microscope (Japan). The electronic microscopy research was conducted to evaluate voluteric density ($\mu m^3/\mu m^3$), surface density (μm2/μm3) and number density (μm0/μm3) of mitochondria, chramatin, granular endoplasmic reticulum (GER), agranular endoplasmic reticulum (AgER), nucleolus (N), lipids (L), lysosomes (Ls), hepatocyte glycogen (G). 50 vision fields of the liver parenchyma shot arbitrarily during the material viewing in the electronic microscope at 3,000 – 20,000 zoom were studied. Volumetric, area and number density of the structures in question were determined on electron diffraction patterns using built-in UTHSCSA ImageTool 3.0 software. The numerical parameters were evaluated per area unit (100 μm2) of the liver parenchyma.

Biochemical Methods. Mitochondrial and cytosolic fractions were obtained using differential centrifugation method. The activity of mitochondrial (mt), cytoplasmic (c) NAD- (nicotine-amidine-dinucleotide) and NADP- (nicotine-amidine-dinucleotide phosphate)-

dependent oxidoreductases: isocitrate dehydrogenases (IDH), malate dehydrogenase (MDH), glucose-6-phosphate dehydrogenase (G-6-PDH), lactate dehydrogenase (LDH) was determined spectrographically at λ=340 nm recording the optical density during the concentration change of NAD/NADP reduced forms in nmole substrate/min per mg protein (M.I. Proxorova, 1982; A.C. Gibb, 2002; A.M. Neyrinck, 2005). Caspase-3 (CPP-32) activity tests were conducted using commercial Caspase-3 Assay Kit manufactured by Sigma on Platr Reader Star-30 KENSTAR spectrophotometer in μm/mg protein. Acid phosphatase voluntary activity in the cytoplasmic fraction was determined by hydrolysis rate of n-nitrophenylphosphate g/min (Merck, Germany) (R. Maciejewski, 2001). Ions concentration was determined using ion-selective method in the cytoplasmic fraction using EasyLyte Calcium instrument in 100 μl of substrate, expressed in meq/l. Protein concentration was determined using modified Lowry method.

Statistical Processing. The critical significance level was adopted equal to $p=0.05$. Statistical characteristics of the parameters in question were determined: quadratic average, mdian, dispersion. Pearson $\chi 2$, Fisher-Snedecor distribution σ were found; they were used to determine the possibility of applying parametrical and non-parametrical analysis methods (Student criterion, Wilcoxon homogeneity - U, Pearson $\chi 2$). To identify the contingency degrees of the indicators in question, inter-species differences Pearson rank-correllation coefficient (r), Mann-Whitney test, multivariate analysis of variance (MANOVA) were applied.

RESERCH FINDINGS AND DATA DISCUSSION

Liver Responsiveness and Flexibility at Organ Level

Intact fishes', amphibias' and reptiles' liver is a tubular structure gland. However, in birds, like in mammals, two rows of short trabecules of hepatocytes are formed. Due to the poorly developed organ stroma support function in fishes, amphibias and reptiles is performed by intercellular structures and collagenic fibers in Disse space. Pigment cells are found in all the animals in question, except mammals (Figure 1, 2, 3, 4). Pigment cells originating from Kuppfer cells may form melano-macrophagal centres. Their main part is centralization, deposition of endogenic and ectogenic substances with subsequent destruction, detoxication and re-use thereof. In pigment granules the presence melanin, haemosiderin originating from ferritin and being a component part of haemoglobin decomposition products is marked.

Melanin is a powerful antioxidant, it protects cells from free oxygen radicals and acts similarly to superoxide dismutase – an enzyme performing this function in unpigmented tissues. After hyperthermia in amphibians and mammals hepatic sinusoidal dilatation and hyperemia is marked in the central vein zone (Figure 5, 6, 7), radial location of sinusoid capillaries and hepatocyte trabecules is disarranged. After hyperthermia in all aminal groups vein congestion, portal tract vascular wall and central veins' swelling, collagen fibers' dissociation, interstitial space hydration in the portal tract area with the vascular wall disintegration is marked (Figure 9).

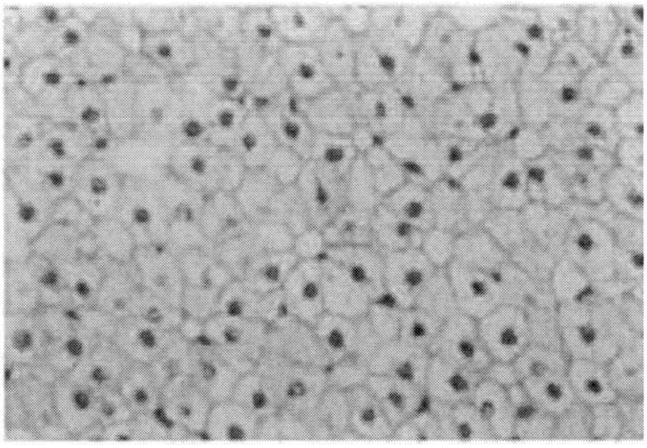

Figure 1. Fishes' liver. Control. Hematoxylin-eosin stain. Zoom 10x40. Liver acinus centrilobular zone.

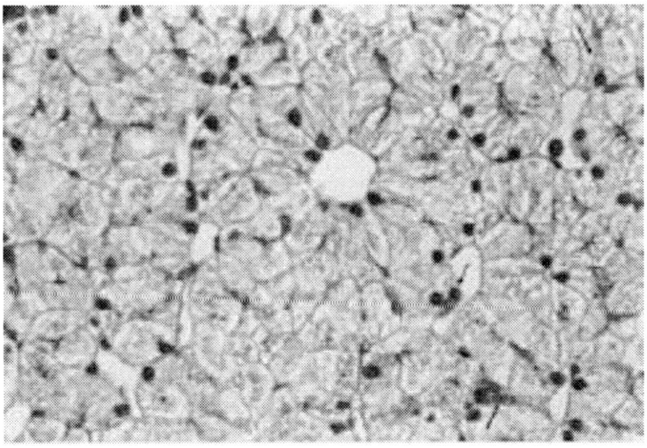

Figure 2. Amphibians' liver. Control. Hematoxylin-eosin stain. Zoom 10x40. Centrilobular zone. Parenchyma tubular structure. Ito cells (arrow). Kuppfer cell in the sinusoid lumen (arrow).

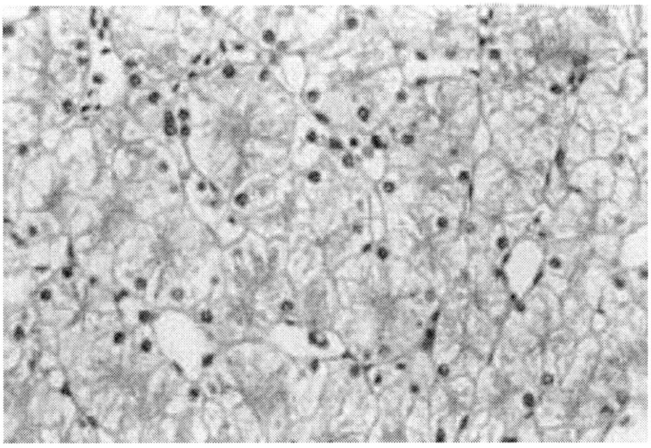

Figure 3. Reptiles' liver. Control. Hematoxylin-eosin stain. Zoom 10x40. Liver acinus centrilobular zone. Ito cell (arrow).

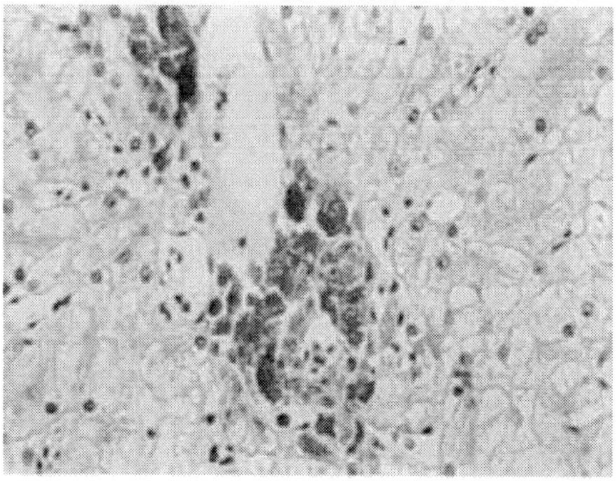

Figure 4. Reptiles' liver. Control. Van Gieson stain. Zoom 10x40. Hepatoportal tract with melano-macrophagal centres.

Due to the organ hemodynamics compromise in all animals acinus architectonics decomplexation – spatial and topographic approximation of portal tract and central veins is observed (Figure 5, 8, 9, 12). Due to the organ hemodynamics instability, researchers identify microcirculation disturbances determine hypoxia development, especially for the hepatocytes in the central veins' zone and, consequently, autointoxication of the liver cells with the influx of reactive metabolites into the acinus system. Hepatocytes' response is manifested in the cytolysis, pycnosis, marginal and demilune shape chromatin location, and nucleus fragmentation (Figure 5, 6, 7, 10, 11). In birds' and mammals' liver periportal zone hepatocytes with swollen and "emptied" cytoplasm occur, which is a symptom of carbohydrate metabolism error (Figure 9, 11). In all the animals' mosaicism is observed in the acinus damage. After hyperthermia in all the fish, amphibia and reptile species in question prominent arteriola diameter reduction is observed. Venule diameter in amphibians, birds and mammals increases and in reptiles - reduces. Central vein diameter in amphibians and reptiles reduces whereas in birds it increases. Microcirculatory bloodstream terminal link expansion provides enhanced blood outflow from the acinus thus minimizing the adverse effect of hypoxia and autointoxication. Lymphatic vessel diameter increases in reptiles, birds and mammals and reduces in fishes.

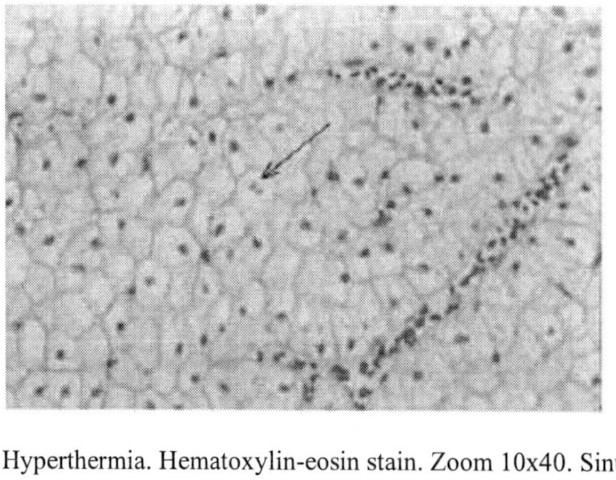

Figure 5. Fishes' liver. Hyperthermia. Hematoxylin-eosin stain. Zoom 10x40. Sinusoidal hyperemia.

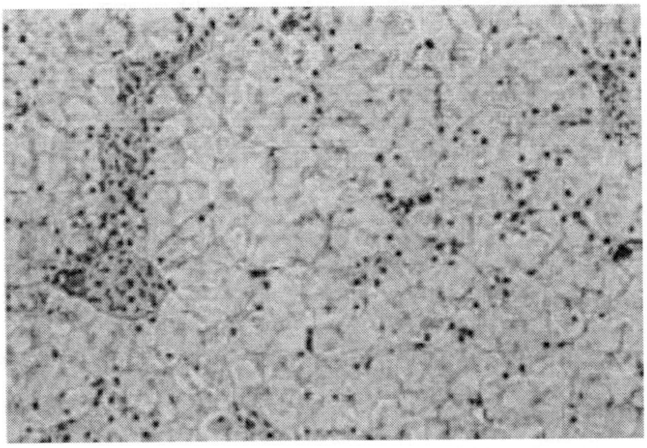

Figure 6. Amphibias' liver. Hyperthermia. Hematoxylin-eosin stain. Zoom 10x20. Sinusoidal hyperemia.

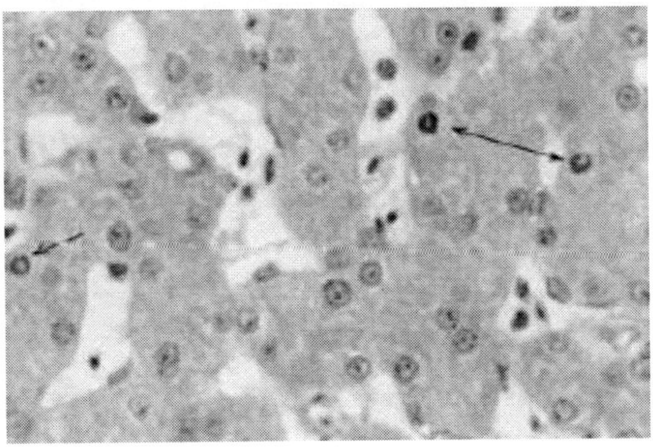

Figure 7. Reptiles' hepatocytes. Hyperthermia. Hematoxylin-eosin stain. Zoom 10x40. Hepatocytes in apoptosis state (arrows).

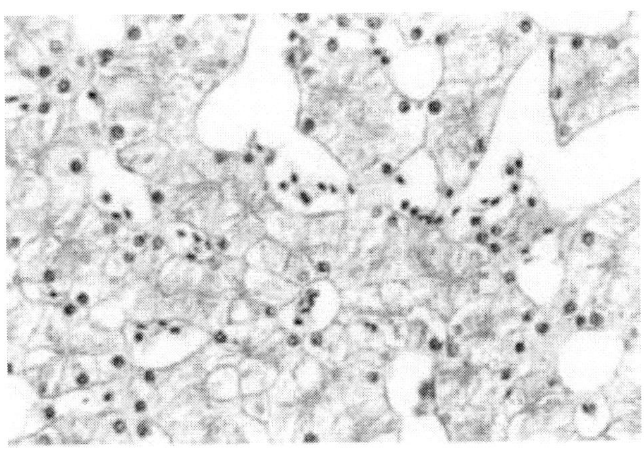

Figure 8. Reptiles' hepatocytes. Hyperthermia. Hematoxylin-eosin staining. Zoom 10x20. Sinusoidal dilation. Liver cytoarchitectonics disorder.

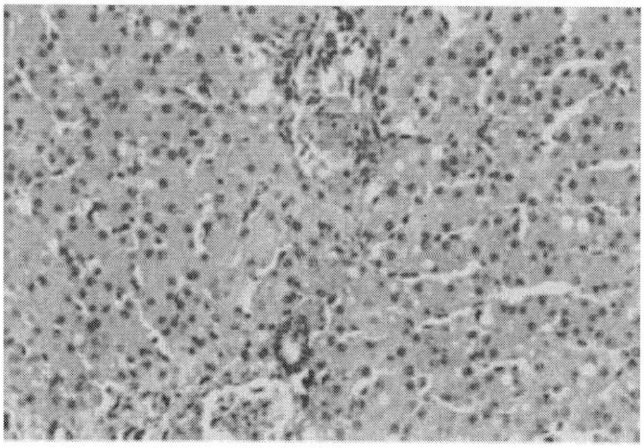

Figure 9. Birds' liver. Hyperthermia. Hematoxylin-eosin stain. Zoom 10x40. Portal tracts' approximation. Venal hyperemia. Sinusoidal dilation. Cytoplasm and hepatocyte nuclei vacuolization.

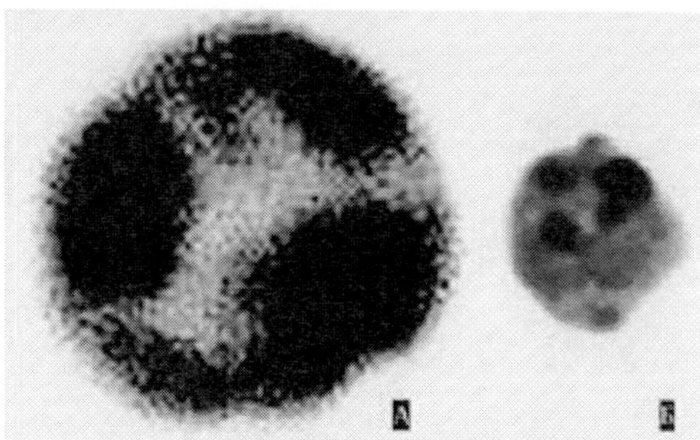

Figure 10. Nucleus in apoptosis state. A). Birds' liver. Hyperthermia. Hematoxylin-eosin stain. Zoom 10x200. B). Birds' liver. Zoom 10x100.

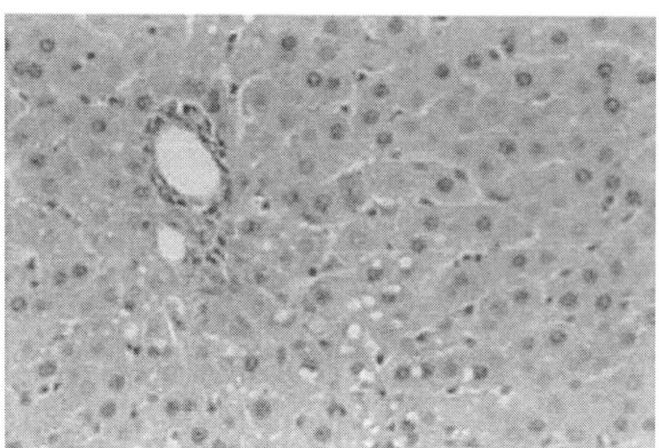

Figure 11. Mammals' liver. Hyperthermia. Hematoxylin-eosin stain. Zoom 10x40. Hepatocytes' nuclei and cytoplasm degeneration (vacuolization).

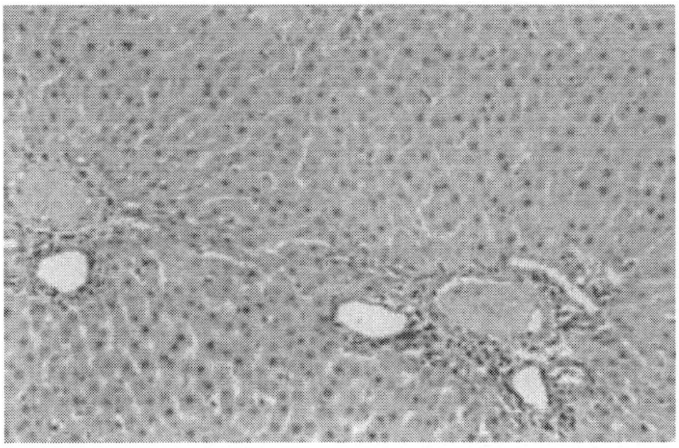

Figure 12. Mammals' liver. Hyperthermia. Hematoxylin-eosin stain. Zoom 10x20. Portal tracts' approximation. Venal hyperemia.

The lymphatic vessel expansion reflects an intensive transcapillary exchange process and may be related to the extracellular water mobilization due to compensatory redistribution of interstitial fluid and activation of metabolism level, drain-immunological function and lymph circulation in the liver, ion balance equalization between intracellular and extracellular spaces (J.I. Borodin, 2000).

Significant changes of the biliary passage internal diameter are found in reptiles and birds. But morphometric changes are directed oppositely: in reptiles the lumen diminution whereas in birds the lumen dilation is observed. Hyperthermia one hour after the effect causes body weight reductions in fishes by 8%, in birds – by 4% and in mammals by 14%.

Liver Responsiveness and Flexibility at Tissular/Intertissular Level

Evolutionally formed spatial and topographic system of liver cytocommunications ensures the functioning of the organ in animals with different thermoregulation system; hereby key role is played by the peculiarities of the liver blood supply in acinus metabolic zones. In this connection, peculiarities in stromal/parenchymal relationships of cytotypes and in hepatocyte polyploidization/ proliferation are found (E.I. Antonova, 2008).

Thus in intact liver of mammals, fishes and reptiles binucleate hepatocytes are uniformly distributed within acinus. In amphiobias the distribution is characterized by venulo-portal gradientt, whereas in birds – by portovenular gradient. On the three zones the total binucleate hepatocyte percentage in fishes, reptiles and birds amounts to 10%, in amphibians – 12%. The largest share of binucleate hepatocytes is found in mammals – 24%.

A peculiar feature of reptiles is the presence of 4-nucleate hepatocytes both in control and experimental animals. After hyperthermia the identified increase of the binucleate hepatocytes' number provides the initiation of the mononucleate hepatocyte future clone potential source zones in the portal tract area (Figure 14) in amphibians and birds - in central lobular zone, and in reptiles and fishes – in perivenular zone (S. Gupta, 2000; I.V. Uryvajeva, 2001; G.A. Sakuta, 2005).

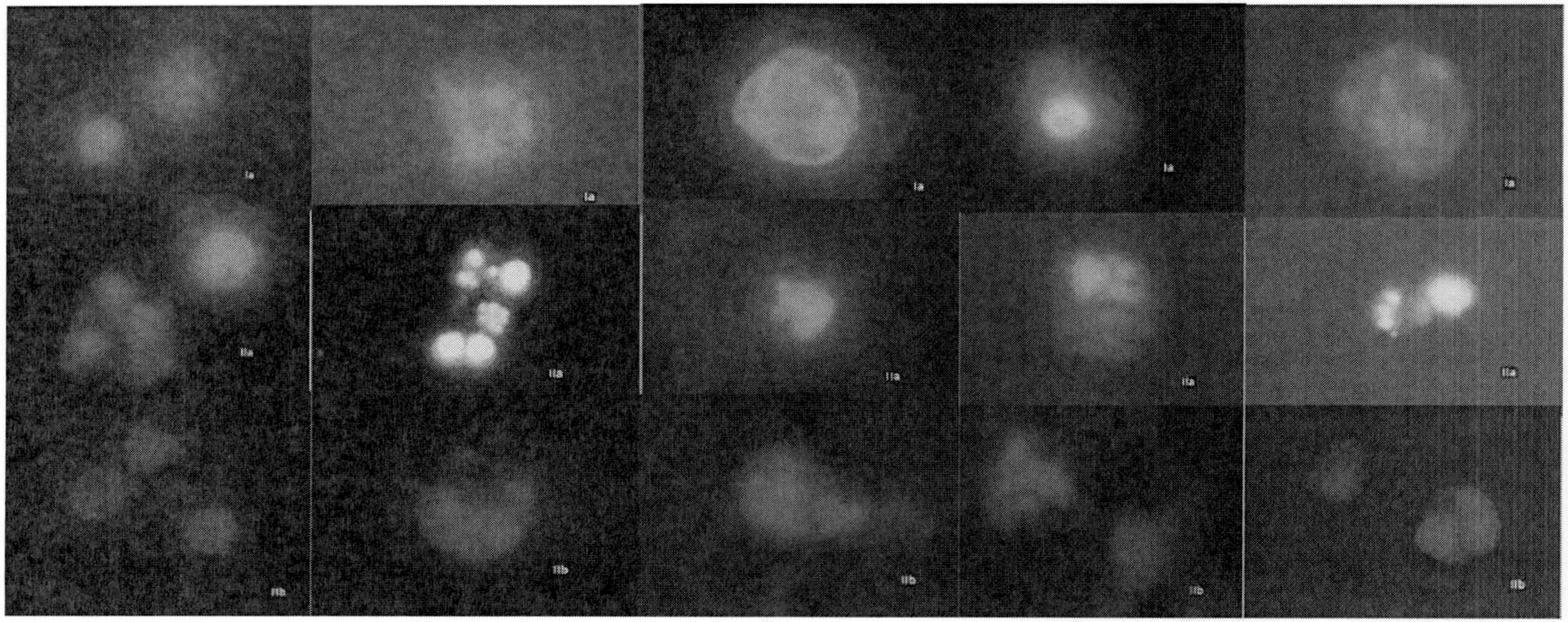

Figure 13. I. Control. II. Hyperthermia. (a) Hoechst 33342 stain, Nucleus pycnosis and fragmentation in apoptosis. (b) Propidium iodide (PI) stain. Zoom 10x40. Necrosis – chromatin random fragmentation, nucleus membrane disintegration. 1. Fish hepatocytes. 2. Amphibian hepatocytes. 3. Reptile hepatocytes. 4. Bird hepatocytes. 5. Mammal hepatocytes.

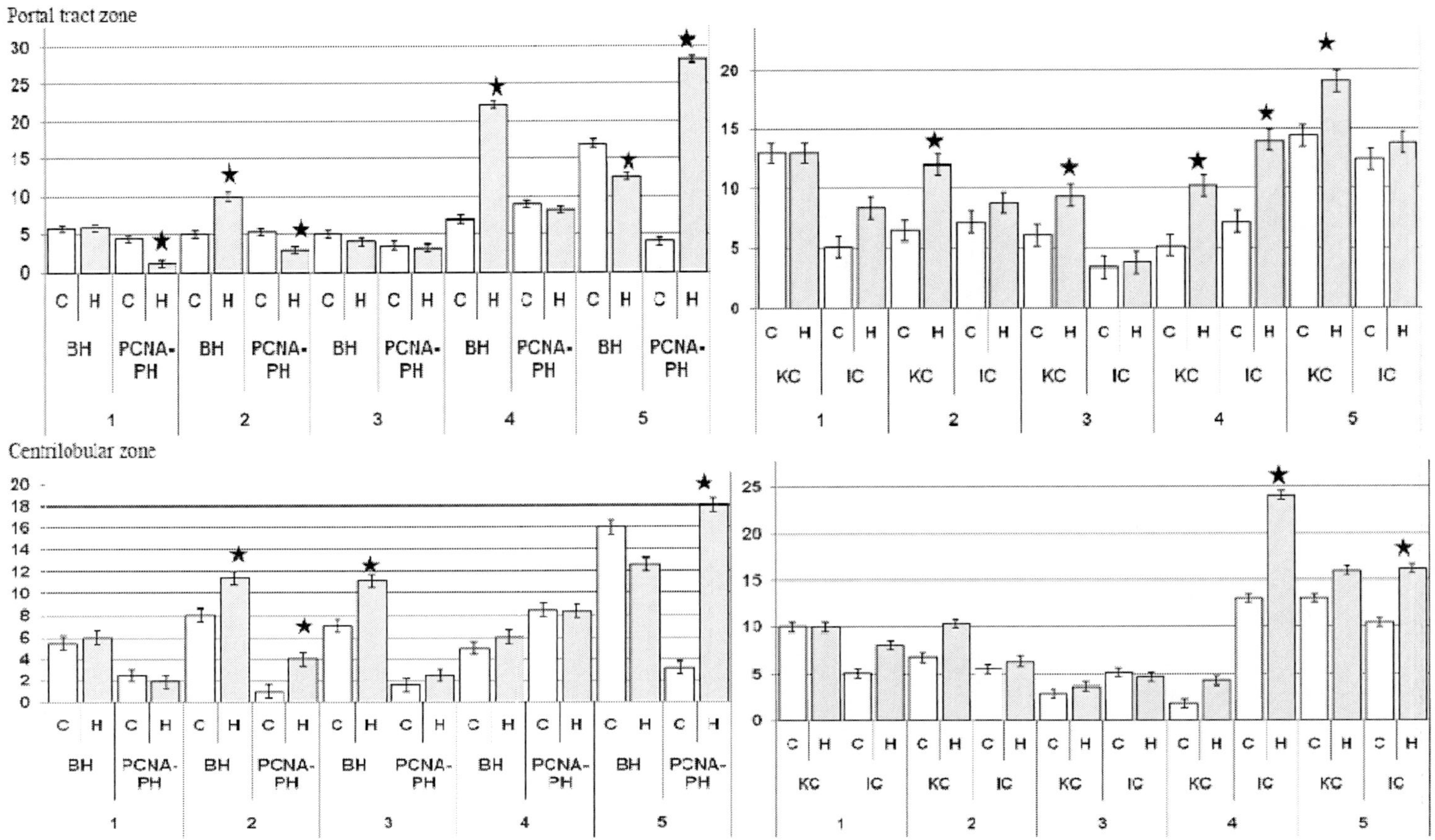

Figure 14. (Continued)

Central vein zone

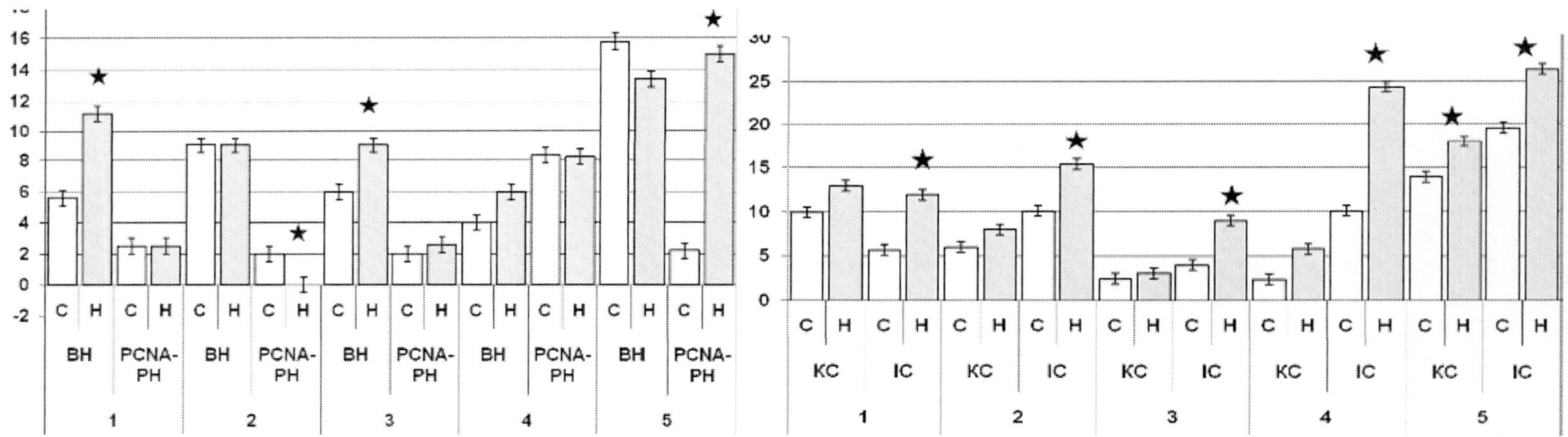

Figure 14. Dynamics of the liver stromal-parenchymal cytotypes ratio in the normal state and after hyperthermia. BH–binucleate hepatocytes, C–control, H – hyperthermia, PCNA-PH – PCNA-positive hepatocytes, KC–Kuppfer cells, IC–Ito cells. 1-fishes; 2-amphibians; 3-reptiles; 4-birds; 5-mammals. * p=0,05.

On average in three zones the post-hyperthermia number of binucleate hepatocytes in fishes is 14%, in amphibias – 20%, in reptiles – 16%, in birds and mammals – 22%.

The largest number of PCNA-positive hepatocytes in intact animals is found in the liver of birds – 11%, in fishes this percentage is 5%, in other animal groups - 4%. The localization of this hepatocyte population in the acinus space is characterized by portovenular gradient. An exception is the birds' group in which the number of PCNA-positive hepatocytes is equal in all the acinus groups (Figure 14, 24).

After overheating the number of PCNA-positive hepatocytes in reptiles and birds matches the control data and in fishes and amphibias it reduces in the periportal zone which manifests the inhibition of poluploidization processes. In amphibians PCNA-positive hepatocytes in the central veins area are not found at all but in the central lobular zone their number increases (Figure 14, 16, 26).

Consequently, in the aforementioned animal groups, in order to prevent the damaging effect of hyperthermia consequences onto DNA structure (and, consequenctly, the genes' structure) in the course of evolution mechanisms were formed which, depending on the liver acinus zone, restrict the fast growth of tetraploid hepatocytes' number (S.P. Otto, 2007). Besides, the tetraploid cells tend to activate p53, cell-cycle block development in G1-stage, and, eventually - to apoptosis activation which we found in every animal group (N.J. Ganem, 2007).

Mammals are the only group in which in all the acinus zones the number of PCNA-positive hepatocytes increases to 28% (Figure 14). Demand for PCNA found in all the organisms obviously reflects the activation of not only proliferative but also destructive processes related to DNA damage and activation of different apoptosis pathways (R.M. Douglas, 2003; J. Frampton, 2006). The increase in PCNA-positive hepatocytes may also be induced by Ito cells' cytokines' synthesis which is confirmed by the increase of stromal cells found in all the animals of this population (C. Schmidt, et al., 2004, 2006; J. Frampton, 2006; P.L. Andersen, 2008). It allows us to assume that the change of metabolic adaptation strategy on biochemical level may be reduced to the setup of new relationships between interconnected tissue-specific metabolic cycles in the animal species in question. Key factors here are organism genotype peculiarities and metabolic system condition during the impact. The identified functional parameters of the compensatory response to the damage enable judging of the initial tissue rersponses' – apoptosis speed and pathways – dynamics as well as proliferative processes corresponding to the hepatocyte death level in the species (A. Gandillet, 2003; A.F. Karapetjan, 2006).

In evolution, thermoresistivity, cell tolerance to oxidative stress and hypoxia as a way of the species' conservative adaptation is in many aspects determined by polyploidy (O.V. Anatskaya, 2000; L.J. Borkin, 2001). Consequently, at the expense of acinus-zone-differentiated increase in the number of binucleate and PCNA-positive hepatocytes, ploidy is increased which characterizes hepatocytes' high functional "capacity". And the identified reduction of Krebs cycle oxidoreductases' activity is apparently related to the activation of hepatocyte polyploidization along the ways of forming binucleate forms in all groups of animals, except mammals. Redundant mitosis-ready tetraploidal hepatocytes and hepatocytes in the DNA-synthesis phase and ready for mitosis process before the stressing time could, obviously, become the source of binucleate hepatocytes (G.G. Avtandilov, 2004). Considering the fact that the time of mitotically inert hepatocytes' existence corresponds to the individual animal's life time, due to the death of a part of the liver parenchyma the hepatocytes can manifest virtually unlimited reproductive capacity. Hereby the regeneration rate is high which enables considering hepatocytes as a unipotent stem cell (I.V. Uryvajeva, 2001; N. Fausto, 2001, 2003). The increased number of binucleate hepatocytes also may be attributed to the urgent reparative regeneration and provides recovery of intercellular adhesions with Ito cells for their activation termination and transformation into myofibroblasts.

In intact animals peroxidase-positive Kuppfer cells are uniformly distributed in the acinus in amphibias and mammals. Predominance of organ-specific macrophagocytes is observed in periportal zone in fishes, reptiles and birds. In fishes desmin-positive Ito cells are uniformly distributed within the acinus. In amphibians and mammals the largest number of desmin-positive Ito cells is found in the central veins zone; in reptiles – in the central lobular zone; in birds – both in the central lobular zones and in the central veins' zone (Figure 14, 15, 19, 20, 22, 25).

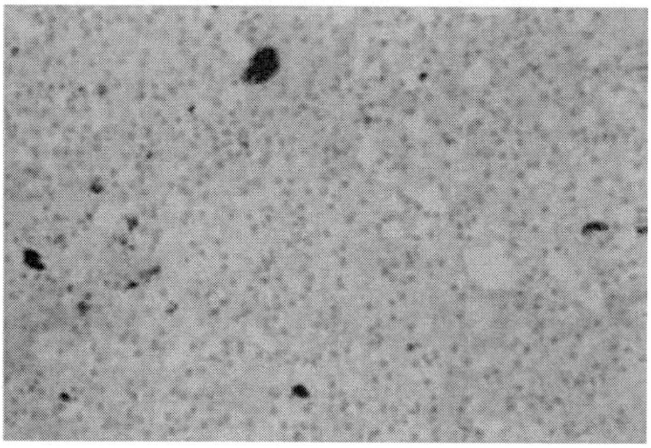

Figure 15. Fishes' liver. Control. Kuppfer cell histochemical peroxidase staining. Zoom 10x20. Central lobular zone.

Figure 16. Fishes' liver. Hyperthermia. Zoom 10x20. PCNA antibidies stain. Proliferating hepatocytes.

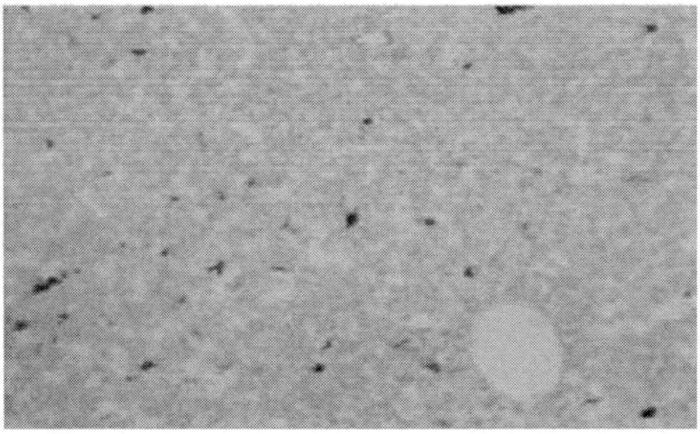

Figure 17. Fishes' liver. Hyperthermia. Kuppfer cell histochemical peroxidase-positive staining. Zoom 10x40. Central vein zone.

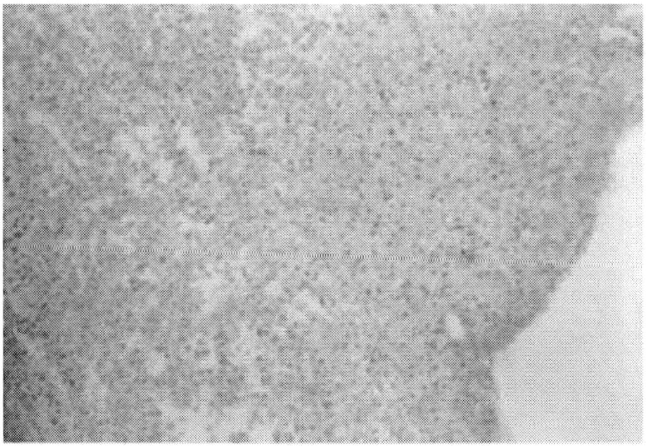

Figure 18. Fishes' liver. Hyperthermia. Zoom 10x400. Desmin antibody stain.

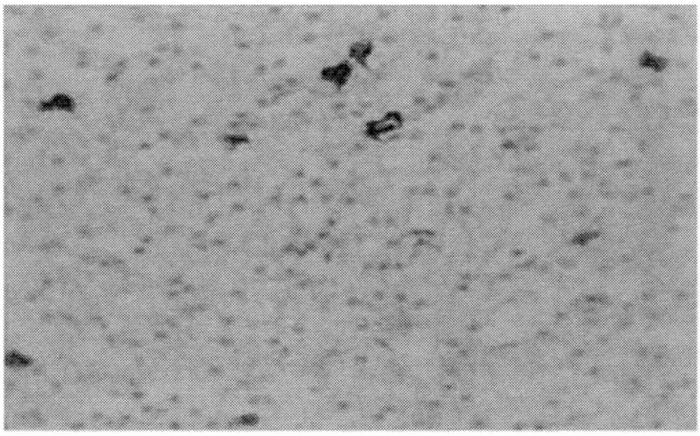

Figure 19. Amphibians' liver. Control. Kuppfer cell histochemical peroxidase staining. Zoom 10x20. Central lobular zone.

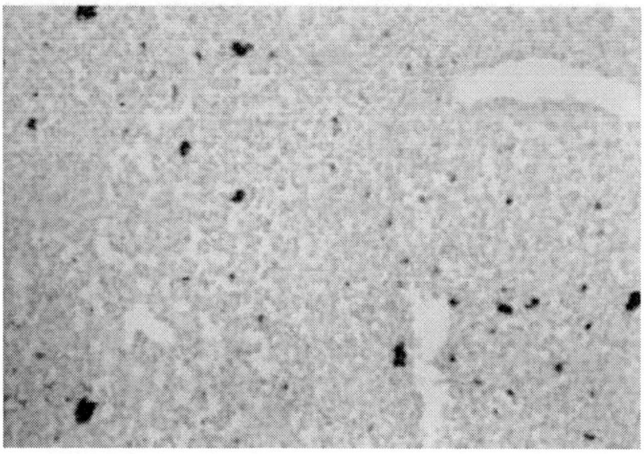

Figure 20. Reptiles' liver. Control. Kuppfer cell histochemical peroxidase staining. Zoom 10x10.

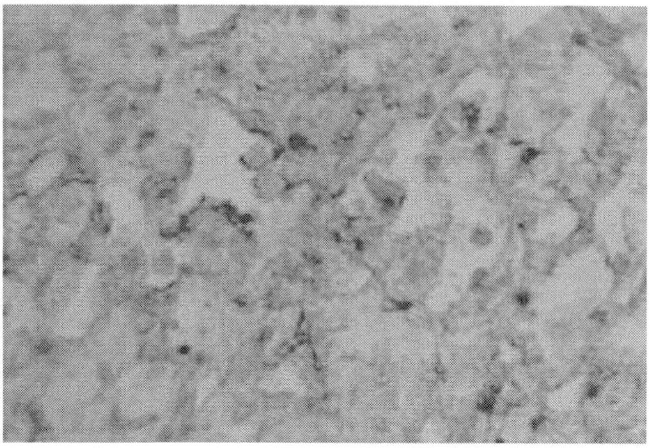

Figure 21. Reptiles' liver. Hyperthermia. Zoom 10x40. Desmin antibody stain.

Figure 22. Birds' liver. Control. Zoom 10x20. Desmin antibody stain.

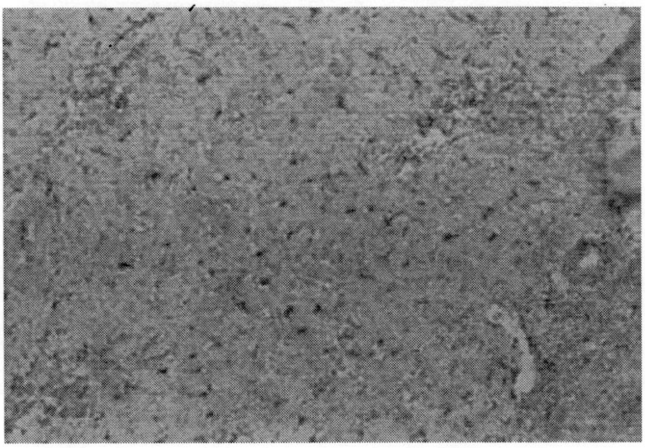

Figure 23. Birds' liver. Hyperthermia. Zoom 10x20. Desmin antibody stain.

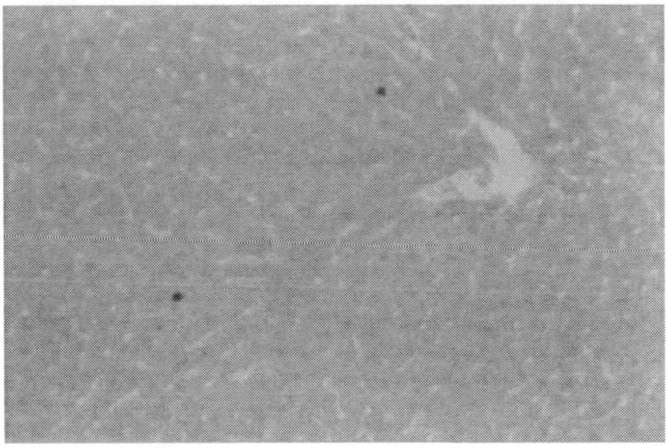

Figure 24. Mammals' liver. Control. Zoom 10x20. PCNA antiobody stain.

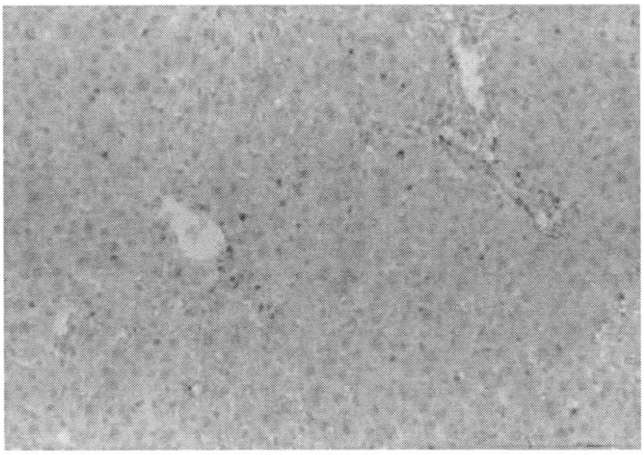

Figure 25. Mammals' liver. Control. Zoom 10x20. Desmin antibody stain.

Figure 26. Mammals' liver. Hyperthermia. Zoom 10x20. PCNA antiobody stain.

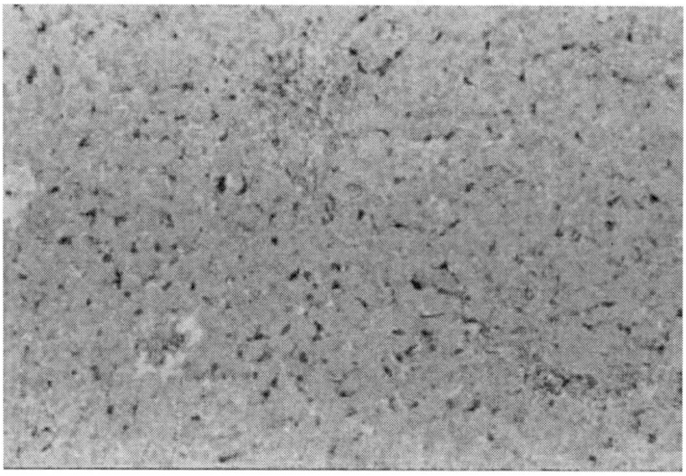

Figure 27. Mammals' liver. Hyperthermia. Zoom 10x20. Desmin antibody stain.

Post-hyperthermic reduction of live hepatocyte number in the pervenular zone is found in all the animal groups; in amphibians and mammals – also in the central lobular zone; and in reptiles, birds and mammals – in the periportal zone. In reptiles, unlike other animals, a higher number of live hepatocytes is observed in the acinus central lobular zone, apparently, due to binucleate forms' increase. In fishes, reptiles, birds and mammals the share of Hoechst-positive hepatocytes amounts to 62-70% of the total number of dead hepatocytes. The largest number of of atopic phenotype dead hgepatocytes is ibserved in birds (Figure 13).

The number of organ-specific macrophages in fishes after overheating corresponds to the control value. However, the development of qualitative (not quantitavtive) changes in this cell group cannot be ruled out (P.A. Knolle, 2000, 2003). In amphibians, reptiles, birds and mammals the increased number of peroxidase-positive Kuppfer cells in the portal tract area is observed. In mammals the number of organ-specific macrophages grows in the central veins' zone as well (Figure 14, 17).

The dynamics of desmin-positive Ito cells' quantity and topography after overheating is manifested in the increase of their quantity in the central veins' zone in all the anumal groups; and , additionally, in the central lobular zone in mammals. In the birds' group the number of desmin-positive Ito cells in all the acinus zones increases (Figure 14, 18, 21, 23, 27). The found Ito cells' activation may be triggered by the termination of contact inhibition with damaged hepatocytes, active oxygen forms and "wound hormones" of the damaged hepatocytes, activation of the lipids' peroxidation. In this case the Ito cell number increase and their activation is, obviously, due to the transition of desmin-negative cells into desmin-positive ones; and the cells which previously did not express this protein or had it in minimum quantities can start its synthesis in case of the liver damage (H. Senoo, 2000, 2004; G. Carpino, 2004; G. Schlaf, 2004; M.R. Alison, 2004; R.G. Wells, 2005; M.H. Ismail, 2009).

In our experiment as early as one hour after the hyperthermic impact the first symptoms of Ito cells' myofibroblastic transformation are manifested; we observed this myofibroblastic transformation in the electronic microscope. If Ito cells' activation is considered from the cascade mechanism standpoint, it may be concluded that at the early stage of the liver post-hyperthermic regeneration Ito cells undergo only stages 1 and 2 – Ito cells' proliferation stimulation with the damaged hepatocytes and triggering of Ito cells' transformation into myofibroblasts using Kuppfer cells' cytokines (R. Issa, 2001; C. Schnabel, 2004). Hepatocyte membrane damage determined the release of substances triggering Ito cell proliferation and cytokines' synthesis by the Ito cells. Particularly, the Transforming Growth Factor -α (TGF) which induces proliferation both of Ito cells and hepatocytes (autocrinically) (A.P. Kijasov, 2002; D.A. Mann, 2002).

The activated Ito cells, in their turn, provide fast rearrangement of the extra-cellular matrix components which is an important condition of the liver regeneration at the earliest stages thereof. The activated Ito cells paracrinically determine Kuppfer cells' activation as well, the latter having stimulating activity related to Ito cells (R. Bataller, 2004, 2005; S.K. Meurer, 2005; A.M. Figueiredo-Fernandes, 2006; C. Schmidt, 2006). Besides, Kuppfer cells act as non-specific immunity components' activators which initiates the liver regeneration early stages; Kuppfer cells also regulate lipids' and glucose liver metabolism (A. Neyrinck, 2004; M. Rentsch, 2005).

Activated Kuppfer cells produce a number of biologically active substances which may cause toxic damage of the hepatocytes (N. Milosevic, 2000; C. Ju, 2002; H. Kono, 2002) and also stimulate proliferative response of the hepatocytes and non-parehchymal cells by triggering the liver regeneratin process (Y. Matsuki, 2002). In particular, the increased number of desmin-positive Ito cells may be an indirect result of TGF-β effect which exerts a pronounced effect onto Ito cells' protein-synthesis function which is manifested through enhanced desmin synthesis. If there is no liver damage, Ito cells secrete anti-inflammatory cytokine IL-10 which reduces Kuppfer cells' activity level.

The found changes in the liver blood-circulation system morphology, xenobiotics delivered with the bloodstream also result in Kuppfer cells' activation (H. Kono, 2002). Consequently, para- and autocrinical activation of the stromal cytotypes in many ways promotes metabolic shifts followed by the implementation of tolerant or resistant metabolism strategy in the animal groups in question (V.I. Kulinskij, 1992).

Liver Responsiveness and Flexibility at Cellular Level (Hepatocytes)

Maximum hepatocyte nucleus DNA chromatin FI (standard units) is found in intact animals in amphibians and mammals group. The second group is formed by fishes and reptiles and the third group - by birds (Figure 28). But, according to Mann-Whitney test (at $p=0.05$), no significant differences were found in the integral optical density between the first and the second groups of aminals.

According to DI DNA amphibias may be classified as Group 1, mammals – as Group 2, fishes and reptiles – as Group 3 (Figure 28), birds – as Group 4.

FM level reduces in the following direction: mammals→amphibians→fishes→reptiles→birds.

DM level is equal in ectothermal animals but lower than in mammals and higher than in birds. Based on DM level comparative analysis (Mann-Whitney test), no significant differences were found in reptiles/fishes and amphibias/fishes groups. Maximum FM and DM difference was found in amphibias and fishes.

DP is higher in amphibian hepatocyte nuclei and lower in fishes, mammals, reptiles and minimum in birds. FP diminishes in the following sequence amphibians →reptiles→fishes→mammals→birds.

By TP of both chromatin fractions tha largest nucleus area is found in amphibians, then fishes' group follows, and then →reptiles→mammals→birds. No meaningful differences in TP were found only in reptiles/mammals group, whereas no meaningful FM differences were observed in mammals/fishes group (Mann-Whitney test).

Maximum hepatocyte number in G0-G1 stage (2n2c) is found in intact fishes, which is due to the minimum number of hypodiploidal hepatocytes and hepatocytes in S and G2-M – cell cycle stages. In birds and reptiles the reverse trend is observed (Figure 28). However, in the reptile group the largest number and in mammals and fishes – the smallest number of hypodiploidal hepatocytes is found.

After hyperthermia according to MANOVA ($p=0.05$) analysis in amphibians'/birds' group DI decrease and FI, DM, T_SR increase is observed and FM increases only in birds. In accordance with the indicators above DP, TP decreases. According to the indicators above increase of the number hepatocytes in cell cycle G0-G1-stage, reduction thereof in S- and G2-M-stage is observed which may result from metabolic depression development (Figure 28). The nucleus area reduces due to eu- and heterochromatin compaction level increase but simultaneously the decondensation to a larger extent affects DNA in heterochromatin.

In fishes'/reptiles'/mammals' group the reverse trend is observed – DI in fishes does not manifest meaningful differences compared with the control values, whereas in reptiles and mammals it decreases. FI and TI in fishes increases and in reptiles and mammals - decreases. Consequently, DP increase is marked in all the three animal groups. Consequently, in fishes TP is decreased due to FI level reduction resulting from FM value increase.

In reptiles and mammals TP increase is determined by the reduction of DI and FI levels resulting from the decrease of DM and FM values. Therefore, in fishes/reptiles/mammals group the number of G2-M-stage hepatocytes in fishes decreases and in reptiles and mammals - increases (Figure 28).

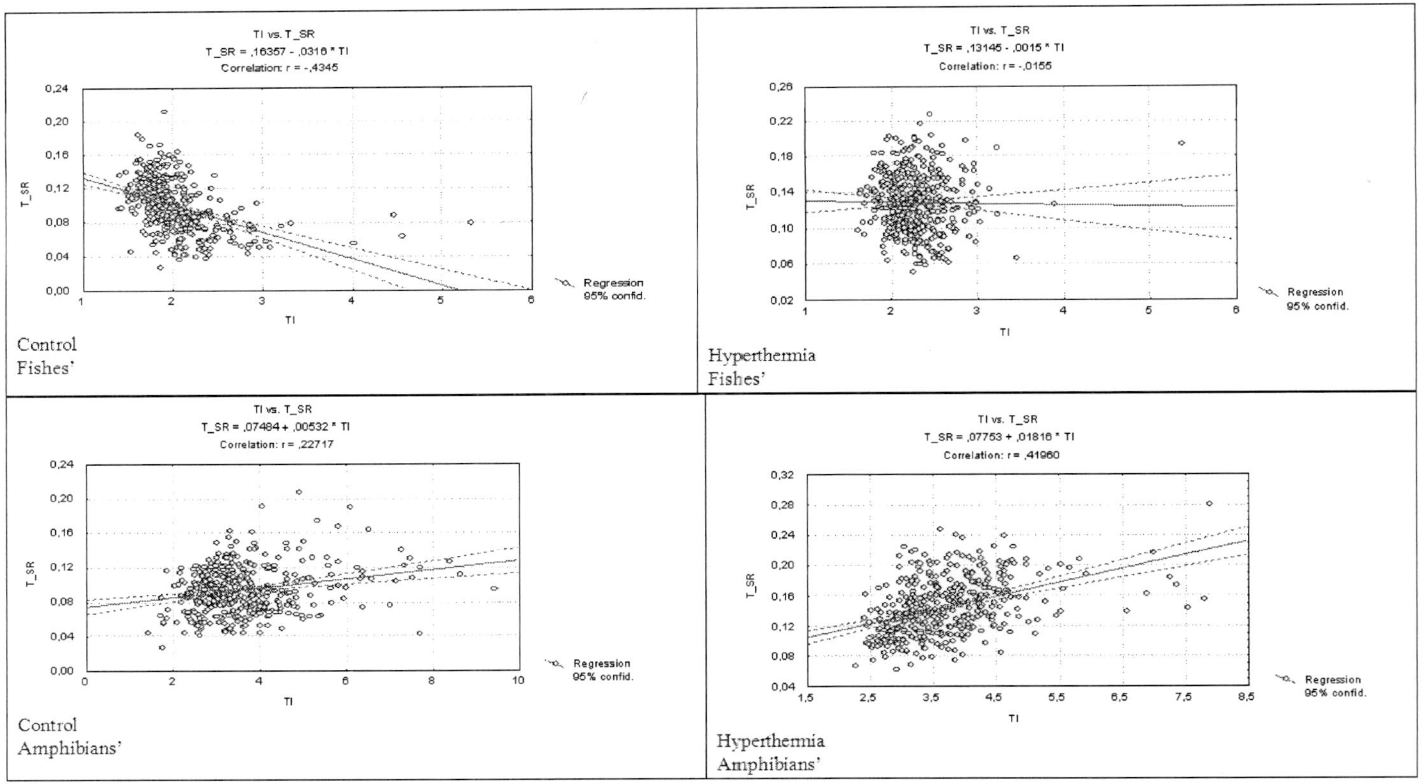

Figure 28. (Continued)

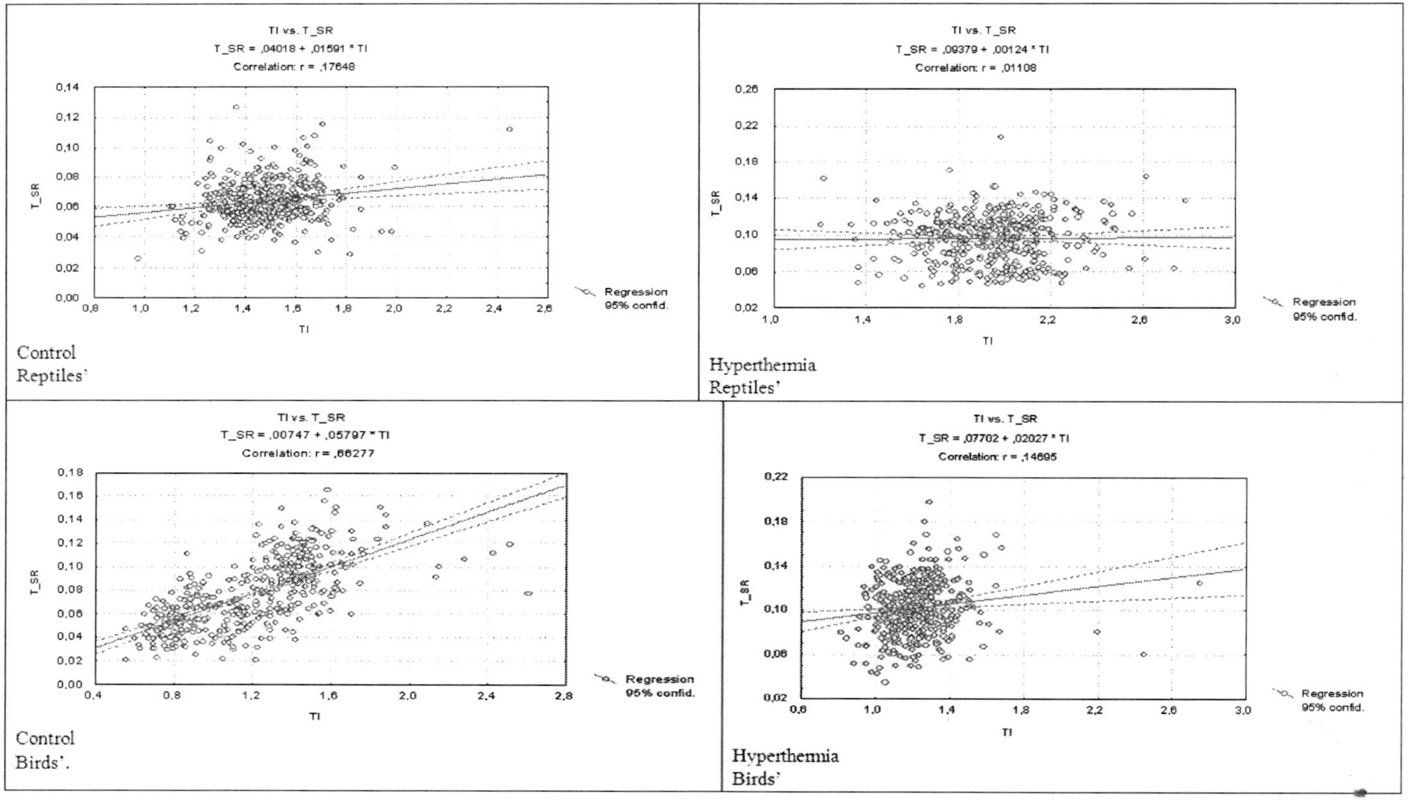

TI vs. T_SR
T_SR = ,04018 + ,01591 * TI
Correlation: r = ,17648
T_SR
TI
Regression
95% confid.
Control
Reptiles'
TI vs. T_SR
T_SR = ,09379 + ,00124 * TI
Correlation: r = ,01108
T_SR
TI
Regression
95% confid.
Hyperthermia
Reptiles'
TI vs. T_SR
T_SR = ,00747 + ,05797 * TI
Correlation: r = ,66277
T_SR
TI
Regression
95% confid.
Control
Birds'.
TI vs. T_SR
T_SR = ,07702 + ,02027 * TI
Correlation: r = ,14695
T_SR
TI
Regression
95% confid.
Hyperthermia
Birds'

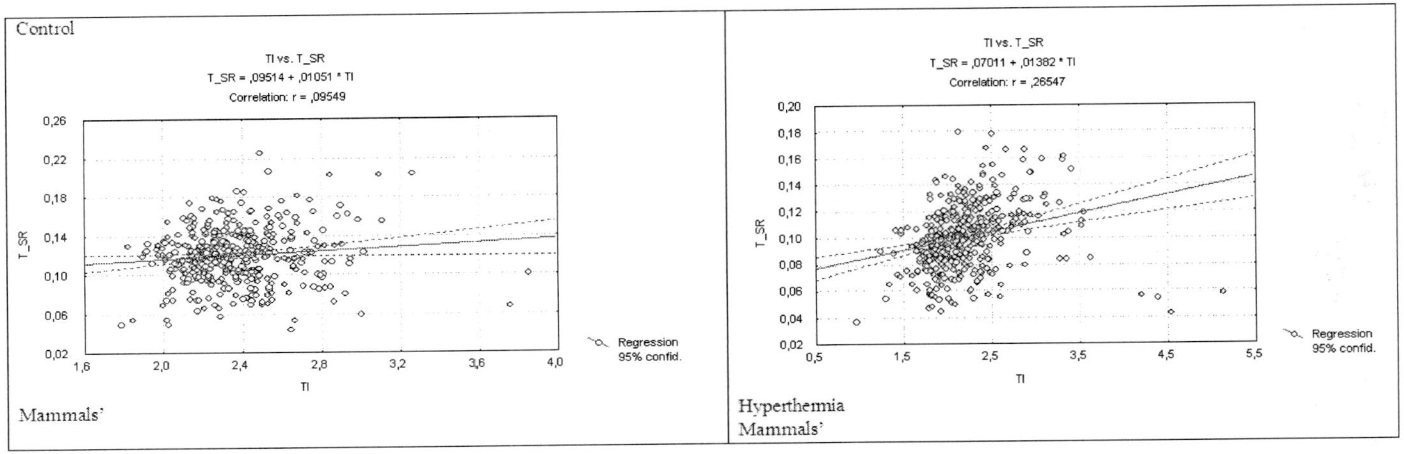

Figure 28. Two-dimensional nuclei distribution in TI – DNA integral optical density and T_RS – hepatocyte DNA average optical density (multivariate analysis of variance (MANOVA), at p=0.05).

In fishes and reptiles the number cell cycle S-stage hepatocytes increases. In reptiles and mammals functional activation of hepatocyte nuclei is observed which is manifested through establishing connection between optical density level and increase in their distribution area in both the chromatin fractions. The identified chromatin decondensation takes place against the background of the increase in [Ca2+]i cytoplasmatic ion concentration (S. Memedula, 2003). Increased hepatocyte nuclei volume in reptiles and mammals may result from the functional swelling with the change of transcriptional activity which causes optimized activity of the enzymes coded by nuclear and mitochondrial genome, provides hepatocytes' adaptation to the changed gas conditions which is a key parameter to establish correlation between chromatin status and the organ metabolic activity within the framework of metabolism strategies' switching (T. Miyahara, 2000; W.G. Muller, 2001; P.O. Vardevanjan, 2001; J.R. Chubb, 2002; J. Frampton, 2006; S.P. Otto, 2007; M.R. Bennett, 2008; N.S. Kenneth, 2008).

Hypodiploid hepatocyte number increases in all the animal groups, to a lerger extent in birds, to a smaller extent – in fishes and reptiles. Hereby a larger increase is marked propidium-iodide-positive hepatocytes. An exception here is the group of amphibians. The changes identified influence apoptosis/necrosis ratio indicators: in mammals the largest reduction is observed – down to 1.2; in reptiles and birds the ratio also reduces and amounts to 1.8 and 1.4, respectively. Apoptosis/necrosis ratio in birds is virtually unchanged. Only in amphibians the ratio increases from 0.6 to 0.8 which is due to a larger increase of Hoechst-positive hepatocytes number. The largest surpass of Hoechst-positive hepatocyte number over propidium-iodide-positive hepatocyte number found in birds, both in control individuals and after overheating is, obviously, related to the availability of biochemical flight adaptations, genome size, breathing physiology peculiarities and metabolic adaptation strategy.

Therefore, different DNA content, its status determines the implementation of various metabolic ways during the adaptation to hyperthermia, selection/switching of different metabolic adaptation strategies in animals: resistant (amphibians and birds) and tolerant (fishes, reptiles and mammals) (L.J. Borkin, 2001, 2004; D.A. Petrov, 2002; A.E. Vinogradov, 2004; T.R. Gregory, 2004; T. Cavalier-Smith, 2005; E.I. Antonova, 2009).

HEPATOCYTE RESPONSIVENESS AND FLEXIBILITY AT SUB-CELLULAR LEVEL

Analysis of hepatocyte structure ultrafine differences in intact animals demonstrated that from fishes to mammals the mitochondrias' volumetric density (Vv) increases. Mitochondrias' number density (Nv) is higher in amphibians and in mammals it is the lowest (Figure 30, 33). The largest number density of mitochondrias in ectothermal animals is a mechanism of adaptation to natural hypoxia. The large number of mitochondrias in ectothermal animals is a mechanism of adaptation to the natural hypoxia and re-oxygenation when an increased functional load on mitochondrias is observed which is also a source of their damage. The type of chondriome presented as single minor mitochondrias is obviously the earliest stage of hepatocytes' program aimed at the protection of the entire cell from functionally damaged mitochondrias and preventing other mitochondrias' damage. Mitochondria membrane surface density (Sv) is higher in reptiles, birds and mammals, as opposed to fishes and amphibians (Figure 29, 30, 31, 32, 33, 34, 35, 36, 37, 38).

The largest nucleolus volume is found in mammals and the lowest – in amphibians. Granular endoplasmic reticulum Vv is higher in fishes, amphibians and mammals, the lowest value is marked in reptiles.

The largest lipidic drop Vv is observed in reptile group hereby Nv is higher in mammals' and reptiles' hepatocytes. Minimum Nv, like lipidic drop Vv, is observed in birds. Maximum glycogen Vv values are found in birds, minimum – in mammals (Figure 29, 30, 31, 32, 33). Maximum lysosome Vv is found in birds, minimum – in reptiles and mammals. However, the largest lysosome Nv is marked in fishes, amphibians and mammals, minimum – in birds and reptiles.

Based on the inter-group comparative analysis (Wilcoxon criterion at p=0.05) it was found that significant differences in:

- mitochondrias' Nv are observed in amphibians vs. all groups of animals;
- mitochondrias' Sv – in amphibians vs. mammals, in fishes vs. mammals, reptiles and birds;
- mitochondrias' Vv is higher in endothermal animals and it significantly differs from ectothermal animals' values;
- lipide Vv is significantly different in reptiles vs. mammals and fishes;
- lipide Nv differs in the group of mammals, reptiles and fishes;
- glycogen Vv differences are found is amphibians vs. mammals; in fishes vs. reptiles as well as between the group of mammals and reptiles;
  lysosome Sv is significantly different in reptiles and birds; in reptiles and mammals;
- lysosome Vv – in birds vs. fishes;
- significant lysosome Nv changes are found only between the groups of mammals and reptiles;
- granular endoplasmic reticulum Vv significantly differs in amphibians vs. reptiles and mammals; in reptiles vs. fishes; in mammals vs. birds.

After hyperthermia in all the animals' groups hepatocyte ultrastructural polymorphism which reflects different degrees of their response to hyperthermia is found; this different response degree is determined by the different functioning rhythm, biological maturity, position in the acinus system, organism evolution status in the taxonomic system and, consequently by the level of hyperthermia resistivity (T. Cavalier-Smith, 2005; J. Das, 2006). Thus in fish hepatocytes (Figure 29, 30, 31, 32, 33) a significant increase of mitochondias' Vv and Sv is marked but Nv decreases which is obviously due to the merging thereof into a single reticulum.

In amphibians' hepatocytes (Figure 30) only Sv increases.

In reptiles and birds (Figure 32) not only Sv but also Nv increases.

In reptiles like in birds and amphibians alongside with Nv increase mitochondrias' Vv and giant mitochondrias' formation is marked.

Mitochondrias' merging provides enhanced transport of energy substrates into the mitochondrias, enables the use thereof as "intracellular cable" transmitting energy as membrane potential from hepatocyte periphery where the level oxygen required for breathing is relatively high to the central part of hepatocytes where the demand in oxidizing metabolites and ATP is especially high (C.T. Taylor, 2008).

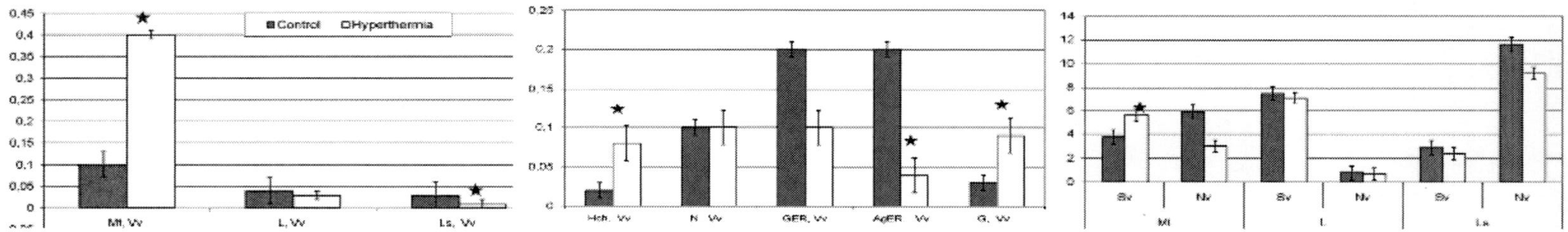

Figure 29. Ultrastructural indicators of Fish hepatocytes in the normal condition and after hyperthermia.

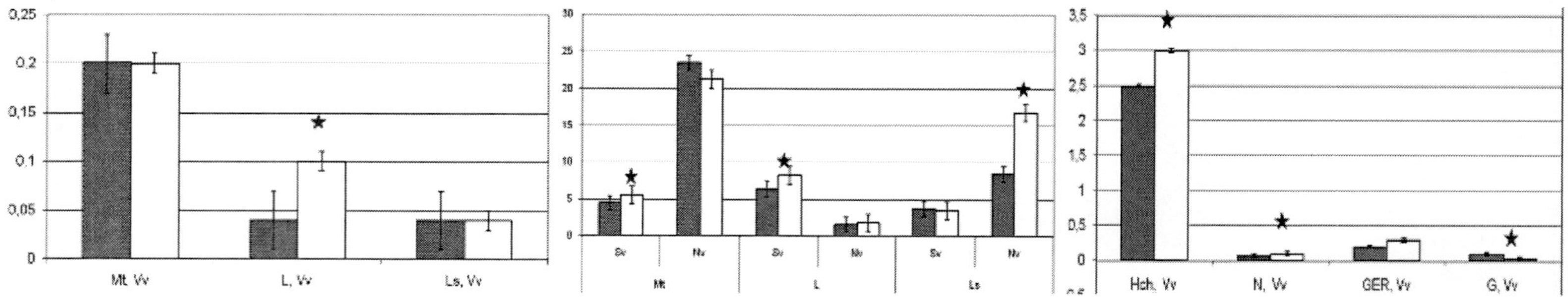

Figure 30. Ultrastructural indicators of Amphibian hepatocytes in the normal condition and after hyperthermia.

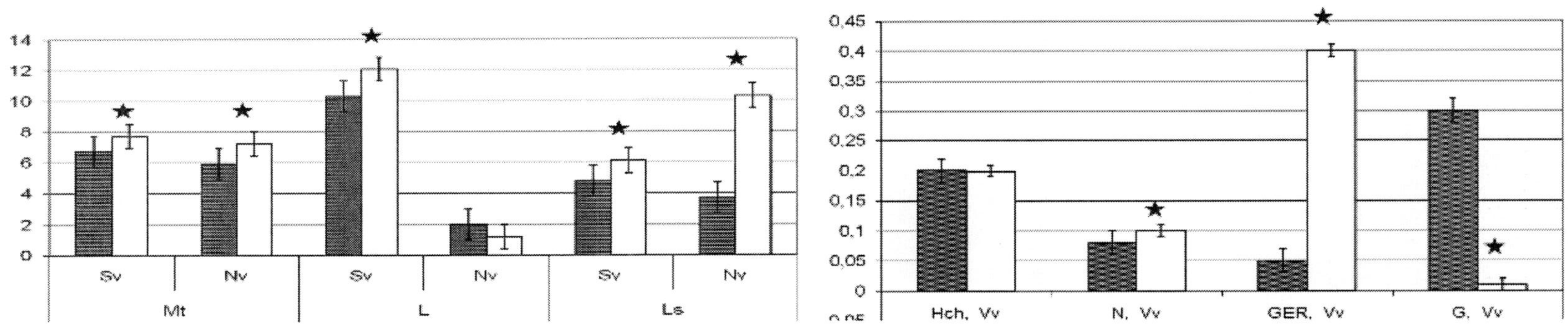

Figure 31. Ultrastructural indicators of Reptile hepatocytes in the normal condition and after hyperthermia.

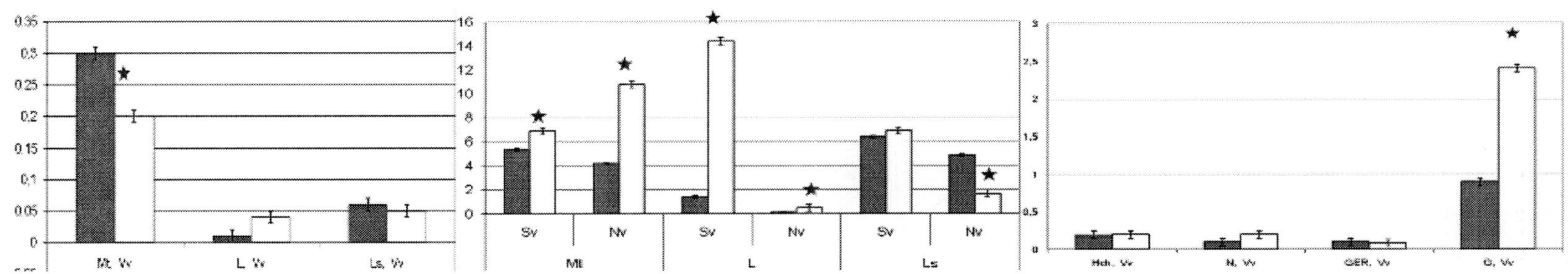

Figure 32. Ultrastructural indicators of bird hepatocytes in the normal condition and after hyperthermia.

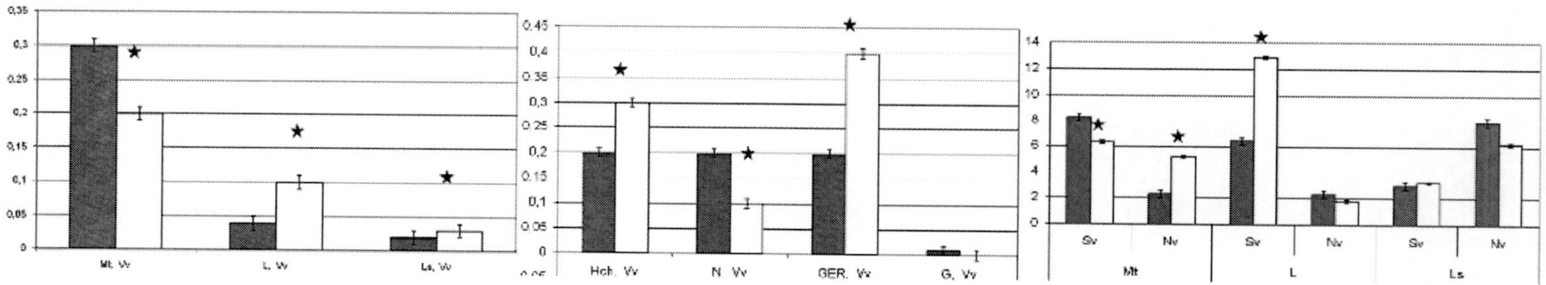

Figure 33. Ultrastructural indicators of Mammal hepatocytes in the normal condition and after hyperthermia.

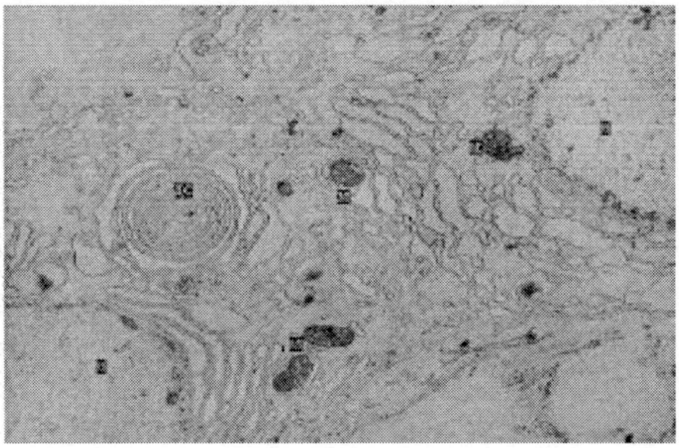

Figure 34. Fish hepatocytes. Control. Zoom x16000. Agranular endoplasmic reticulum forms dactyloscopic structures (DS).

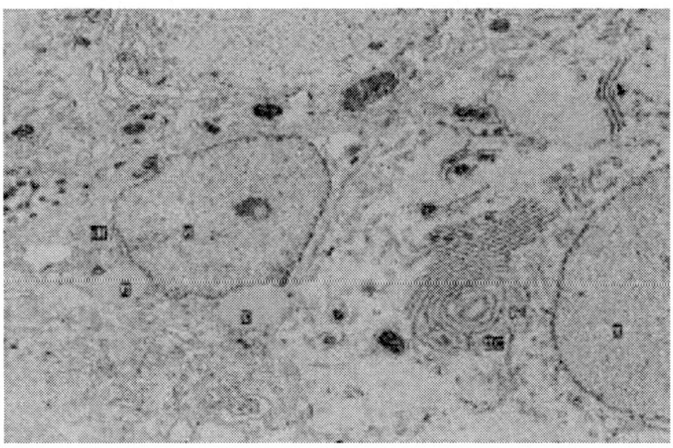

Figure 35. Fish hepatocytes. Control. Zoom x5000. Ito cells (Ic) and lipid drops (L) are situated between the hepatocytes.

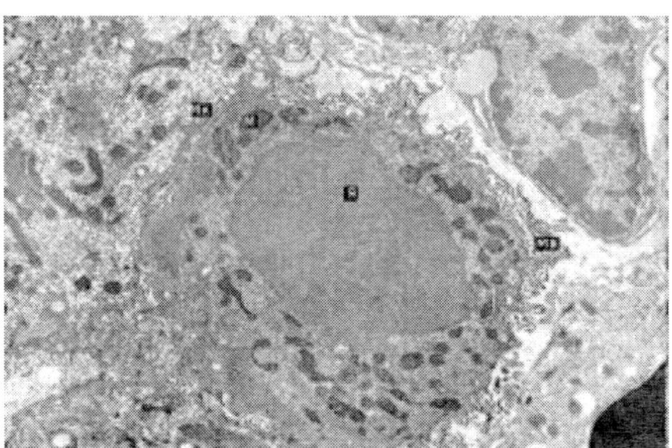

Figure 36. Amphibian hepatocytes. Control. Zoom x3500. Intracellular canaliculi, mitochondrian heterogeneity.

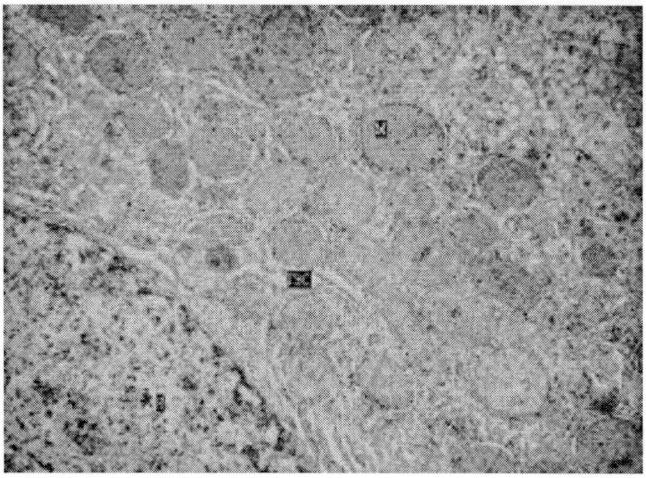

Figure 37. Bird hepatocytes. Control. Zoom x5000. Topographic contact of mitochondrias and granular endoplasmic reticulum.

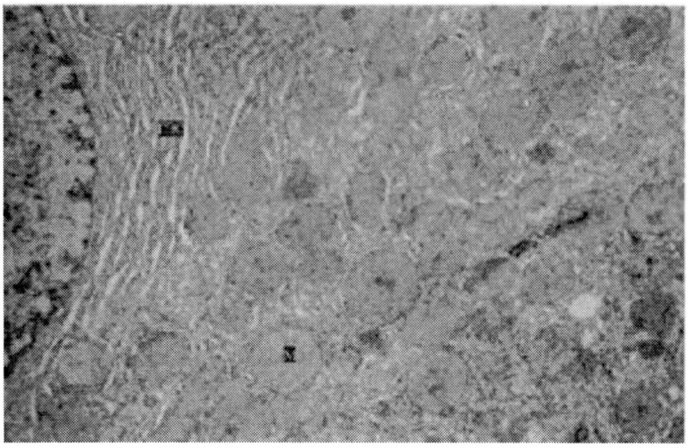

Figure 38. Mammal hepatocytes. Control. Zoom x5000. Powerful pack of granular endoplasmic reticulum adjacent to the nucleus. Morphologic heterogeneity of mitochondrias in a close topographic contact with the agranular endoplasmic reticulum canaliculi.

As opposed to other groups, in mammals mitochondrias' Vv and Sv reduction with only Nv increase is found. Mitochondrias' hyperplasis in mammals may reflect, on the one hand, increased proliferative activity of hepatocytes and on the other hand – increased functional load. Simultaneously mitochondrial fragmentation may result from both breathing inhibition and increased production level of active oxygen forms generated by the respiratory chain of the damaged mitochondrias. Thus, in mammals the protection program against mitochondrias with damaged functions found in intact ectothermal animals is implemented (E.F. Lušnikov, 2001; E.L. Xolmuxamedov, 2008; J. E. Flanigan, 2004; J. M. Lee, 2006).

It should be pointed out that inside the same hepatocyte in all the animal groups in question increased Vv and Nv both in lysosomes and in mitochondrias was found (Figure 41), which points out at the overlapping of thanatogenic sub-programs inside the hepatocytes (S.Ja. Porskurjakov, 2002; L.E. Bakejeva, 2006).

Lysosome Nv increase in amphibians and reptiles (Figure 29, 30, 31, 32, 33), Vv in mammals (as opposed to birds and fishes) reflects, on the one hand, hepatocytes' increased synthetic activity and on the other hand – initiation of autophagia processes which is linked with energy consumption for which reason a stand-alone mitochondrias' population manifests the symptoms of maximum functional activity (D.R. Green, 2004; S. Rodriguez-Enriquez, 2004; M.E. Guicciardi, 2004; B. Winchester, 2005; H. Erdal, 2005).

The changed lipids' utilization intensity is demonstrated through an increased concentration of lipids in all the animal groups' hepatocytes. Activation of glycogen utilization processes as the main energy source is marked in amphibians, reptiles and mammals (Figure 29, 30, 31, 32, 33). Increase in lysomes Nv, Vv and Sv is correlated with both reduced glycogen Vv and increased lipids' Vv. An exception here is the birds' group in which glycogens' and lipids' value ratio manifests the reverse trend. Glycogen Vv is increased in birds and fishes.

It should also be considered that in the fish liver glycogenolysis enzymes' activity is lower and glycogen, even in fasting period is consumed slowly.

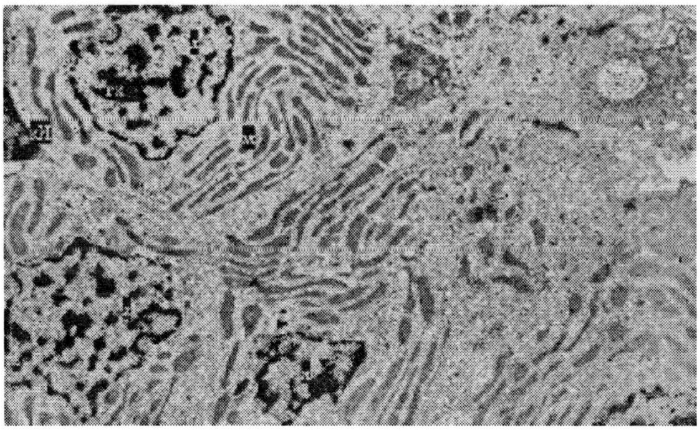

Figure 39. Amphibian hepatocytes. Hyperthermia. Zoom x4000. Festoon-shaped nuclei with pronounced heterochromatin component, nucleolus absence. Ito cell.

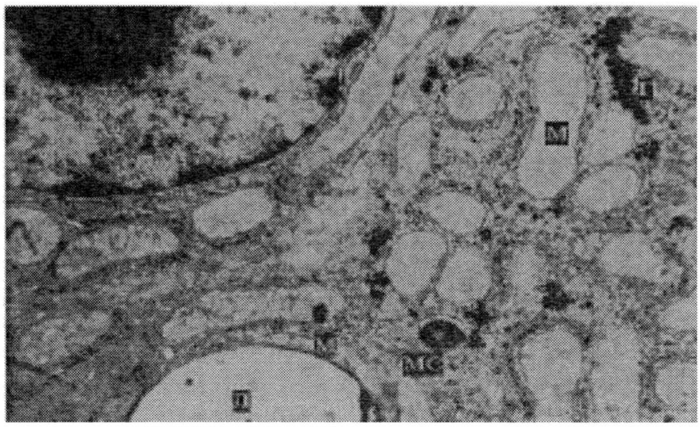

Figure 40. Bird hepatocytes. Hyperthermia. Zoom x5000. Topographic contact of lipid drops and mitochondrias. Glycogen fields.

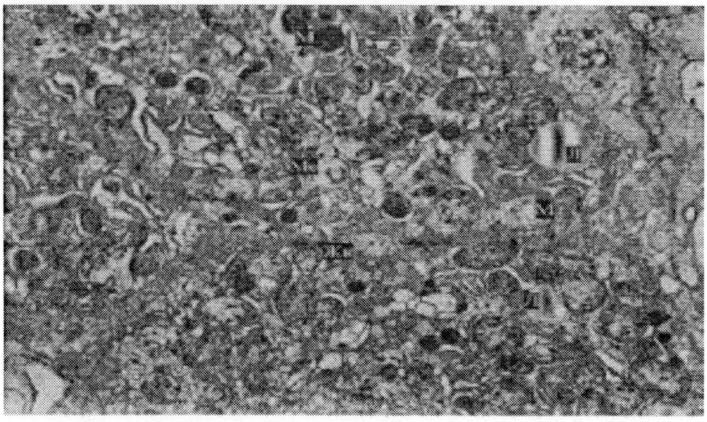

Figure 41. Mammal hepatocytes. Hyperthermia. Zoom: x5000. Lipid drops, myelin-like structures, heterogeneity of mitochondrias' morphology.

Apparently, after hyperthermia the main energy source alongside with glycogen are lipids. This is confirmed in all the animals through a close topographic contact of mitochondrias with lipid (Figure 40) drops which reflects the activation of ATP-dependent process of fatty acids' β-oxidation using mitochondrial enzymes (M.M. Kalašnikova, 2003).

In amphibians and reptiles the lipid drops' volume increase amounts to maximum 30% whereas in birds and mammals lipid drops' Sv, Vv and Nv significantly increases. The lysosome Nv decrease in birds (Figure 32), decrease in hepatocyte acid phosphatase free activity morpho-biochemically determine the largest surpass of the number of Hoechst-positive hepatocytes over that of propidium-iodide-positive ones.

Based on the data obtained lipogenesis/glycogtenesis (Vl/Vg) volumetric ratio may be found. Thus lipid and glycogen Vv ratio in control fishes is 57/43, respectively, after hyperthermia - 25/75. In amphibians the control-found ratio of 80/20 is retained after the overheating.

In control group reptiles lipid and glycogen Vv ratio is 50/50, and after overheating it amounts to 95/5. In birds control-found inclusion ratios of 1/99 are retained which is due to very fast metabolic depression activation, like in amphibians. In mammals the control-identified ratio of 82/18 after overheating is presented only by lipids which is due to the complete glycogen utilization (S. Mitev, 2005).

Nucleus chromatin (Figure 39) structure manifests adaptive re-organization in accordance with the hepatocytes' demand in RNA and protein synthesis thus providing alignment between transcription and translation level. Thus, in fishes nucleolus segregation is manifested which may reflect protein synthesis inactivation at the level of pre-ribosome sub-units assembly inhibition. Simultaneously this nucleolus structure occurs in intact fishes' hepatocytes. It may be a morphological equivalent of the implementation of evolutionally formed hyperthermia resistivity mechanisms reflecting a lower metabolism level in normal condition with subsequent transcription modulation, (C.R. Currens, 2002). The change in the hepatocytes' nucleolus functional morphology in amphibians and mammals, as opposed to birds. reflects the inhibition of rRNA biosynthsis processes.

After hyperthermia in all the animal groups the Ito cell populations are singled out: with increased lipid contents; with the manifestation of activation and transformation into myofibroblasts symptoms (H.A. Singh, 2006). Kuppfer cells' activation is manifested through the accumulation of damaged cells' cytoskeletal structures in the phagosomes, and through the large quantity of the captured cell detrit.

The manifested peculiarities in post-hyperthermia hepatocytes' structure, mosaicism and type non-uniformity of the response of adjacent hepatocytes and even organelles inside the same hepatocyte is the reflection of the law of intermittent activity of the functioning structures (G.N. Kryžanovskij, 1973; K. Imai, 2000; A. J. Koster, 2003; J. Das, 2006; E.I. Antonova, 2007, 2008).

Peculiarities of Liver Oxidizing Processes, Metabolic Mechanisms

Analysis of "metabolism intensity compensation" as a component of different metabolic adaptation strategies in the comparative row of intact animals revealed that more intensive catabolism of substates with oxidoreductases and supply of respiratory chain with reduced equivalents at the level of IDHmt (NAD)-responses is observed in mammals, the lowest enzyme activity is found in reptiles (Figure 42, 43, 44, 45, 46). The highest IDHmch (NADP) activity is found in amphibians and reptiles and, to a lower degree, - in mammals, which reflects the peculiarity of amphibians' and reptiles' organism – availability of natural hypoxia and, consequently, a higher level of energy-aimed regulation of tri-carboxylic acid cycle oxidoreductases, metabolism coordination and integration. The highest MDHmt(NAD) activity level is found in the group of mammals and amphibias, the lowest – in reptiles. Reactions catalyzed by mitochondrian MDHmt(NADP) flow more actively in amphibians and birds and, to a lower extent – in mammals. The activity of oxidoreductases' cytoplasmatic forms manifests dynamics similar to that of mitochondrias (Figure 42, 43, 44, 45, 46).

Maximum LDH activity is found in fishes, minimum – in amphibians. The ratio of mitochondrian oxidoreductases/LDH activity in fishes amounts to 40/60; in amphibians - 87/13; in reptiles - 36/64; in birds - 43/57 and in mammals - 53/47 which reflects the trend of energetic metabolism in these animal groups. Free activity of acid phosphatase and caspase-3 is by an order of magnitude higher in birds than in other animals, this activity has minimum value in amphibians and fishes.

G6PD activity in all the animals is similar, except mammals in which the activity is by factor of 5 lower.

Concentration of [Ca2+]i ions in hepatocyte cytosol is equal in all the animals. [K+]i ions are to a larger degree found in fishes and mammals whereas the largest number of [Na+]i ions is found in mammals and the smallest number – in reptiles.

The identified primary complex of metabolic modifications and metabolic pathways associations after hyperthermia is manifested in the activation of the tricarboxylic acids' cycle early stage due to the increased activity of mitochondrian NAD-dependent IDH in the group of fishes and reptiles (Figure (Figure 42, 43, 44, 45, 46). Simultaneously, in the birds' group the enzyme activity is reduced which, in all probability, determines the decrease of generation of the reduced equivalents for the respiratory chain, decrease of NAD+/NAD(N) and NADP+/NADPN ratio, substrate catabolism level in the cycle of tricarboxylic acids. The increased activity of mitochondrian NADP-dependent IDH is found only in mammals, whereas in amphibians and mammals the enzyme activity is reduced.

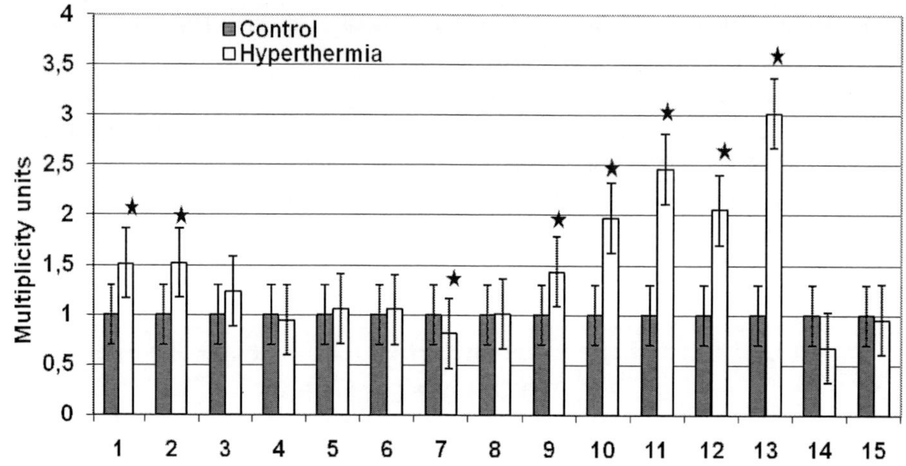

Figure 42. Fish liver enzymes' activity.

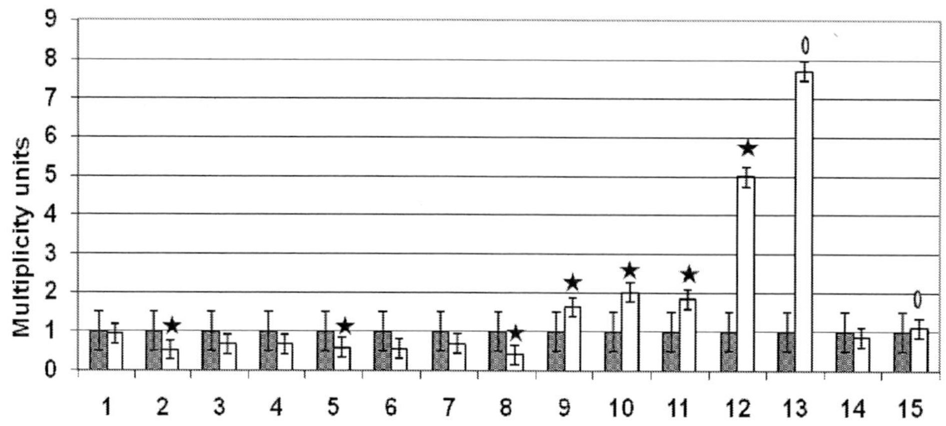

Figure 43. Amphibian liver enzymes' activity.

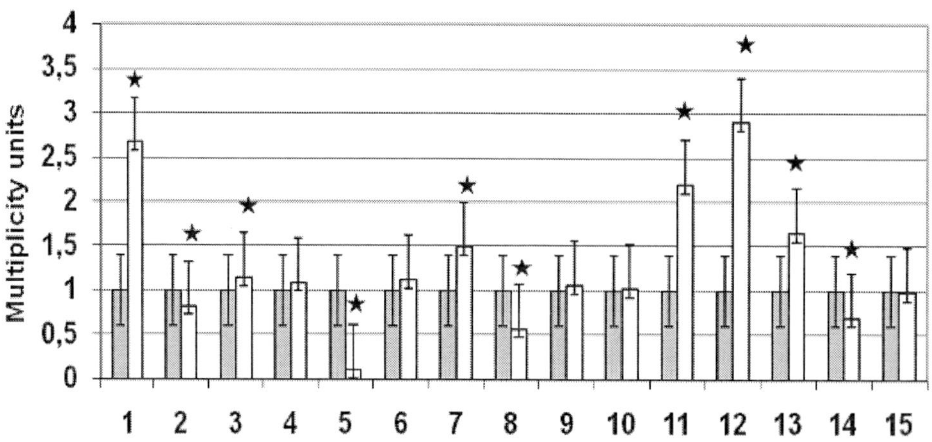

Figure 44. Reptile liver enzymes' activity.

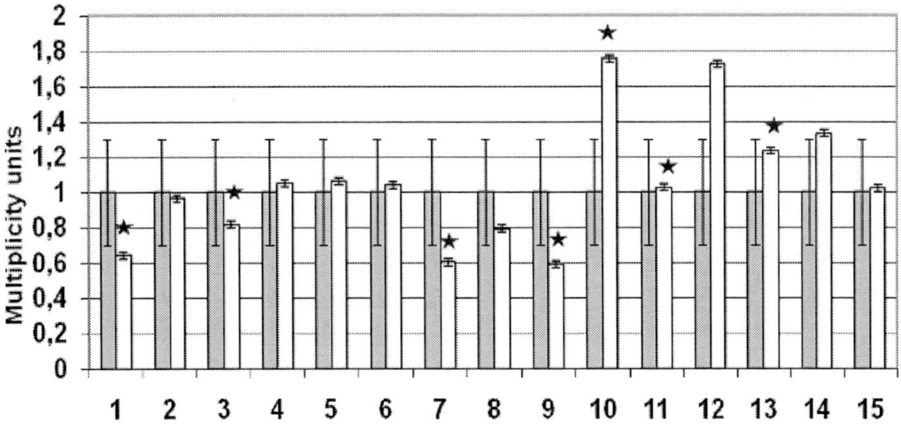

Figure 45. Bird liver enzymes' activity.

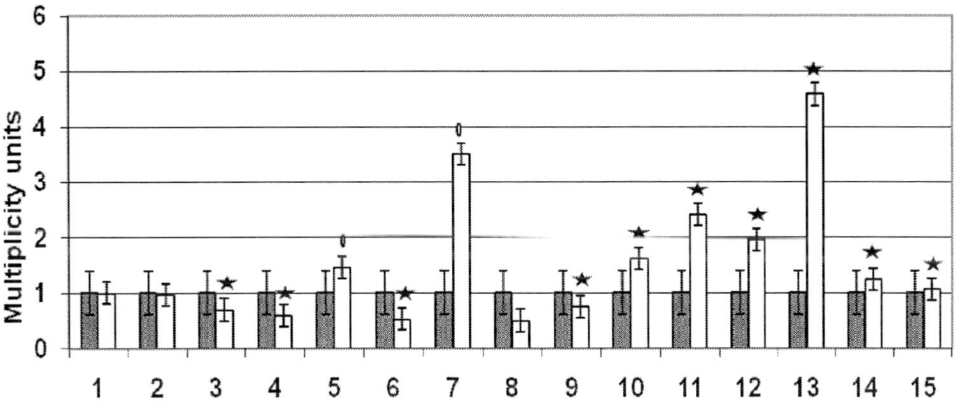

Note: In Nmol substrate per mg protein – 1–IDHNADm; 2–IDH(NAD)c; 3–MDH(NAD)m; 4–MDH(NAD)c; 5–IDH(NADP)m; 6–IDH(NADP)c; 7–MDH(NADP)m; 8–MDH(NADP)c; 9-NAD/NADP; 10 G6PD; 11–free activity of acid phosphatase; 12- Caspase-3 in μm/mg protein; 13–LDH in μmol/l; 14–[K]i; 15–[Na]i. At p=0.05 Student and Wilcoxon criterion.

Figure 46. Mammal liver enzymes' activity.

Mitochondrian MDH(NAD) activity is increased in reptiles; and reduced in birds' and mammals' group. Mitochondrian MDH(NADP) in reptiles and mammals also increases, apparently, this is related to the increased lipids' number in hepatocytes (H. Eisler et al, 2004; E.T. Sadowska, 2005). In fishes and birds the enzyme activity is inhibited which results in the reduced β-oxidation rate of fatty acids.

It is quite possible the identified increased activity of energy-production reactions of tri-carboxylic acids' cycle in fishes, reptiles and mammals is due to the fact that the process of lipid peroxidation provides the flow of a number of enzymatic reactions in the tri-carboxylic acids' cycle, in hypoxia conditions. Simultaneously, the reduced activity of mitochondrian oxidoreductases of tri-carboxylic acids' cycle is a safeguarding mechanism which influences the keeping of the integrity of hepatocytes and structures thereof, reduced procuction of active oxygen forms also determining the ratio modulation at biochemical level of hepatocytes' apoptosis pathways (R.G. Boutilier, 2000; J. St-Pierre, 2000; J.E. Flanigan, et al., 2004; G.

Kroemer, et.al, 2005; G. Balogh, 2005; T. Yorimitsu, 2005; E-L. Eskelinen, 2005; M. Mandi, 2006; L.E. Bakejeva, 2006; Y. Zhang, 2007; S. Luo, 2007; N. Kourtis, 2008).

The activity of NAD-dependent IDHc in cytosol fraction of the fish liver increases and in amphibias and reptiles it reduces. The activity of NADP-dependent IDHc significantly reduces in mammals. Activity of MDH(NADPc) in amphibians and reptiles reduces (Figure 42, 43, 44, 45, 46) which, obviously, results in the inhibition of the early gluconeogenesis stages, proliferative processes and detoxication.

In re-oxygenation period G6PD in many ways determines the enhancement of reparative-proliferative processes in all the animal species in question: in fishes, amphibias, birds and mammals. Increased activity of phosphogluconate pathway is, in all probability, induced by the lipid peroxidation, high concentrations of glucose-6-phosphate as glucose conversion product and glycogen reserves mobilization indicator with the reduction of volumetric density thereof in the group of amphibians, reptiles and mammals. Generation of reduced equivalents in the phosphogluconate pathway reactions is the main source for the functioning of glutathione reductase/glutathione peroxidase anti-oxidant system as well as sythesis of pentose nucleotides and fatty acids (P.W. Hochachka, et.al., 2001; S. Filosa, 2003; G. Tozzi, 2006; L. Agius, 2008).

LDH activity increases which indicates the possible intensification of anaerobic pathway of oxidation in the liver. To a larger extent LDH activation is observed in the group of amphibians, mammals, fishes and, to a smaller extent in the groups of birds and reptiles (Figure 42, 43, 44, 45, 46). Thus, against the background of the reduced activity of tricarboxylic acids' cycle oxidoreductases in the amphibian group the anaerobic glycolysis becomes the only ATP source in stress conditions.

Against the background of the increased number of [Ca2+]i ions in the cytosol fraction of amphibian hepatocytes an increased number of [Na+]I ions; and in reptiles – reduction of [K+]i ions concentration is found; in mammals both the concentration of [K+]i, and [Na+]I ions is increased; in birds the concentration of only [Ca2+]i is increased. The ion concentration changes identified lead to the development of potential-dependent ion channels, differentiated disoprder of the hepatocytes ionic and osmotic homeostasis.

In its turn, the increased concentration of [Ca2+]i ions in cytosol determined the implementation of different hepatocyte apoptosis pathways which is confirmed by the increase of acid phosphatase and caspase-3 free activity (S.J. Porskurjakov, 2002; K. Paszty, 2002, 2005; N. Itano, 2003; T. Yorimitsu, 2006; S. Kumar, 2007; L. Galluzzi, 2007, 2008). Increased caspase-3 activity has nearly become a synonym of cell death (B. McLaughlin, et al. 2003; K. Kivinen, 2005; S. A. Lakhani, 2006). Thus, in intact fishes caspase-3 activity and acid phosphatase free activity are equal, after hyperthermia the enzymes' activity increases more than twofold. In amphibians caspase-3 activity increases to a larger extent. In mammals and reptiles acid phosphatase and caspase-3 free activity increases. In birds the acid phosphatase free activity does not change after overheating, whereas caspase-3 activity increases. Reduced number density of lysosomes in birds and low acid phosphatase free activity, in all probability, determines the security of glycogen against the background of pronounced metabolic depression.

Therefore, the hypercatabolic metabolism strategy (resistant) implemented on the basis of post-overheating stress reaction is found in reptiles, mammals and, to a smaller extent, in fishes. Phylogenetically more ancient – hypometabolic- reaction, tolerant hypobiosis (tolerant) is found in amphibians and birds. Metabolic foundation of this strategy is reduced catabolism, energy consumption and oxygen consumption, effectors here are adenosine, serotonin (V.I. Kulinskij, 1992; I.A. Volčegorskij, 2000; A. Makarieva, 2006; R-P. Sabat, 2006). Metabolic depression in intact ectothermal animals has natural foundation when hypoxic state is changed by hypertoxic state due to reoxygenation (J.B. Pritchard, 2002; M. Hermes-Lima, 2002; D.C. Jackson, 2004). However, in birds hypometabolism develops which is manifested at the level of both mitochondrial and cytosolic oxidoreductases. It highlights the provision that in evolution new possibilities/ functions arise on the basis of the previously achieved level due to the activation of evolutionally older mechanisms of hepatocytes' functioning and energy-supply. In its turn, the identified activation of oxidoreductases of tricarboxylic acids' cycle and cytosol in reptiles is due to the fact that being descendants of ancient amphibians, reptiles, particularly, tortoises, started two major branches of the present-day higher vertebrates – birds and mammals. Tortoise order descends directly from cotylosaurs and, consequently, their position is the closest to the reptile-like ancestors of mammals. Consequently, it is possible to suggest that in extreme conditions endothermal animals may activate poikilothermal thermal regulation mechanisms and vice versa.

The identified primary metabolic reaction of hepatocytes to hyperthermia action in all the animal species in question, is, as per "oxygen compliance" strategy, provided by the differentiated reduction of activity of tricarboxylic acids' cycle oxidoreductases, intensity of energy-consuming processes in the cytoplasm, ensuring metabolic ATP sources security and hepatocytes' integrity. The identified changes in the activity of the enzymes in question, according to "minimum energy consumption" principle solve two problems – metabolism adaptation to hyperthermia and protection against oxidative stress in re-oxygenation period (N.D. Ozernjuk, 2000, 2004; R.G. Boutilier, et al., 2000; V.I. Lushchak, 2001; P.W. Hochachka, et al., 2002; G.N. Somero, 2002; B.G. Lovegrove, 2003; C.T. Taylor, 2008; E.I. Antonova, 2010). The metabolic strategy trends identified are implemented through overlapping of cellular compensatory reactions and response reaction cascade of higher biological hierarchy levels (N.D. Charles, 2005).

Conclusion

In the comparative row of vertebrates with different thermoregulation system species specificity of the liver acinus functional morphology, liver responsivity and flexibility complying with the hierarchic pattern of the living systems organization after a short-term hyperthermia action is detected. Multi-parametric principle of the liver responsivity and flexibility studies in the normal condition and after hyperthermia enabled identifying the trend in which the peculiarities of the liver structures're-organization peculiarities ensuring "survival" and subsequent reparative regeneration stages were formed. Our studies revealed:

1. Liver acinus histological structure in the comparative row of intact animals changes from tubular type (fishes, amphibias, reptiles) to tubular-trabecular type (birds) and trabecular type (mammals). Number density of liver acini, volumetric density of mitochondrias, hepatocyte apoptosis/necrosis ratios increase from ectothermal animals to endothermal ones. In the hepatocyte ultrastructure species specificity of the nucleolus structure, volumetric,

surface and number indicators of mitochondrias, granular endoplasmic reticula, lysosomes, glycogen and lipids is identified.

In the evolution polyplodization due to acytolkinetic mitosis is the leading form of hepatocytes' physiological regeneration. And, if in mammals, reptiles and fishes this hepatocyte population dominanation zones are not present, in birds this is portal tract zone, and in amphibias – central veins' zone. Hereby the number of PCNA-positive hepatocytes in birds in all acinus zones is by an order of magnitude higher. In other animal groups their number is smaller with the predominant localization in the periporal zone.

The number of desmin-positive Ito cells grows from ectothermal animals to endothermal ones; whereas the maximum number of organ-specifi macrophages is found in fishes and mammals.

In intact fishes and amphibias NAD- and NADP-dependent mitochondrial oxidation paths are expressed to an equal degree; in reptiles and birds a higher activity of NADP-dependent dehydrogenases and in mammals – NAD-dependent dehydrogenases is found. The highest activity of hepatocyte apoptosis biochemical markers is found in birds and the lowest activity – in amphibias and fishes.

2. Single hyperthermia in "amphibias/birds" and "fishes/reptiles" animal groups causes species-specific changes of the bloodstream morphometrical indicators, increased number of stromal liver cells with the change of the topography thereof within the acinus; changes of cell cycle indicatrors, state and proportions of hepatocyte chromatin fractions', change of physiological regeneration sources and hepatocyte death pathways.

Thus, hyperthermia determines the change of hepatocyte cell cycle structure: in "amphibias/birds" group the number of hepatocytes in the cell cycle G0-G1-stage increases and in the "fishes/reptiles/ mammals" group the increase is observed in S- and G2-M-stages. The number of hypodiploid hepatocytes in all the animal groups increases, the main death pathway is necrosis/autophagy.

Hepatocytes' ultrastructural changes are of mosaic and oppositely-directed nature. Volumetric, number and surface density of the mitochondrias, nucleolus, granular endoplasmic reticula, lysosomes as well as lipids/glycogen volumetric ratios change. Activation of the liver stromal cells manifested as the increase of their number, transformation of Ito cells into myofibroblasts and increase of autophagolysosomes and cell debris fragments in the Kuppfer cells is marked. Depending on the development of hypo- or hypermetabolic responses in the animal species in question, these changes reflect the increase of the number of binucleate or PCNA-positive hepatocytes or both the hepatocyte populations.

In reptiles hyperthermia causes increased activity of NAD- and NADP-containing mitochondric oxidoreductases; in mammals – increased activity of NADP-containing mitochondric oxidoreductases and in fishes - NAD-containing mitochondric oxidoreductases which reflects hypercatabolism development. In birds and amphibias the activity of NAD- and NADP-dependent mitochondric oxidoreductases is inhibited which reflects hypometabolism development. Species differences demonstrate cytosol oxidoreductases' activity, ions' concentration, activities of biochemical markers of hepatocytes' proliferation and death pathways.

REFERENCES

Agius, L. Glucokinase and molecular aspects of liver glycogen metabolism [Text] / L. Agius // *Biochem*. 2008. 414. P. 1–18.

Alison, M.R. Hepatic stem cells: from inside and outside the liver? [Text] / M.R. Alison, P.Vig, F. Russo, B.W. Bigger, E. Amofah, M. Themis, S. Forbes // *Cell. Prolif*.- 2004.- Vol.37.-№ 1.- P.1-21.

Anackaja, O.V. Struktura kletočnoj populjacii gepatocitov ptencovyx i vyvodkovyx ptic s različnoj massoj tela [Text]: XIII Vserossijskij simpozium Struktura i funkcija kletočnogo jadra /O.V. Anackaja, A.E. Vinogradova, N.N. Smirnova//*Citologija*.-2000.- vol. 42, No.3.- p. 257.

Andersen, P. L. Eukaryotic DNA damage tolerance and translesion synthesis through covalent modifications of PCNA [Text] / P.L. Andersen, Xu Fang, Xiao Wei // *Cell. Research*.- 2008.- № 18.- P. 162–173.

Antonova, E.I. Mexanizmy kratkosročnoj teplovoj kompensatorno-prisposobitel'noj reakcii pečeni ryb vida Carassius aurata gibelio [Text]/E.I. Antonova//*Sibirskij ekologičeskij žurnal*.-2010.- No.1.- pp.79-87.

Antonova, E.I. Rannije, reparativnyje, sročno realizujemyje reorganizacii subkletočnyx struktur ptic vida Columbia livia posle gipertermii [Text] / E.I. Antonova// *Morfologičeskije vedomosti*.-2007.- No. 3-4. pp. 5-8.

Antonova, E.I. Stromal'no-pareximnyje projavlenija regeneracii pečeni ryb Carassius auratus gibelio posle gipertermii [Text]/ E.I. Antonova//*Morfologičeskije vedomosti* -2008.- No. 1-2.- pp.8-11.

Antonova, E.I. Strtegija strukturno-metaboličeskoj adaptacii pervičnoj kompensatorno-prisposobitel'noj reakcii gepatocitov Columbia livia posle gipertermii [Text] / E.I. Antonova//*Sibirskij ekologičeskij žurnal*.-2009.- No. 3.- pp. 413-422.

Antonova, E.I. Ul'trastrukturnyje projavlenija pervičnoj kompensatorno-prisposobitel'noj reakcii gepatocitov životnyx s različnoj sistemoj termoreguljacii posle vozdejstvija gipertermii [Text]/ E.I. Antonova //*Morfologija*.-2008.- No. 4.- pp.24-28.

Avtandilov, G.G. Plodoimetričeskaja diagnistika funkcional'nogo sostojanija gepatocitov v oblasti vozdejstvija plazmennogo luča [Text]/G.G. Avtandilov, M.D. Bajramkulov//*Naučno-praktičeskij žurnal dlja klinicistov* -2004.- No. 1.- pp. 2-6.

Bakejeva, L.E. Ul'trastruktura mitoxondrij pri apoptoze [Text]/L.E. Bakejeva//XXI Rossijskaja konferencija po elektronnoj mikroskopii: Sbornik naučnyx trudov.- *Černogolovka*, 2006.- p. 206.

Balogh, G. The hyperfluidization of mammalian cell membranes acts as a signal to initiate the heat shock protein response [Text] / G. Balogh, I. Horváth, E. Nagy, Z. Hoyk, S. Benkõ, O. Bensaude, L. Vígh // *FEBS*. – 2005. – Vol. 272. – P. 6077-6086.

Bataller, R, Hepatic stellate cells as a target for the treatment of liver fibrosis [Text] / R. Bataller, D.A. Brenner // *Semin. Liver Dis*. – 2001, № 21.- P. 437–451.

Bataller, R. Liver fibrosis [Text] / R. Bataller, D.A. Brenner // *Clin. Invest*. - 2005. –Vol. 115, №2. - P. 209–218.

Battle, M.A. Hepatocyte nuclear factor 4alpha orchestrates expression of cell adhesion proteins during the epithelial transformation of the developing liver [Text] / M.A. Battle, G. Konopka, F. Parviz // *Proc. Natl. Acad. Sci. USA*.- 2006.- № 103.- P. 8419–24.

Bennett, M.R. Metabolic gene regulation in a dynamically changing environment [Text] / M.R. Bennett, W.L. Pang, N.A. Ostroff, B.L. Baumgartner, S. Nayak, L.S. Tsimring, J. Hasty // *Nature*.- 2008. – Vol. 454.- P.1119-1122.

Borkin, L. J On distribution of diploid, triploid and tetraploid green toads (Bufo viridis complex) in south-eastern Kazakhstan [Text] / L. J. Borkin, V. K. Eremchenko, N. Helfenberger, A. M. Panfilov, J. M. Rosanov // *Russ. J. Herpetol.* - 2001. - Vol. 8- №1.- P. 175-178.

Borkin, L.J. Mass occurrence of polyploid green frogs (Rana esculenta complex) in Eastern Ukraine [Text] / L.J. Borkin, A.V. Korshunov, G.A. Lada // *Russian Journal of Herpetology*. – 2004. – Vol.11, №3. – P. 194–213.

Borodin, Ju.I. Problemy limfodetoksikacii i limfosanacii [Text]/Ju.I. Borodin//*Problemy eksperimental'noj i kliničeskoj limfologii: Trudy NIIKEL SO RAMN*. - Novosibirsk, 2000. - vol. 8. - pp. 5-9.

Boutilier, R. G. Surviving hypoxia without really dying [Text] / R. G. Boutilier // Comparative Biochemistry and Physiology - *Part A: Molecular and Integrative Physiology*.- 2000. -V.126, №4. -P.481-490.

Carpino, G. Activated hepatic stellate cells in liver cirrhosis. A morphologic and morphometrical study [Text] / G. Carpino, A. Franchitto, S. Morini, S. G. Corradini, M. Merli, E. Gaudio // *Anat Embryol*. – 2004. – Vol. 109, № 4. – P. 225-238.

Cavalier-Smith, T. Economy, speed and size matter: evolutionary forces driving nuclear genome miniaturization and expansion [Text] / T. Cavalier-Smith // *Ann. Bot.* (Lond).- 2005.-Vol.95, №1.-P.147-75.

Chang, Y.-M. Thermal acclimation of vetabolic rate may be seasonally dependent in the subtropical anuran latouche's frog (Rana latouchii, Boulenger) [Text] / Y.-M. Chang, P.-C. L. Hou // *Physiological and Biochemical Zoology*. - 2005. – Vol. 78, № 6. – P. 947-955.

Charles, N. D. Hydra and the evolution of apoptosis [Text] / N. D. Charles, N. Schmidt, M. Schade, B. Pauly, O. Alexandrova, A. Böttger // *Integr. Comp. Biol*. – 2005. – Vol. 45. – P. 631-638.

Chen, S. C. Hyperthermic pretreatment decreases microvascular protein leakage and attenuates hypotension in anaphylactic shock in rats [Text] / S. C. Chen, T. S. Lu // *Microvasc. Res*. – 2001. – Vol. 61, № 2. – P. 152–159.

Chubb, J. R. Chromatin motion is constrained by association with nuclear compartments in human cells [Text] / J. R. Chubb, S. Boyle, P. Perry, W. A. Bickmore // *Curr. Biol.* - 2002. - Vol. 12. - P. 439-445.

Currens, C. R. Effects of temperature and time of day on the resting metabolic rates of paedomorphic and metamorphic mole salamanders, Ambystoma talpoideum [Text] / C. R. Currens, P.H. Niewiarowski, H.H. Whiteman // *Copeia*. – 2002.– P. 489-495.

Das, J. The role of mitochondrial respiration in physiological and evolutionary adaptation [Text] / J. Das // *Bio. Essays*. – 2006. – Vol. 28, №9. – P. 890-901.

Ding, H. Effects of Kupffer cell inhibition on liver function and hepatocellular activity in mice [Text] / H. Ding, R. Peng, E.Reed, Q.Q. Li // *Int. J. Mol. Med*.- 2003.-Vol.12, №4.- P.549-557.

Ding, H. Influence of Kupffer cells on hepatic signal transduction as demonstrated by second messengers and nuclear transcription factors World [Text] / H. Ding, J.A. Huang, J.Tong , X.Yu, J.P. Yu // *Gastroenterol.* 2003.-Vol.9, №11.-P. 2519-2522.

Douglas, R.M. Genetic Models in Applied Physiology Invited Review: Effect of oxygen deprivation on cell cycle activity: a profile of delay and arrest [Text] / R. M. Douglas, G.G. Haddad // *J. Appl Physiol*.-2003.-Vol.94.-P. 2068-2083.

Eisler, H. Starvation for an essential amino acid induces apoptosis and oxidative stress in yeast [Text] / H. Eisler, K.H. Frolich, E. Heidenreich // *Exp. Cell Res*. - 2004. – Vol. 300. - P. 345-353.

Erdal, H. Induction of lysosomal membrane perme- abilization by compounds that activate p53-independent apop- tosis [Text] / H. Erdal, M. Berndtsson, J. Castro, U. Brunk, M.C. Shoshan, S. Linder // *Proc. Natl. Acad. Sci. USA*.-2005.-Vol.102.-P. 192–197.

Eskelinen, E-L. Autophagy can promote both cell survival and cell death [Text] / E-L. Eskelinen, D. Jekyll, M. Hyde // *Cell Death and Differentiation*.-2005.-Vol.12. - P.1468–1472.

Fausto, N. Liver regeneration: from laboratory to clinic [Text] / N. Fausto // *Liver Transpl*. – 2001. – Vol. 7, № 10. – P. 835-844.

Fausto, N. The role of hepatocytes and oval cells in liver regeneration and repopulation [Text] / N. Fausto, J.S. Campbell // *Mech. Dev*.- 2003.-Vol.120. - P.117-130.

Figueiredo-Fernandes, A. M. Temperature and gender influences on the hepatic stroma (and associated pancreatic acini) of Nile tilapia, Oreochromis niloticus (Teleostei, Cichlidae): A stereological analysis by light microscopy [Text] / A. M. Figueiredo-Fernandes, A. A. Fontaнnhas-Fernandes, M. A. Reis-Henriques, E. Rocha // *Morphology*.-2006.-Vol.267, №2. - P. 221-230.

Filosa, S. Failure to increase glucose consumption through the pentose-phosphate pathway results in the death of glucose-6-phosphate dehydrogenase gene-deleted mouse embryonic stem cells subjected to oxidative stress [Text] / S. Filosa, A. Fico, F. Paglialunga, M. Balestrieri, A. Crooke, P. Verde, P. Abrescia, J. M. Bautista, G. Martini // *Biochem*. – 2003. – Vol. 15. – P. 935-943.

Flanigan, J. E. Metabolic depression and sodium-potassium ATPase in the aestivating frog, Neobatrachus kunapalari [Text] / J. E. Flanigan, M. Guppy // *Journal of Comparative Physiology B: Biochemical, Systemic, and Environmental Physiology*. – 2004. – Vol. 167, № 2. P. 135 145.

Frampton, J. Postreplication repair and PCNA modification in schizosaccharomyces pombe [Text] / J. Frampton, A. Irmisch, C. M. Green, A. Neiss, M. Trickey, H. D. Ulrich, K. Furuya, F. Z. Watts, A. M. Carr, A. R. Lehmann // Originally published as MBC in Press. – 2006. - Vol. 17, №7. – P. 2976-2985.

Galluzzi, L., Kroemer, G. Necroptosis: A Specialized Pathway of Programmed Necrosis [Text] / L. Galluzzi, G. Kroemer // *Cell*.- 2008. - Vol. 135, №7. – P. 1161-1163.

Galluzzi, L. Cell death modalities: classification and pathophysiological implications [Text] / L. Galluzzi, M.C. Maiuri, I. Vitale, H. Zischka, M. Castedo, L. Zitvogel, G. Kroemer // *Cell Death and Differentiation*. - 2007.- Vol.14.- P.1237–1243.

Gandillet, A. Hepatocyte ploidy in normal young rat [Text] / A. Gandillet, E. Alexandre, V. Holl, C. Royer, P. Bischoff, J. Cinqualbre, P. Wolf, D. Jaeck, L. Richert // *Comp. Biochem. Physiol. a Mol. Integr. Physiol*. – 2003. – Vol. 134, № 3. - P. 665-73.

Ganem, N. J., Pellman D. Limiting the Proliferation of Polyploid Cells Cell [Text] / N. J. Ganem, D. Pellman // *Cell*. – 2007. – Vol.131, № 3. – P. 437-440.

Gibb, A. C. Functional morphology and biochemical indices of performance: is there a correlation between metabolic enzyme activity and swimming performance? [Text] / A.

C. Gibb, K. A. Dickson // *Integrative and Comparative Biology*. – 2002. – Vol. 42, № 2. – P. 199-207.

Green, D.R The pathophysiology of mitochondrial cell death [Text] / D.R. Green, G. Kroemer // *Science*. – 2004. –Vol. 305. – P. 626–629.

Gregory, T. R. Insertion-deletion biases and the evolution of genome size [Text] / T. R. Gregory // *Gene*. - 2004. –Vol. 7. – P. 15-34.

Gregory, T.R. Genome size is not correlated positively with longevity in ẽshes (or homeotherms) [Text] / T.R. Gregory // *Exp. Gerontol*.- 2004.-Vol.39.- P. 859-860.

Gregory, T.R. Macroevolution, hierarchy theory, and the C-value enigma [Text] / T.R. Gregory // *Paleobiology*. - 2004.- Vol. 30. – P.179-202.

Gregory, T.R. Variation across amphibian species in the size of the nuclear genome supports a pluralistic, hierarchical 624 Genomes and evolution [Text] / T.R. Gregory // *Current Opinion in Genetics and Development*.- 2004.- Vol. 14. – P.620-626.

Guicciardi, M.E. Lysosomes in cell deat [Text] / M.E. Guicciardi, M. Leist, G.J. Gores // *Oncogene*. -2004. - Vol.23. - P.2881–2890.

Gupta, S. Hepatic polyploidy and liver growth control [Text] / S. Gupta // *Semin. Cancer Biol*. - 2000. – Vol. 10. – P. 161-171.

Hermes-Lima, M. Animal response to drastic changes in oxygen availability and physiological oxidative stress [Text] / M. Hermes-Lima, T. Zenteno-Savin // *Comp. Biochem. Physiol*. – 2002. – Vol. 133. – P. 537–556.

Hochachka, P. W. *Biochemical adaptation: Mechanism and process in physiological evolution* [Text] / P. W. Hochachka, G. N. Somero // New York: Oxford University Press, 2002. P. 1-32.

Hochachka, P. W. Mechanism, origin, and evolution of anoxia tolerance in animals [Text] / P. W. Hochachka, P. L. Lutz // *Comp. Biochem. Physiol*. - 2001. – Vol. 130. – P. 435-459.

Imai, K. Storage of lipid droplets in and production of extracellular matrix by hepatic stellate cells (vitamin A-storing cells) in Long-Evans cinnamon-like colored (LEC) rats [Text] / K. Imai, M. Sato, N. Kojima, M. Miura, T. Sato, T. Sugiyama, K. Enomoto, H. Senoo // *Anat. Rec*.- 2000. - Vol. 258, № 4. - P. 338-348.

Ismail, M.H., Pinzani, M. Reversal of liver fibrosis [Text] / M.H. Ismail, M. Pinzani // *Gastroenterol*. – 2009. – Vol.15, №1. – P. 72-79.

Issa, R. Apoptosis of hepatic stellate cells: involvement in resolution of biliary fibrosis and regulation by soluble growth factors [Text] / R. Issa, E. Williams, N. Trim, T. Kendall, M. J. Arthur, J. Reichen, R. C. Benyon, J. P. Iredale // *Gut*. – 2001. – Vol. 48. – P. 548-557.

Itano, N. Cell spreading controls endoplasmic and nuclear calcium: A physical gene regulation pathway from the cell surface to the nucleus [Text] / N. Itano, S.Okamoto, D.Zhang, S. A. Lipton, E. Ruoslahti // *PNAS*. - 2003. - Vol.100, № 9.-P. 5181-5186.

Jackson, D. C. Acid-base balance during hypoxic hypometabolism: selected vertebrate strategies [Text] / D. C. Jackson // *Resp. Phys. Neur*. – 2004. – Vol. 141. – P. 273–283.

Ju, C. Protective role of Kupffer cells in acetaminophen-induced hepatic injury in mice [Text] / C. Ju ,T. P. Reilly , M. Bourdi , M.F. Radonovich, J.N. Brady, J.W.George., L. R. Pohl // *Chem. Res. Toxicol*.- 2002.-Vol.15, №12.- P.1504-1513.

Kalašnikova, M.M. Ul'trastrukturnyje aspekty prisposobitel'nyx osobennostej kletok pečeni pozvonočnyx [Text]/M.M. Kalašnikova.- Moscow: IPEE, 2003.- 241 p.

Karapetjan, A. F. O regeneratornom potenciale pečeni ozërnoj ljaguški Rana Ridibunda posle častičnoj gepaekomii [Text]/A.F. Karapetjan, K.A. Dživanjan// *Citologija*.- 2006.- vol. 48, No. 4.- pp. 346-354.

Kenneth, N. S. Regulation of gene expression by hypoxia [Text] / N. S.Kenneth, S. Rocha // *Biochem.* - 2008.- Vol. 414.- P.19–29.

Kijasov, A.P. *Sovremennyje texnologii morfologičeskix issledovanij* [Text]/A.P. Kijasov.- Kazan: KGMU.-2000.-38 p.

Kivinen, K. Caspase-3 is required in the apoptotic disintegration of the nuclear matrix [Text] / K. Kivinen, M. Kallajoki, P. Taimen // *Exp. Cell Res.* – 2005. – Vol. 311, № 1. – P. 62-73.

Knolle, P. A. Control of immune responses by savenger liver endothelial cells [Text] / P. A. Knolle, A. Limmer // *Swiss Med. Wkly.* – 2003. – Vol. 133, № 37-38. – P. 501-506.

Knolle, P. A. Local control of the immune response in the liver [Text] / P.A. Knolle, G. Gerken // *Immunol. Rev.* - 2000. – Vol. 174. – P. 21-34.

Kohji, O. Erythrophagocytosis by Liver Macrophages (Kupffer Cells) Promotes Oxidative Stress, Inflammation, and Fibrosis in a Rabbit Model of Steatohepatitis [Text] / O. Kohji, K. Kohji, F. Hideki // *Am. J. Pathol.*- 2007.- Vol.170, №3.-P. 967–980.

Kono, H. Functional heterogeneity of the kupffer cell population is involved in the mechanism of gadolinium chloride in rats administered endotoxin [Text] / H. Kono, H. Fujii, M. Asakawa, M. Yamamoto, A. Maki, M. Matsuda, I. Rusyn, Y. Matsumoto // *Surg. Res.* -2002.- Vol.106, №1. - P. 179-187.

Koster, A. Electron microscopy in cell biology: integrating structure and function [Text] / A. Koster, J. Klumperman // *Nature Reviews Molecular Cell Biology*.-2003.- №4.- P. 6–10.

Kourtis, N. Autophagy and cell death in model organisms [Text] / N. Kourtis, N. Tavernarakis // *Cell Death and Differentiation*.- 2008.- №3. -P. 56-62.

Kroemer, G. Classification of cell death: recommendations of the Nomenclature Committee on Cell Death [Text] / G. Kroemer, W. S. El-Deiry, P. Golstein, M. E. Peter // *Cell Death and Differentiation*.-2005.- Vol.12.- 1463–1467.

Kryžanovskij, G.N. Bioritmy i zakon strukturno-funkcional'noj vremennoj diskretnosti biologičeskix processov [Text]/G.N. Kryžanovskij // *Biologičeskije ritmy v mexanizmax kompensacii narušennyx funkcij*.- Moscow, 1973.- pp.20-34.

Kulinskij, V.I. Dve adaptacionnyje strategii v neblagoprijatnyx uslovijax – rezistentnaja i tolerantnaja. Rol' gormonov i receptorov [Text]/V.I. Kulinskij, I.A. Ol'xovskij// *Uspexi soveremennoj biologii*.-1992.-No. 6. – pp. 697-714.

Kumar, S. Caspases and their many biological functions [Text] / Kumar S. // *Cell Death and Differentiation*. - 2007.- Vol.14.- P.1–2.

Lakhani, S. A. Caspases 3 and 7: key mediators of mitochondrial events of apoptosis [Text] / S. A. Lakhani // *Science*. - 2006. – Vol.311. – P. 847-851.

Lee, J. M. The epithelial–mesenchymal transition: new insights in signaling, development, and disease [Text] / J. M. Lee, S. Dedhar, R. Kallur, W. E. Thompson. // *JCB* - 2006. - Vol. 172, № 7. – P. 973-981.

Liu, C.-C. Glucocorticoids reduce interleukin-1β concentration and result in neuroprotective effects in rat heatstroke [Text] / C.-C. Liu, C.-H. Chien, M.-T. Lin // *Physiol.* – 2000. – Vol. 527. - P. 333–343.

Lovegrove, B. G. The influence of climate on the basal metabolic rate of small mammals: a slow-fast metabolic continuum [Text] / B. G. Lovegrove // *Comp. Physiol.* – 2003. – Vol. 173, № 2. – P. 87-112.

Luo, S. Atg5 and Bcl-2 provide novel insights into the interplay between apoptosis and autophagy [Text] / S. Luo, D.C. Rubinsztein // *Cell Death and Differentiation*.- 2007.- Vol.14.- P.1247–1250.

Lushchak, V. I. Oxidative stress and antioxidant defense in goldfish Carassius auratus during anoxia and reoxygenation [Text] / V. I. Lushchak, L. P. Lushchak, A. A. Mota // *Am. Physiol. Regul. Integr. Comp. Physiol.* - 2001.- Vol.- 280.- P. 100−107.

Lušnikov E.F.. *Gibel' kletki* (apoptoz) [Text]/E.F. Lušnikov, A.Ju. Abosimov.- Moscow: Medicina, 2001. -192 p.

Maciejewski, R. Acid phosphatase activity in different organs as a marker of acute pancreatitis [Text] / R. Maciejewski, M. Kopieniak, B. Madej, A.Sek, A. Dabrowski // *Ann. UMCS. D.* -2001.- Vol. 56.- P. 357-361.

Makarieva, A. Dimensions Blackwell Publishing Ltd - and temperature independence of the minimal metabolic norms bolstering life {discharges} [Text] / A. Makarieva, V.Gorshkov, I.-L.Liá, S. L. Chown // *Functional Ecology*.- 2006.- № 20.- P. 83-96.

Mandi, M. Ca2+ release triggered by NAADP in hepatocyte microsomes [Text] / M. Mandi, B. Toth, G. Timar, J. Bak // *Biochem.* – 2006. – Vol. 395. – P. 233–238.

Mann, D.A. Transcriptional regulation of hepatic stellate cell activation [Text] / D.A. Mann, D.E. Smart // *Gut*.- 2002.- Vol.50.- P.891-896.

Matsuki, Y. Soluble Fas gene therapy protects against Fas-mediated apoptosis of hepatocytes but not the lethal effects of Fas-induced TNF-alpha production by Kupffer cells [Text] / Y. Matsuki , L. Li , H.C. Hsu , P.A.Yang, R. Zheng // *Cell Death Differ*.- 2002.- Vol. 9, №6.- P. 626-635.

McKechnie, A. E. Partitioning of evaporative water loss in white-winged doves: plasticity in response to short-term thermal acclimation [Text] / A. E. McKechnie, B. O. Wolf // *Exp. Biol.* – 2004. – Vol. 207. – P. 203-210.

McLaughlin, B. Caspase 3 activation is essential for neuroprotection in preconditioning [Text] / B. McLaughlin // *Proc. Natl. Acad. Sci.* – 2003. – Vol. 100. – P. 715-720.

Memedula, S. Sequential Recruitment of HAT and SWI/SNF Components to Condensed Chromatin by VP16 [Text] / S. Memedula, A. S. Belmont // *Curr. Biol.* - 2003. – Vol. 13. – P. 241-246.

Metody bioximičeskix issledovanij [Text]/ Ed. by M.I. Prokhorova.-Leningrad: University, 1982.- 380 p.

Meurer, S. K. Identification of endoglin in rat hepatic stellate cells: new insights into transforming growth factor beta receptor signaling [Text] / S. K. Meurer, L. Tihaa, B. Lahme, A. M. Gressner, R. Weiskirchen // *Biol. Chem.* – 2005. - Vol. 280, № 4. – P. 3078-3087.

Milosevic, N. Lead stimulates intercellular signalling between hepatocytes and Kupffer cells [Text] / N. Milosevic, P. Maier // *Pharmacol.* – 2000. – Vol. 401, № 3. – P. 317-328.

Mitev, S. Effect of the acclimation to high environmental temperature on the activity of hepatic glycogen phosphorylase (a+b and a), liver glycogen content and blood glucose level in rats [Text] / S. Mitev, S. Dinevska-Kovkarovska, B.Miova // *Thermal Biology*.- 2005.- Vol.30, №8.- P. 563-568.

Miyahara, T. Peroxisome proliferator-activated receptors and hepatic stellate cell activation [Text] / T. Miyahara, L. Schrum, R. Rippe, S. Xiong, H. F. Yee, K. Motomura, F. A. Anania, T. M. Willson, H. Tsukamoto // *Biol. Chem.* – 2000. – Vol. 275, № 46. – P. 5715-5722.

Muller, W.G. Large-scale chromatin decondensation and recondensation regulated by transcription from a natural promoter [Text] / W. G. Muller, D. Walker, G. L. Hager, J. G. McNally // *Cell. Biol.* - 2001. – Vol. 154. - P. 33-48.

Muraca, M. Hepatocyte transplantation as a treatment for glycogen storage disease typeIa. [Text] / M. Muraca, G.Gerunda, D. Neri, M.T.Vilei, A.Granato, P. Feltracco, M. Meroni, G.Giron, A.B.Burlina // *Lancet.*- 2002.- Vol.359.- P.317-318

Neyrinck, A. M. Insight into the involvement of Kupffer cell-derived mediators in the hepatoprotective effect of glycine upon inflammation: study on rat precision-cut liver slices [Text] / A. M. Neyrinck, S. Margagliotti, N. M. Delzenne // *Inflamm. Res.* – 2005. – Vol. 54, № 3. – P. 106-112.

Neyrinck, A. Modulation of Kupffer cell activity: physio-pathological consequences on hepatic metabolism [Text] / A. Neyrinck // *Bull. Mem. Acad. Med. Belg.* – 2004. – Vol. 159, № 5-6. – P. 358-366.

Otto, S.P. The Evolutionary Consequences of Polyploidy [Text] / S.P. Otto // *Cell.* – 2007. – Vol.131, №3. – P. 452-462.

Ozernjuk, N.D. Ontogentičeskije temperaturnyje adaptacii fermentov poikilotermnyx životnyx [Text]/N.D. Ozernjuk //*Uspexi soveremennoj biologii.*-2004.-vol. 124, No. 6. - pp. 89-98.

Ozernjuk, N.D. *Temperaturnyje adaptacii* [Text]/N.D. Ozernjuk.-Moscow: izdatel'stvo MGU, 2000. -205 p.

Paszty, K. Plasma membrane Ca2+ATPase isoform 4b is cleaved and activated by caspase-3 during the early phase of apoptosis [Text] / K. Paszty, A. K. Verma, R. Padanyi, A. G. Filoteo, J. T. Penniston, A. Enyedi // *Biol. Chem.* – 2002. – Vol. 277. – P. 6822–6829.

Paszty, K. The caspase-3 cleavage product of the plasma membrane Ca2+-ATPase 4b is activated and appropriately targeted [Text] / K. Paszty, G. Antalffy, A. R. Penheiter // *Biochem.* – 2005. – Vol. 391. – P. 687–692.

Petrov, D. A. Mutational equilibrium model of genome size evolution [Text] / D. A. Petrov // *Theor. Popul. Biol.* – 2002. – Vol. 61. – P. 531–544.

Porskurjakov, S.Ja. Nekroz – aktivnaja forma programmirujemoj kletočnoj gibeli [Text]/ S.Ja. Porskurjakov //*Bioximija.*-2002.- vol.67, issue 4.-pp. 468-487.

Pritchard, J. B. Comparative models and biological stress [Text] / J.B. Pritchard // *Am. J. Physiol. Regul. Integr. Comp. Physiol.*- 2002.- Vol.283.- P. 807-809

Rentsch, M. Benefit of Kupffer cell modulation with glycine versus Kupffer cell depletion after liver transplantation in the rat: effects on postischemic reperfusion injury, apoptotic cell death graft regeneration and survival [Text] / M. Rentsch, K Puellmann, S. Sirek, I. Iesalnieks, K. Kienle, T. Mueller, U. Bolder, E. Geissle, K.W. Jauch, A. Beham // *Transpl. Int.*- 2005.- Vol.18, №9.- P.1079-1089.

Rice, A. N. Heating and cooling rates of eastern diamondback rattlesnakes, Crotalus adamanteus [Text] / A. N. Rice, T. L. Roberts IV, M. E. Dorcas // *Thermal Biology.* – 2006. – Vol. 31, №6. – P. 501-505.

Rodriguez-Enriquez, S. Role of mitochondrial permeability transition pores in mitochondrial autophagy [Text] /S. Rodriguez-Enriquez, L. He, J.J. Lemasters // *Int. J. Biochem. Cell Biol.* 2004.- Vol.36.- P.2463-2472.

Sabat, R.-P.Water and energy economy of an omnivorous bird: Population differences in the collared Sparrow (Zonotrichia capensis) [Text] / R.-P.Sabat ,G. Cavieres , C. Veloso , M. Canals // *Comparative Biochemistry and Physiology - Part A: Molecular and Integrative Physiology*.- 2006.- Vol. 144, № 4.- P. 485-490.

Sadowska, E.T. Genetic correlations between basal and maximum metabolic rates in a wild rodent: consequences for evolution of endothermy [Text] / E.T. Sadowska, M. K. Labocha, K. Baliga, A. Stanisz, A.K. Wroblewska, W. Jagusiak, P. Koteja // *Evolution Int. J. Org. Evolution*. -2005.- Vol.59, №3.- P.672-681.

Sakuta, G.A. Kletočnyje mexanizmy regeneracii cirrotičeski izmenennoj pečeni krys. Vilijanije častičnoj gepaektomii na proliferaciju, poliploidizaciju i gipertrofiju gepatocitov [Text]/G.A. Sakuta, B.N. Kudrjavcev//*Citologija.*–2005- vol. 47, No 10.- pp. 379-387.

Schlaf, G. Upregulation of fibronectin but not of entactin, collagen IV and smooth muscle actin by anaphylatoxin C5a in rat hepatic stellate cells [Text] / G. Schlaf, M. Schmitz, I. Heine, T. Demberg, H. L. Schieferdecker, O. Gotze // *Histol. Histopathol.* – 2004. – Vol. 19, № 4. – P. 1165-1174.

Schmidt, A. Changes in body temperatures in king penguins at sea: the result of fine adjustments in peripheral heat loss? [Text] / A. Schmidt, F. Alard, Y. Handrich // *Am. J. Physiol. Regul. Integr. Comp. Physiol*.- 2006.- Vol. 291.- P. 608-618.

Schnabel, C. Expression of cytosolic and membrane associated tissue transglutaminase in rat hepatic stellate cells and its upregulation during transdifferentiation to myofibroblasts in culture [Text] / C. Schnabel, I. Sawitza, C.G. Tag, B. Lahme, A.M. Gressner, K. Breitkopf // *Hepatol. Res.* – 2004. – Vol. 28, № 3. – P. 140-145.

Scholz, H. Adaptational responses to hypoxia [Text] / H. Scholz // *Am. J. Physiol. Regul. Integr. Comp. Physiol*.- 2002.- Vol. 282.- P. 1541-1543.

Senoo, H. Storage of lipid droplets in and production of extracellular matrix by hepatic stellate cells (vitamin A-storing cells) in Long-Evans cinnamon-like colored (LEC) rats [Text] / H. Senoo // *Anat. Rec.* -2000. – Vol.258, №4.- P.338-48.

Senoo, H. Structure and function of hepatic stellate cells [Text] / H. Senoo // *Med. Electron. Microsc.* - 2004.- Vol. 37.- P. 3-15.

Singh, H. A. Functional Module of Yeast Mediator That Governs the Dynamic Range of Heat-Shock [Text] / H. A. Singh, A. M. Erkine, S. B. Kremer, H. M. Duttweiler, D. A. Davis, J. Iqbal, R. R. Gross, D. S. Gross // *Gene Expression Genetics*. – 2006. - Vol. 172. – P. 2169-2184.

Škurupij, V.A.. Ul'trastruktura kletok pečeni pri stresse [Text]/V.A. Škurupij.- Novosibirsk: *Nauka*. Sibirskoje otdelenije, 1989.- 144 p.

Somero, G. N. Thermal Physiology and Vertical Zonation of Intertidal Animals: Optima, Limits, and Costs of Living Integrative and Comparative [Text] / G. N. Somero // *Biology*. – 2002. - Vol. 42, № 4. - P. 780-789.

St-Pierre, J. Metabolic depression and enhanced O2 affinity of mitochondria in hypoxic hypometabolism [Text] / J. St-Pierre, G. J. Tattershall, R. G. Boutilier // *Physiol. Regul. Integr. Comp. Physiol.* – 2000. - Vol. 279. - P. 1205-1214.

Talbot, H. Waterman Evolutionary challenges of extreme environments (part 2) [Text] / H. Talbot // *Experimental Zoology: Molecular and Developmental Evolution*. – 2001. – Vol. 291, №2. - P. 130–168.

Taylor, C.T. Mitochondria and cellular oxygen sensing in the HIF pathway [Text] / C.T. Taylor // *Biochem*.- 2008.- Vol. 409.- P.19-26.

Tozzi, M. G. Pentose phosphates in nucleoside interconversion and catabolism [Text] / M. G. Tozzi, M. Camici, L. Mascia, F. Sgarrella, P. L. Ipata // *FEBS*. – 2006. - Vol. 273. - P. 1089-1101.

Uryvajeva, I.V. Replikativnyj potencial gepatocitov I stvolovyje kletki pečeni [Текст]/I.V. Uryvajeva //Izvestija RAN. *Serija Biologija*.-2001.-No. 6.-pp.728-737.

Vardevanjan, P.O. Svjaz' elektrokinetičeskix svojstv izolirovannyx jader gepatocitov krys s ix funkcional'noj aktivnost'ju [Text]/P.O. Vardevanjan, S.A. Gonjan, S. G. Tiracujan //*Biofizika*.- 2001.- vol. 46, No. 2. - pp. 271-274.

Vinogradov, A. E. A spiral of DNA: importance of to be the Rich collector of dust [Text] / A. E. Vinogradov // *Nucleic. Acids. Res*. – 2003. – Vol. 31. – P. 1838-1844.

Vinogradov, A. E. Isochores and specificity of a fabric [Text] / A. E. Vinogradov // *Nucleic. Acids. Res*.- 2003.- Vol.31.- P.5212-5220.

Vinogradov, A. E. Relationship of hepatocyte ploidy levels with body size and growth rate in mammals [Text] / A. E. Vinogradov, O. V. Anatskaya, B. N. Kudryavtsev // *Genome*. – 2001. – Vol. 44, № 3. – P. 350-360.

Vinogradov, A.E. Genome size and extinction risk in vertebrates [Text] / A. E. Vinogradov // *Proc. R Soc. Lond B Biol. Sci*.- 2004.- Vol.271.- P.1701-1705.

Vinogradov, A.E. Test of complexity of a genome [Text] / A. E. Vinogradov // *A science*.- 2004.- Vol.304.- P.389-390.

Vinogradov, A.E. The dimension of a genome and inspissation of a chromatin in vertebrate animals [Text] / A. E. Vinogradov // *Chromosoma*.- 2005.- Vol. 113.-P. 362-369.

Volčegorskij, I.A. Eksperimental'noje modelirovanije i laboratornaja ocenka adaptivnyx reakcij organizma [Text]/I.A. Volčegorskij, I.I. Dolgušin, O.L. Kolesniokov, V.E. Cejlikman.– Chelyabinsk, 2000.- 167 p.

Warren, P. P. Calculating Climate Effects on Birds and Mammals: Impacts on Biodiversity, Conservation, Population Parameters, and Global Community Structure [Text] / P. P. Warren, S. Budaraju, W. E. Stewart, N. Ramankutty // *American Zoologist*. – 2000. – Vol. 40, № 4. – P. 597-630.

Wells, R. G. The role of matrix stiffness in hepatic stellate cell activation and liver fibrosis [Text] / R. G. Wells // *Clin. Gastroenterol*. – 2005. – Vol. 39, № 2. – P. 158-161.

Winchester, B. Lysosomal metabolism of glycoproteins [Text] / B. Winchester // *Glycobiology*. -2005.- Vol. 15, № 6. - P. 1–15.

Xolmuxamedov, E.L. Rol' mitoxondrij v obespečenii normal'noj žiznedejatel'nosti i vyživanija kletok mlekopitajuščix [Text]: *Avtoreferat dissertacii na soiskanije učënoj stepeni doktora biologičeskix nauk*: 03.00.02/E.L. Xolmuxamedov.- Pushchino, 2008.- 35p.

Yorimitsu, T. Autophagy: molecular machinery for self-eating [Text] / T. Yorimitsu, D.J. Klionsky // *Cell Death and Differentiation*.- 2005.- №12.- P.1542–1552.

Yorimitsu, T. Invokes {produces} Klionsky DJ autophagy [Text] / T. Yorimitsu, E. R. Strain // *Biol. Chem*. – 2006. –Vol. 281, №40. – P. 299-304.

Zhang, Y. The role of autophagy in mitochondria maintenance: characterization of mitochondrial functions in autophagy-deficient S. cerevisiae strains. [Text] / Y. Zhang, H. Qi, R. Taylor, W. Xu, L. F. Liu, S. Jin // *Autophagy*. – 2007. – Vol. 3. – P. 337-346.

In: Hyperthermia: Recognition, Prevention and Treatment ISBN 978-1-61942-275-9
Editor: Xing-Hua Gao and Hong-Duo Chen, 165-198

Chapter 5

Hyperthermia Induced by the Psychostimulants Methamphetamine, MDMA, and Cocaine

Katsutoshi Shioda, Makiko Kasai and Koichi Nisijima
Department of Psychiatry,
Jichi Medical University, Shimotsuke-shi,
Tochigi-Ken, Japan

Abstract

The abuse of psychostimulant drugs, including methamphetamine (METH), 3,4-methylenedioxymethamphetamine (MDMA), and cocaine, is popular in many parts of the world, especially among younger generations. These psychostimulant drugs cause various acute and chronic adverse effects, including hyperthermia, arrhythmia, tachycardia, and agitated delirium, of which hyperthermia is one of the most life threatening. Numerous studies have reported that hyperthermia induces multiorgan failure, rhabdomyolysis, disseminated intravascular coagulation, and even mortality. However, pharmacological treatment for psychostimulant-induced hyperthermia has not yet been established. We have reported MDMA- and METH-induced dopamine (DA) and serotonin (5-HT) intracellular concentration increases in the hypothalamus, the thermoregulation center, and that selective D1 and 5-HT2A receptor antagonists attenuated psychostimulant-induced hyperthermia. In this review we describe the pharmacological mechanisms of these three psychostimulants and present the basic research we have performed thus far. In addition, we address pharmacological treatment for hyperthermia induced by these psychostimulants.

Introduction

The abuse of psychostimulants as recreational drugs, including methamphetamine (METH), 3,4-methylenedioxymethamphetamine (MDMA), and cocaine, is widespread worldwide, especially among young people. As abuse of these psychostimulants has become popular, it has become a major global health issue.

Previous studies have shown that these psychostimulants induce acute and chronic adverse effects, including a wide range of physiological changes such as arrhythmias, seizures, liver dysfunction, rhabdomyolysis, hypertension, and hyperthermia. Of these, one of the most life-threatening acute effects of psychostimulants is hyperthermia. Numerous investigators have reported that psychostimulants, including METH, MDMA, and cocaine, induce hyperthermia, which can lead to rhabdomyolysis, disseminated intravascular coagulation, multiorgan failure, and in some cases, death. Therefore, clinicians must keep in mind that an overdose of psychostimulants can induce hyperthermia, which may have serious complications or be fatal. However, there is no established pharmacological treatment for psychostimulant-induced hyperthermia, and development of such treatment is urgently needed.

Body temperature is regulated by the balance between heat gain and heat loss. The central and peripheral thermoregulation systems rely on intricate mechanisms to maintain body temperature within a narrow range. Psychostimulants such as METH, MDMA, and cocaine impair these thermoregulatory mechanisms and induce hyperthermia. In this review, we describe the mechanisms of hyperthermia induced by each of these psychostimulants and discuss potential treatment approaches.

1. Methamphetamine

METH is used as an illegal recreational drug primarily in Asia, including Japan and China, and this abuse has spread to other parts of the world such as North America and Europe. Street names for METH include speed, chalk, ice, crystal, and glass; in the USA, approximately 1.2 million Americans aged 12 and older had abused METH at least once in the year prior to being surveyed in 2009 (National Survey on Drug Use and Health, Substance Abuse and Mental Health Administration web site; http://www.samhsa.gov/). METH is commonly administered by insufflation, ingestion, or smoking, or frequently by intravenous injection. METH has both acute and chronic effects. Chronic effects that have been reported include degeneration of the dopamine (DA) and serotonin (5-HT) neuronal systems in the brain and with long-term use, depression, anxiety, insomnia, aggressiveness, psychosis, and other mental disorders (Darke et al., 2008; Homer et al., 2008; Scott et al., 2007). Ingestion of large doses of METH induces many acute adverse effects such as hyperthermia, hypertension, arrhythmia, tachycardia, and agitation. (Darke et al., 2008; Schep et al., 2010). Hyperthermia is one of the most serious acute adverse effects and can lead to mortality (Chan et al., 1994; Davidson et al., 2001; Ishigami et al., 2003; Kojima et al., 1984).

Pharmacological Mechanism of Methamphetamine

The acute adverse effects of METH are induced by increased concentrations of various neurotransmitters such as monoamines [DA, 5-HT, and noradrenaline (NA)], glutamate (Glu), and nitric oxide (NO) in the brain. The intricate mechanism by which METH increases monoamine concentrations is described as follows (Figure 1) (Barr et al., 2006; Seger, 2010). 1) METH enters the nerve terminals via DA transporter (DAT); 2) once METH enters the neuroterminals, it causes the release of DA into the cytoplasma from the vesicles; 3) METH binds to the DAT and prevents reuptake from the synapses; 4) METH has a monoamine oxidase (MAO)-inhibiting effect; 5) METH blocks the vesicular monoamine transporter (VMAT), which uptakes cytoplasmic DA into the synaptic vesicles; and 6) as the cytoplasmatic concentration of DA increases, METH induces a rapid release of DA to the neuronal synapses via the DAT. The synaptic concentrations of other monoamines are also increased by a similar mechanism via other monoamine transporters. As METH has a high affinity to DAT, METH particularly induces an increase in the intercellular concentration of DA. Previous experimental studies have reported that DA nervous system hyperactivity induces Glu and NO concentration increase in the intercellular (Chen et al., 2003; Nicola et al., 1996). Therefore, Glu and NO increases associated with METH are assumed to result from METH-induced DA nervous system hyperactivity.

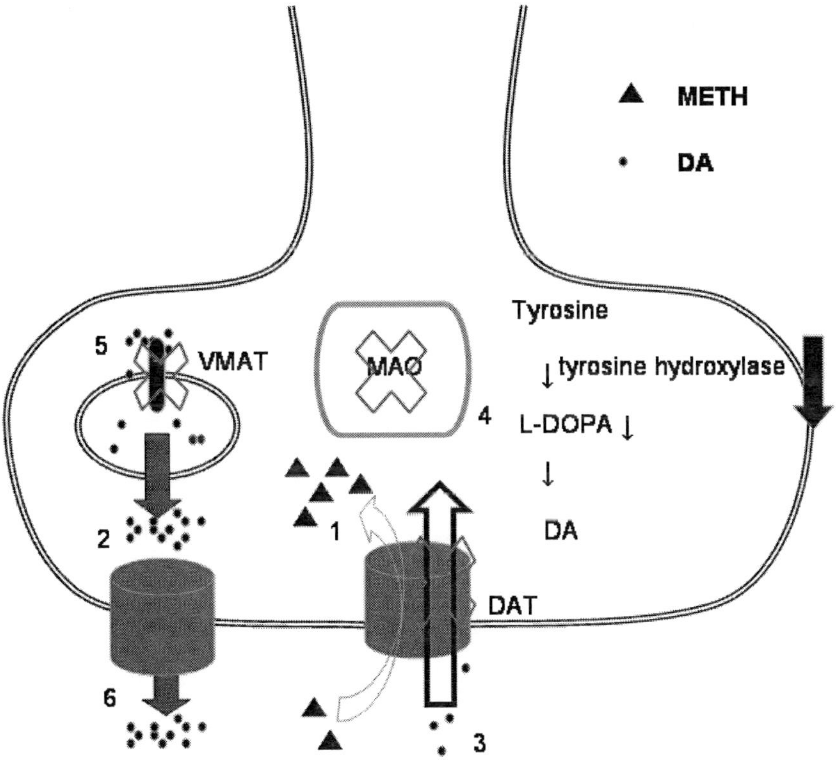

Figure 1. Pharmacological mechanism of the methamphetamine-induced increase in the DA concentration in the synapses. 1) METH enters the nerve terminals via DAT. 2) Once METH enters the neuroterminals, it causes the release of DA into the cytoplasma from the vesicles. 3) METH binds to the DAT and prevents reuptake from the synapses. 4) METH has a monoamine oxidase (MAO)-inhibiting effect. 5) METH blocks the vesicular monoamine transporter (VMAT), which uptakes cytoplasmic monoamine into the synaptic vesicles. 6) As the cytoplasmatic DA concentration increases, METH induces a rapid release of DA to the neuronal synapses via the DAT (Barr et al., 2006; Segar et al., 2010). DAT = dopamine transporter; VMAT = vesicular monoamine transporter.

Dopamine and Methamphetamine-Induced Hyperthermia

Numerous studies have reported METH-induced lethality associated with severe hyperthermia (Chan et al., 1994; Davidson et al., 2001; Ishigami et al., 2003; Kojima et al., 1984). METH-induced hyperthermia has been reported to be caused primarily by increased release of DA, which is well known to be centrally involved in control of thermoregulation (Yamada et al., 1998; Yamawaki et al., 1983). Recently, we demonstrated that subcutaneous injection (s.c.) of METH (10 mg/kg) induced hyperthermia in rats. When the rats were injected with METH, their rectal temperature rapidly increased to >41°C 60 min after the METH injection (Figure 2A).

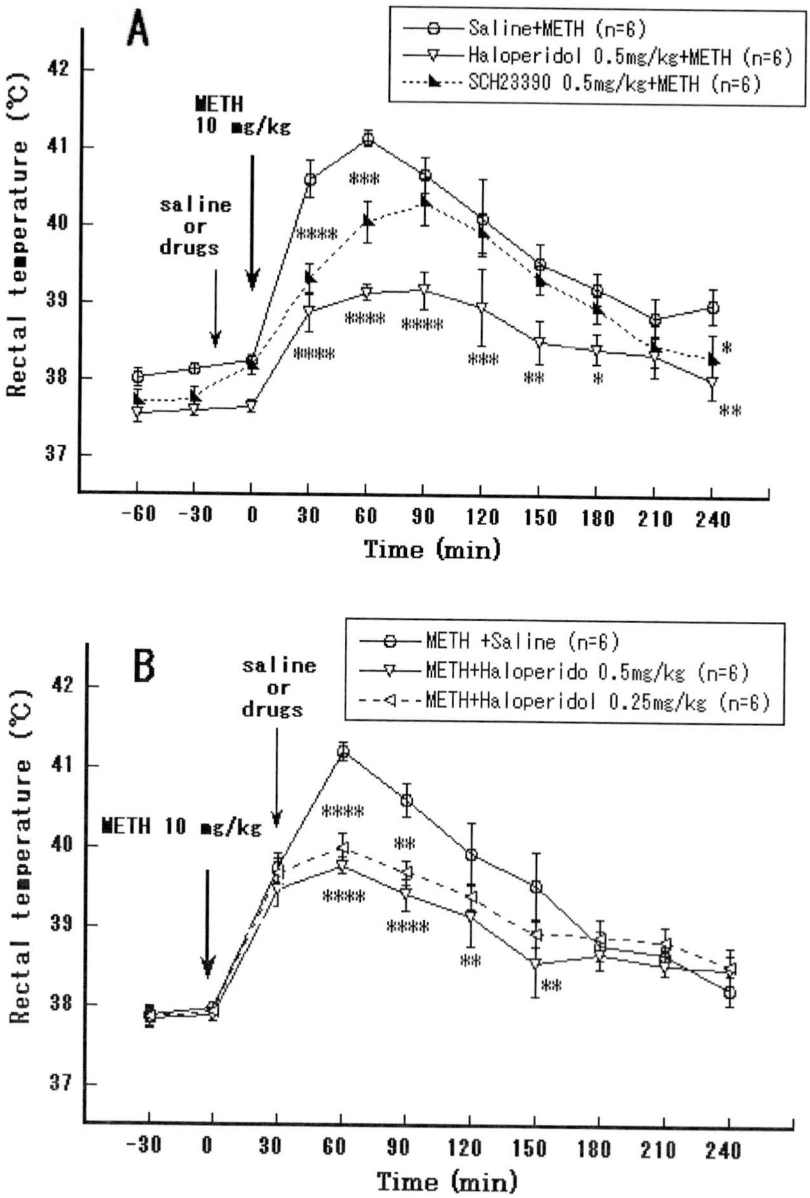

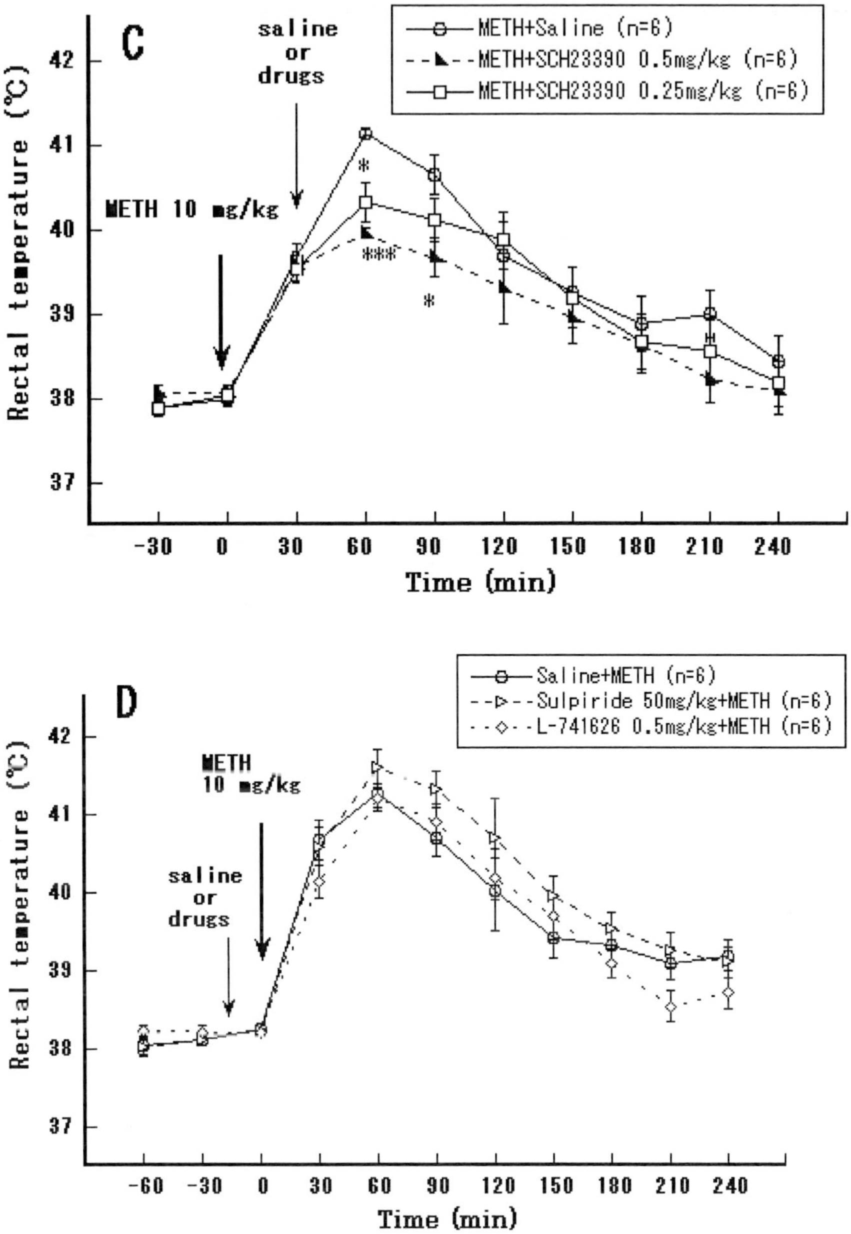

Figure 2. Effects of dopamine receptor antagonists on METH-induced hyperthermia. (A) Saline, haloperidol (0.5 mg/kg), or SCH 23390 (0.5 mg/kg i.p.) was injected into the rats, and 15 min later, METH (10 mg/kg s.c.) was injected. (B) METH (10 mg/kg s.c.) was injected into the rats, and 30 min later, haloperidol (0.25 mg/kg and 0.5 mg/kg i.p.) was injected. (C) METH (10 mg/kg s.c.) was injected into the rats, and 30 min later, SCH 23390 (0.25 or 0.5 mg/kg i.p.) was injected. (D) Saline, sulpiride (50 mg/kg), or L-741626 (0.5 mg/kg) was injected i.p. into the rats, and 15 min later, METH (10 mg/kg s.c.) was injected. Values are presented as means ± standard error of the mean (SEM). Statistical differences among the groups are indicated as follows: *, $p < 0.05$; **, $p < 0.01$; ***, $p < 0.001$; and ****, $p < 0.0001$ [vs. METH + saline; analysis of variance (ANOVA) followed by Fisher's protected least significant difference (PLSD)] (Shioda et al., 2010).

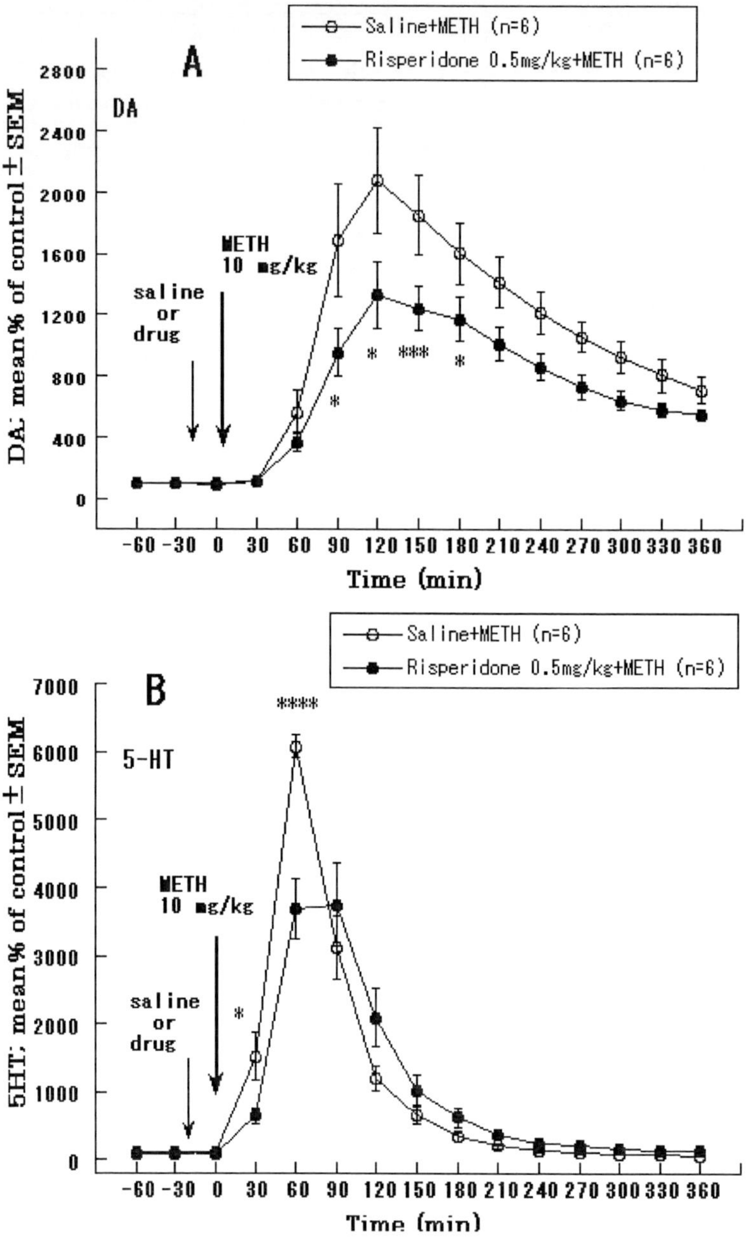

Figure 3. Extracellular levels of (A) DA and (B) 5-HT in the anterior hypothalamus after METH administration and the effect of risperidone pretreatment on these levels. Saline or risperidone (0.5 mg/kg i.p.) were injected into the rats, and 15 min later, METH (10 mg/kg s.c.) was injected. The average DA and 5-HT concentrations before the METH injection (2 measurements performed every 30 min) were used as baseline values (= 100%). The DA and 5-HT values are presented as mean % of control ± SEM. Statistical differences among the groups are indicated as follows: *, $p < 0.05$; **, $p < 0.01$; ***, $p < 0.001$; and ****, $p < 0.0001$ (vs. saline; ANOVA followed by Fisher's PLSD) (Shioda et al., 2010).

Microdialysis demonstrated that the METH injection induced an increase in the level of DA in the rat hypothalamus, which is the central control area for thermoregulation. Injection of METH (10 mg/kg) induced an approximately 20-fold increase in the DA concentration compared to baseline (Figure 3A). In addition, our study showed that haloperidol, a D1 and D2 receptor antagonist, and SCH 23390, a selective D1 receptor antagonist, significantly attenuated and reversed METH-induced hyperthermia (Figure 2A–C), but sulpiride and L-741626, D2 receptor antagonists, did not alter METH-induced hyperthermia (Figure 2D). These results are in agreement with those of Broening et al. (2005), who reported that the D1 receptor blocking effect is more important than D2 receptor blocking for attenuating METH-induced hyperthermia.

Serotonin and Methamphetamine-Induced Hyperthermia

5-HT is also widely known to be involved in the central control of thermoregulation. The 5-HT2A receptor agonist 1-(2,5-dimethoxy-4-iodophenyl)-2-aminopropane (DOI) is known to induce hyperthermia in animals (Mazzola-Pomietto et al., 1995). We found that 5-HT levels in the rat anterior hypothalamus increased 60-fold after injection of METH (10 mg s.c.) using microdialysis (Figure 3B). 5-HT concentrations increased more (60-fold) than did DA concentrations (20-fold) after METH (10 mg/kg) injection (Figure 3B). Previous investigators using microdialysis have also shown that METH increases the extracellular level of DA and 5-HT in several other regions of the rat brain; however, the effects of METH on DA concentrations were greater than those on 5-HT concentrations in the prefrontal cortex (Ago et al., 2007), the striatum, nucleus accumbens, and other areas (Kuczenski et al., 1995; Shimada et al. 1996). As the rat brain regions evaluated in our study are different from the previous studies, the results cannot be directly compared.

In our study (Shioda et al., 2010), the 5-HT2A antagonists ritanserin, ketanserin, and R-96544 significantly attenuated METH (10 mg/kg)-induced hyperthermia, and ritanserin significantly reversed METH-induced hyperthermia (Figure 4A, B), but the 5-HT2B/2C receptor antagonist SB 206553 and the 5-HT2C receptor antagonist SB 242084 did not prevent METH-induced hyperthermia (Figure 4C). Thus, an increase in the intercellular 5-HT concentration and activation of 5-HT2A receptors are assumed to be as important as the DA system in METH-induced hyperthermia.

Glutamate and Methamphetamine-Induced Hyperthermia

Recently, Glu has been assumed to play a role in hyperthermia and neurotoxicity induced by METH, because the Glu receptor antagonists MK801 and MPEP have been reported to attenuate these effects (Bowyer et al., 1994; Gołembiowska et al., 2003)

Using microdialysis, we found that METH induces an increase in the Glu level in the rat anterior hypothalamus (Figure 5). When rats were injected with METH (10 mg/kg s.c.), the Glu concentration immediately increased by about 40%, and it gradually increased by about 90% 210 min after METH administration.

Other investigators have reported that METH injection increases the Glu level in the rat striatum (Nash and Yamamoto, 1992; Ohmori et al., 1996). Although there have been no previous reports analyzing Glu in the anterior hypothalamus, these results of our reports were consistent with these previous results.

Pehek et al. (2006) showed using microdialysis that DOI, a selective 5-HT2A receptor agonist, increased the Glu concentration in the ventral tegmental area (VTA) and that the 5-HT2A receptor antagonist M100907 blocked this increase. Stephans and Yamamoto (1994) showed using microdialysis that haloperidol, a D1 and D2 receptor antagonist, attenuated the METH-induced Glu level increase, and Nicola et al. (1996) reported an electrophysiological study of the potential direct and indirect modulation of Glu release by D1 receptor stimulation. Thus, the METH-induced increase in Glu concentration levels is assumed to be due to hyperactivation of D1 and 5-HT2A receptors by METH.

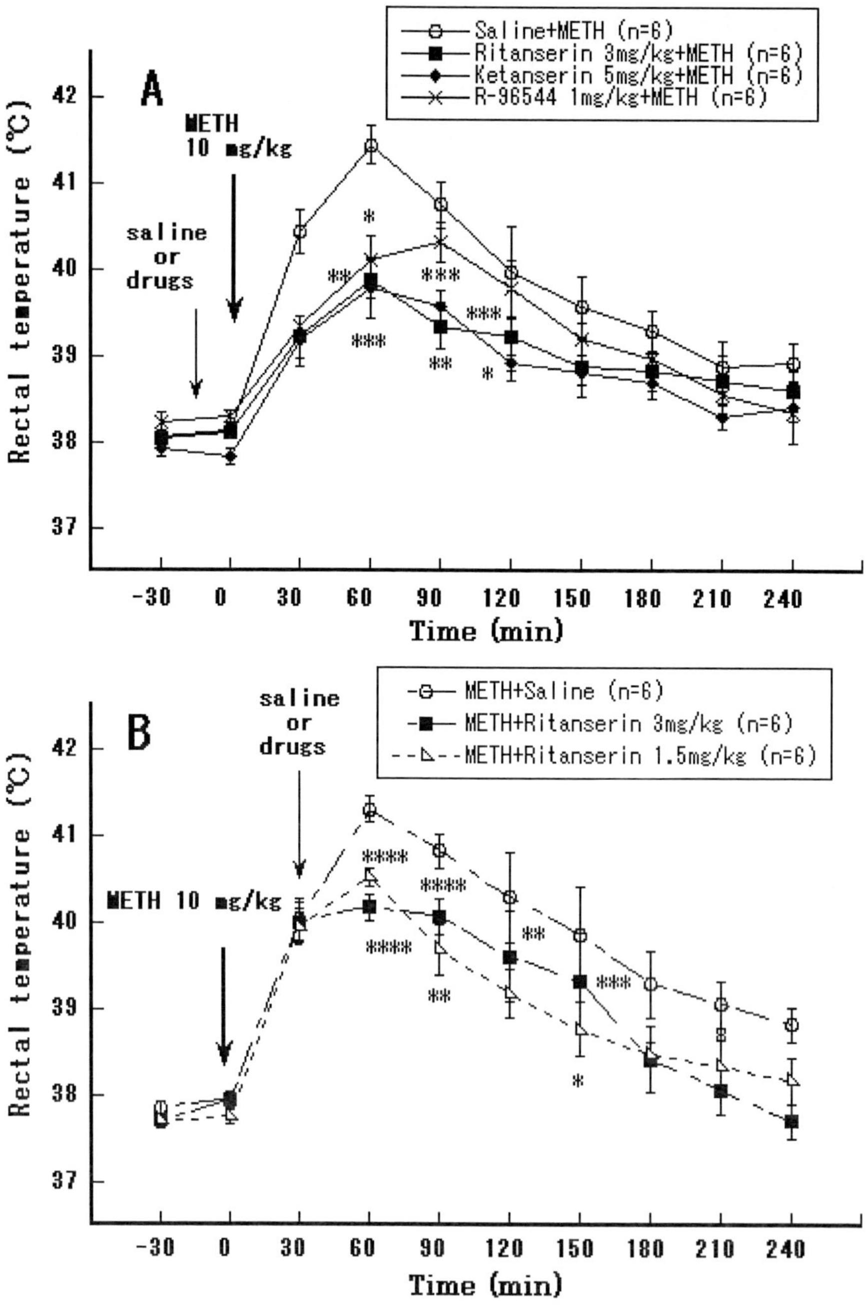

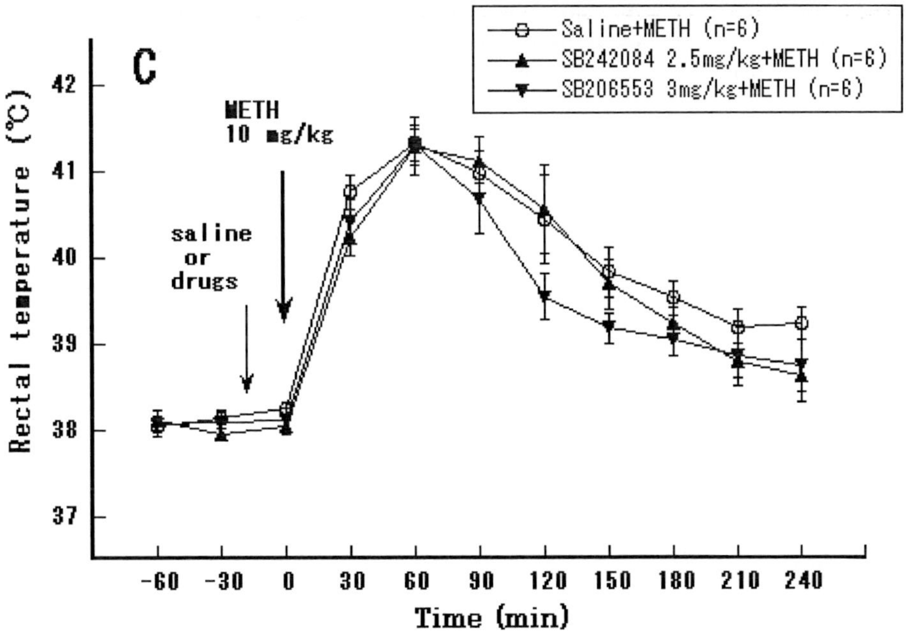

Figure 4. Effects of 5-HT antagonists on METH-induced hyperthermia. (A) Saline, ritanserin (3 mg/kg), ketanserin (5 mg/kg), or R-96544 (1 mg/kg) were injected i.p. into the rats, and 15 min later, METH (10 mg/kg s.c.) was injected. (B) METH (10 mg/kg s.c.) was injected, and 30 min later, ritanserin (1.5 mg/kg or 3 mg/kg i.p.) was injected. (C) Saline, SB 242084 (2.5 mg/kg), or SB206553 (3 mg/kg) were injected i.p. into the rats, and 15 min later, METH (10 mg/kg s.c.) was injected. Values are presented as means ± SEM. Statistical differences among the groups are indicated as follows: *, $p < 0.05$; **, $p < 0.01$; ***, $p < 0.001$; and ****, $p < 0.0001$ [vs. saline; ANOVA followed by Bonferroni/Dunn (A) or Fisher's PLSD (B, C)] (Shioda et al., 2010).

NO and Methamphetamine-Induced Hyperthermia

Recent research suggests that NO is also associated with METH-induced hyperthermia and neurotoxicity. Several studies have reported that NO synthase inhibitors such as 7-nitroindazole (7-NI) and L-NG-nitroarginine methyl ester (L-NAME) prevent hyperthermia and neurotoxicity induced by METH (Abekawa et al., 2001; Callahan and Ricaurte, 1998; Itzhak and Ali, 1996; Itzhak et al, 2000; Sanchez et al., 2003; Sheng et al., 1996). Because NO readily metabolizes into NO2- and NO3-, it is difficult to directly measure the NO concentration using microdialysis. Therefore, we measured the concentration of NO metabolites (NOx; NO2- + NO3-) in the anterior hypothalamus, and found that the level of NOx in the anterior hypothalamus increased after METH administration (Shioda et al., 2010) (Figure 6).

Previous studies have demonstrated that D1 receptor activation induces increased NO efflux and activates NO synthase (Chen et al., 2003; Sammut et al., 2006). Park and West (2009) also reported that NO efflux was elicited in the striatum by systematic administration of the D1 receptor agonist SKF-81297, and that this effect was blocked by the non-specific ionotropic glutamate receptor antagonist kynurenic acid as well as the selective NMDA receptor antagonist 3-phosphonopropyl-piperazine-2-carboxylic acid. These studies demonstrate that the D1 receptor and glutamate receptor activation may be involved in NO efflux and activation of the NO receptor. Hence, METH-induced DA and Glu concentration increases are assumed to promote NO efflux in the hypothalamus.

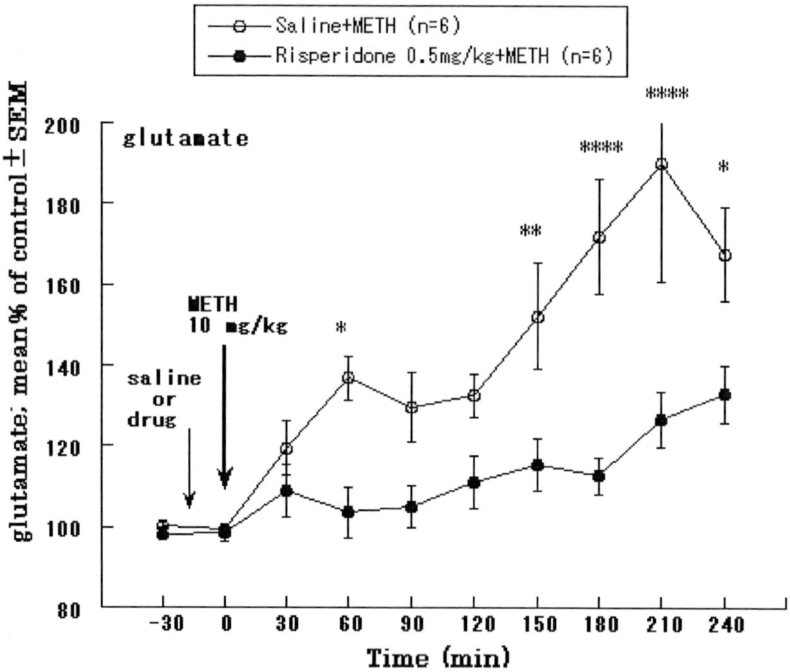

Figure 5. Extracellular levels of glutamate in the anterior hypothalamus after METH administration and the effect of risperidone pretreatment on these levels. Saline or risperidone (0.5 mg/kg i.p.) were injected into the rats, and 15 min later, METH (10 mg/kg s.c.) was injected. The average glutamate concentrations before the METH injection (2 measurements performed every 30 min) were used as the baseline values (= 100%). The glutamate values are presented as mean % of control ± SEM. Statistical differences among the groups are indicated as follows: *, $p < 0.05$; **, $p < 0.01$; ***, $p < 0.001$; and ****, $p < 0.0001$ (vs. saline; ANOVA followed by Fisher's PLSD).

Ambient Temperature and Methamphetamine-Induced Hyperthermia

Several studies have reported that the ambient temperature influences METH-induced hyperthermia (Bowyer et al., 1992; Myles et al., 2008; Xie et al., 2000); high ambient temperatures reinforce METH-induced hyperthermia and low ambient temperatures attenuate the hyperthermia. Myles et al. (2008) reported that when METH (5 mg/kg) was injected into rats at 20, 24, or 28 °C, an ambient temperature of 20 °C significantly decreased the body temperature compared with the body temperature after injection with METH (5 mg/kg) at 24 or 28 °C. However, the mechanism of this effect is not well understood. Recently, Zhang et al. (2011) demonstrated that warm ambient temperatures enhanced responsivity of 5-HT2A receptors in DOI-induced hyperthermia. Therefore, the reinforcement of METH-induced hyperthermia by warm ambient temperatures may be due to 5-HT2A receptor activation.

In addition, Xie et al. (2000) reported that the ambient temperature affects DAT function, and that a high ambient temperature (40 °C) significantly increased DAT-mediated cellular accumulation of METH compared with 34 or 37 °C ambient temperature. These results indicate that DAT activity may play a role in METH-induced hyperthermia at warm ambient temperatures.

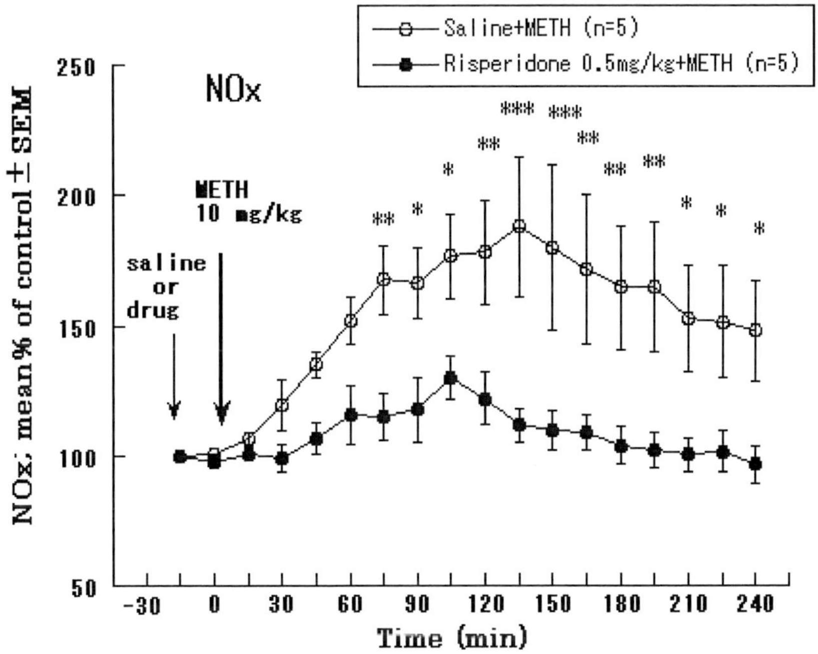

Figure 6. Extracellular levels of NOx in the anterior hypothalamus after METH administration and the effect of risperidone pretreatment on these levels. Saline or risperidone (0.5 mg/kg i.p.) were injected into the rats, and 15 min later, METH (10 mg/kg s.c.) was injected. The average NOx concentrations before the METH injection (2 measurements performed every 30 min) were used as the baseline values (= 100%). The NOx values are presented as mean % of control ± SEM. Statistical differences among the groups are indicated as follows: *, $p < 0.05$; **, $p < 0.01$; ***, $p < 0.001$; and ****, $p < 0.0001$ (vs. saline; ANOVA followed by Fisher's PLSD) (Shioda et al., 2010).

Pharmacological Treatment for Methamphetamine-Induced Hyperthermia

Management of METH-induced hyperthermia has thus far consisted solely of supportive treatment, such as body temperature reduction and fluid replacement. In an animal experiment, the muscle relaxant dantrolene, used as a treatment option for malignant hyperthermia and neuroleptic malignant syndrome (Nisijima et al., 2007), reportedly attenuated METH-induced hyperthermia (Makisumi et al., 1998). However, there are no case reports indicating that dantrolene is effective for METH-induced hyperthermia in humans. The acute adverse effects of METH are assumed to occur primarily due to dopamine receptor hyperactivity. Accordingly, several studies have reported that antipsychotics such as haloperidol (Bowyer et al., 1994; Funahashi et al., 1990), olanzapine (He et al., 2004), and quetiapine (He et al., 2005, 2006) attenuated METH-induced hyperthermia in animals. We have demonstrated the effectiveness of risperidone on hyperthermia and neurotransmitter increases induced by METH administration. Risperidone, an atypical antipsychotic drug, has potent 5-HT2A and D1 receptor antagonistic effects (5-HT2A, Ki = 0.6 nM; 5-HT2B, pKi = 7.67; 5-HT2C, pKi = 8.31; D1, Ki = 75 nM) (Bymaster et al., 1996; Wood et al., 2006), which suggests that risperidone may attenuate METH-induced hyperthermia. When rats were injected with 10 mg/kg METH s.c., the rectal temperature immediately increased to >41 °C. Risperidone (0.5 mg/kg i.p.) administration 15 min prior to METH injection significantly abolished METH-induced hyperthermia (Figure 7A).

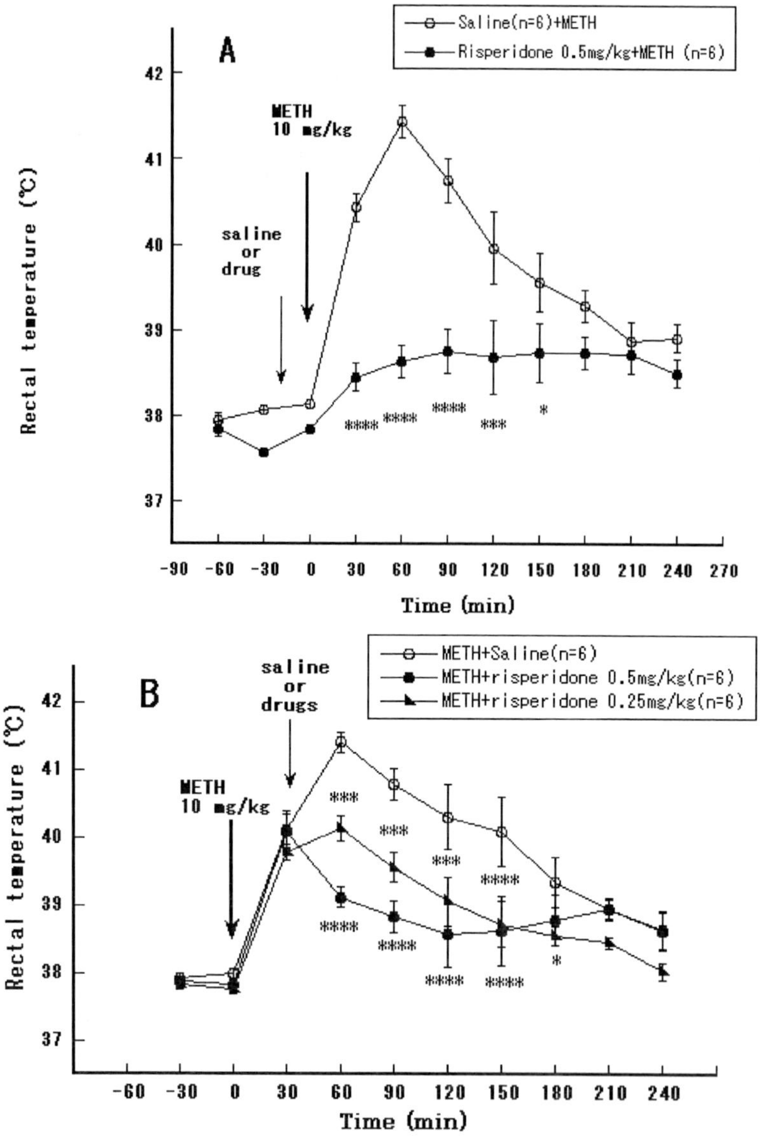

Figure 7. Effect of risperidone on METH-induced hyperthermia. (A) Saline or risperidone (0.5 mg/kg i.p.) were injected into the rats, and 15 min later, METH (10 mg/kg s.c.) was injected. (B) METH (10 mg/kg s.c.) was injected, and 30 min later, risperidone (0.25 and 0.5 mg/kg i.p.) was injected. Values are presented as means ± SEM. Statistical differences among the groups are indicated as follows: *, $p < 0.05$; **, $p < 0.01$; ***, $p < 0.001$; and ****, $p < 0.0001$ (vs. saline; ANOVA followed by Fisher's PLSD) (Shioda et al., 2010).

Furthermore, risperidone (0.25 and 0.5 mg/kg i.p.) administration 30 min after METH injection also significantly suppressed METH-induced hyperthermia (Figure 7B). METH causes hyperthermia mainly by increasing D1 and 5-HT2A receptor activity; accordingly, the potent 5-HT2A and D1 receptor antagonist risperidone appeared to abolish and reverse METH-induced hyperthermia. When METH (10 mg/kg s.c.) was injected, the DA and 5-HT levels in the anterior hypothalamus increased by approximately 20- and 60-fold (Figure 3A, B). Risperidone (0.5 mg/kg i.p.) administration 15 min prior to METH injection (0.5 mg/kg) significantly attenuated the increase in DA and 5-HT levels (Figure 3A, B). Similarly, Porras et al. (2002) reported that the 5-HT2A receptor antagonist SR46349B reduces amphetamine-induced DA release in the striatum and nucleus accumbens. 5-HT2A receptor activation has been reported to increase DA synthesis and release in the rat brain (Huang and Nichols, 1993; Pehek et al., 2006). In addition, O'Dell et al. (1993) observed that SCH23390, a selective D1 receptor antagonist, and eticlopride, a selective D2 receptor antagonist, attenuates METH-induced DA release in the striatum. The findings of these studies are consistent with our results. Attenuation of METH-induced DA release by risperidone pretreatment may be due to the antagonistic effects of risperidone on the 5-HT2A, D1, and D2 receptors. Ritanserin, a selective 5-HT2A antagonist, has been reported to suppress METH-induced 5-HT release in the pre-frontal cortex (Ago et al., 2007). In that study, it was assumed that 5-HT2A receptor activation might lead to an enhancement in METH-induced 5-HT release. The attenuation of the METH-induced increase in 5-HT concentrations by pretreatment with risperidone may be due to the potent 5-HT2A receptor antagonistic effect of risperidone.

Several studies have reported that Glu is involved in METH-induced hyperthermia (Bowyer et al., 1994; Gołembiowska et al., 2003). An increase in the Glu level in the hypothalamus after METH injection was demonstrated using microdialysis (Figure 5). Previous reports have suggested that D1 and 5-HT2A receptors are involved in the increase in Glu concentration (Nicola et al., 1996; Peheek et al., 2006; Stephans and Yamamoto, 1994). The suppression effect of risperidone on the METH-induced increase in the Glu level is likely due to the antagonistic effect of risperidone on 5-HT2A and D1 receptors, and may be one of the pharmacological mechanisms by which risperidone attenuates METH-induced hyperthermia.

Previous studies have also reported that NO is associated with METH-induced hyperthermia and neurotoxicity. Callanhan et al. (1998) and Itzhak et al. (2000) reported that an NO synthesis inhibitor attenuated METH-induced hyperthermia in an animal model. Risperidone attenuates the METH-induced increase in NO efflux in the hypothalamus (Figure 6). Activation of the D1 receptor induces NO efflux and an increase in NO synthesis (Chen et al., 2003; Sammut et al., 2006), and a glutamate receptor antagonist attenuated D1 agonist-induced NO efflux (Park and West, 2009). Attenuation of the METH-induced increase in NOx levels by risperidone treatment is likely due to exertion of an antagonistic effect on the D1 receptor, possibly due to attenuation by risperidone of the METH-induced increase in Glu levels. Attenuation by risperidone of the METH-induced increase in NO levels may be an additional mechanism by which risperidone protects against METH-induced hyperthermia.

In summary, hyperthermia induced by METH appears to be mainly due to DA and 5HT receptor hyperactivity; in particular, D1 and 5-HT2A receptor activation play important roles in METH-induced hyperthermia. The atypical antipsychotic drug risperidone is a potent 5-HT2A and D1 receptor antagonist that attenuates and reverses METH-induced hyperthermia and attenuates METH-induced increases in DA, 5-HT, Glu, and NO in the anterior hypothalamus in rats. Risperidone is used as an atypical antipsychotic drug worldwide and its safety has been established. Therefore, risperidone is a potentially useful drug for treatment of METH-induced hyperthermia. Because warm ambient temperatures facilitate METH-induced hyperthermia, body-cooling therapy is assumed to be effective and may have a synergistic effect with risperidone administration. In this chapter, we focused on the efficacy of risperidone on METH-induced hyperthermia. However, because it has been reported that atypical antipsychotics other than risperidone (olanzapine and quetiapine) also reduce hyperthermia induced by METH (He et al., 2004, 2006), these antipsychotics may also be effective against METH-induced hyperthermia.

2. 3,4-METHYLENEDIOXYMETHAMPHETAMINE (MDMA)

MDMA is a synthetic drug that is related to METH and is usually taken orally as a tablet or capsule. MDMA street names include ecstasy, X, XTC, love drug, etc. MDMA was first synthesized by Merck and Co. in 1914 as an appetite suppressant, and is now a popular illegal recreational drug among youth in North America and Europe. Recently, its abuse has spread to Asia, including Japan and China. In 2009, approximately 2.8 million Americans aged 12 and older had abused MDMA at least once in the year prior to being surveyed (National Survey on Drug Use and Health, Substance Abuse and Mental Health Administration web site : http://www.samhsa.gov/).

An overdose of MDMA induces tachycardia, arrhythmia, hypertension, and rhabdomyolysis (Hall and Henry, 2006). One of the life-threatening adverse effects of MDMA is hyperthermia and the associated 5-HT syndrome (Mueller and Krey, 1998; Parrott, 2002). Although it is estimated that the fatality rate of MDMA is only 0.00167% among users aged 16–24 years (Schifano et al., 2010), MDMA-induced hyperthermia and the related 5-HT syndrome can cause mortality if the peak of the patient's core temperature exceeds 42 °C (Dar and McBrien, 1996; Dowsett, 1996; Mueller and Korey, 1998; O'Connor, 1994).

The Pharmacological Mechanism of MDMA

The acute adverse effects of MDMA are reportedly induced by increases in the intercellular concentrations of various neurotransmitters in the brain. The pharmacological mechanisms of MDMA are similar to those of METH. Although MDMA and METH both induce the release of neurotransmitters, including 5-HT, DA, and NA, and bind to their respective reuptake transporters, MDMA has a stronger affinity for SERT (Crespi et al., 1997; Iravani et al., 2000; Han and Gu, 2006). The mechanisms by which MDMA induces 5-HT concentration increases can be described as follows (Figure 8) (Capela et al., 2009). 1) MDMA, like serotonin (5-HT), is a substrate of SERT and uses the transporter to enter the neuronal terminal, although at high concentrations, it may also enter by diffusion. 2) Once inside, MDMA produces an acute and rapid enhancement in the release of 5-HT from the synaptic vesicles, possibly by entering the vesicles via VMAT, and depletes vesicular neurotransmitter stores via a carrier-mediated exchange mechanism. 3) MDMA binds to SERT and prevents reuptake of 5-HT from the synapses. 4) Monoamine oxidase B (MAO-B), located in the outer membrane of the mitochondria of serotonergic neurons, is the enzyme responsible for 5-HT degradation and its activity is partially inhibited by MDMA. 5) MDMA blocks VMAT, which uptake cytoplasmic monoamine into the synaptic vesicle. 6) Due to the increase in the free cytoplasmatic pool of 5-HT, MDMA promotes rapid release of intracellular 5-HT to the neuronal synapses via SERT. Other monoamine synaptic concentrations are also increased by mechanisms similar to those of METH.

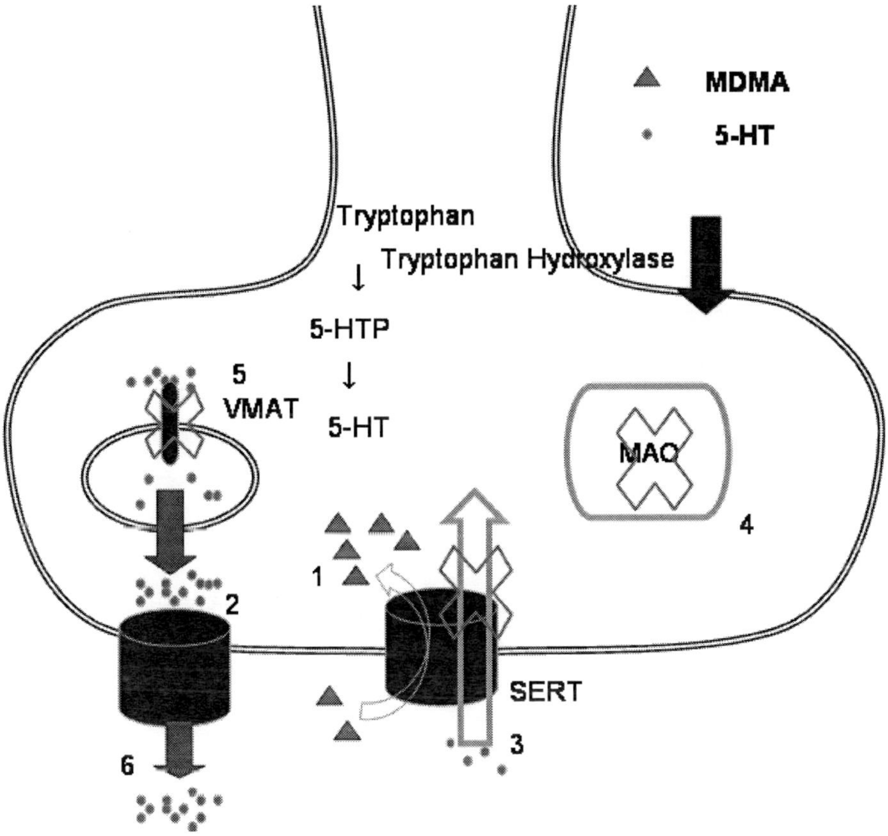

Figure 8. The pharmacological mechanism of action of MDMA at the neuronal serotonergic terminals and synapses. 1) Like serotonin (5-HT), MDMA enters the nerve terminals via SERT. 2) Once inside, MDMA produces an acute and rapid enhancement in the release of 5-HT from the synaptic vesicles into the cytoplasm. 3) MDMA binds to the SERT and prevent reuptake from the synapse. 4) Monoamine oxidase is partially inhibited by MDMA. 5) MDMA blocks VMAT, which uptakes cytoplasmic monoamine into the synaptic vesicles. 6) Due to the increase in the free cytoplasmatic pool of 5-HT, MDMA promotes a rapid release of intracellular 5-HT to the neuronal synapses via reversal of serotonin SERT activity (Capela et al., 2009). SERT = serotonin transporter; VMAT = vesicular monoamine transporter.

Serotonin and MDMA-Induced Hyperthermia

Body temperature is regulated by 5-HT and DA in the anterior hypothalamus (Yamawaki et al., 1983). In previous studies, co-administration of tryptophan, 5-HT, and a monoamine oxidase inhibitor induced hyperthermia (Grahame-Smith, 1971), and the 5-HT2A receptor agonist DOI was found to increase body temperature in animals (Mazzola-Pomietto et al., 1995). In our experiment (Shioda et al., 2008), when rats were injected with MDMA (10 mg/kg s.c.) at 24 °C ambient temperature, their rectal temperature rapidly increased to >41 °C (Figure 9A). We found that pretreatment with the 5-HT2A antagonists ritanserin, ketanserin, and R-96544 significantly attenuated and post-treatment with ritanserin reversed MDMA-induced hyperthermia (Figure 9A, B). These results are consistent with those of Vanattou-Saifoudine et al. (2010). However, the 5-HT2B/2C receptor antagonist SB 206553 and the 5-HT2C receptor antagonist SB 242084 did not prevent MDMA-induced hyperthermia (Figure 9C).

Several studies using microdialysis have demonstrated that MDMA induces 5-HT concentration increases in several brain regions in animals (Baumann et al., 2008; Gough et al., 2002; Mechan et al., 2002), although no previous studies have been done of the hypothalamus. We evaluated MDMA-induced 5-HT and DA concentration increases in the hypothalamus, and 5-HT levels increased approximately 50-fold compared to baseline after injection of MDMA (10 mg/kg) (Figure 10B).

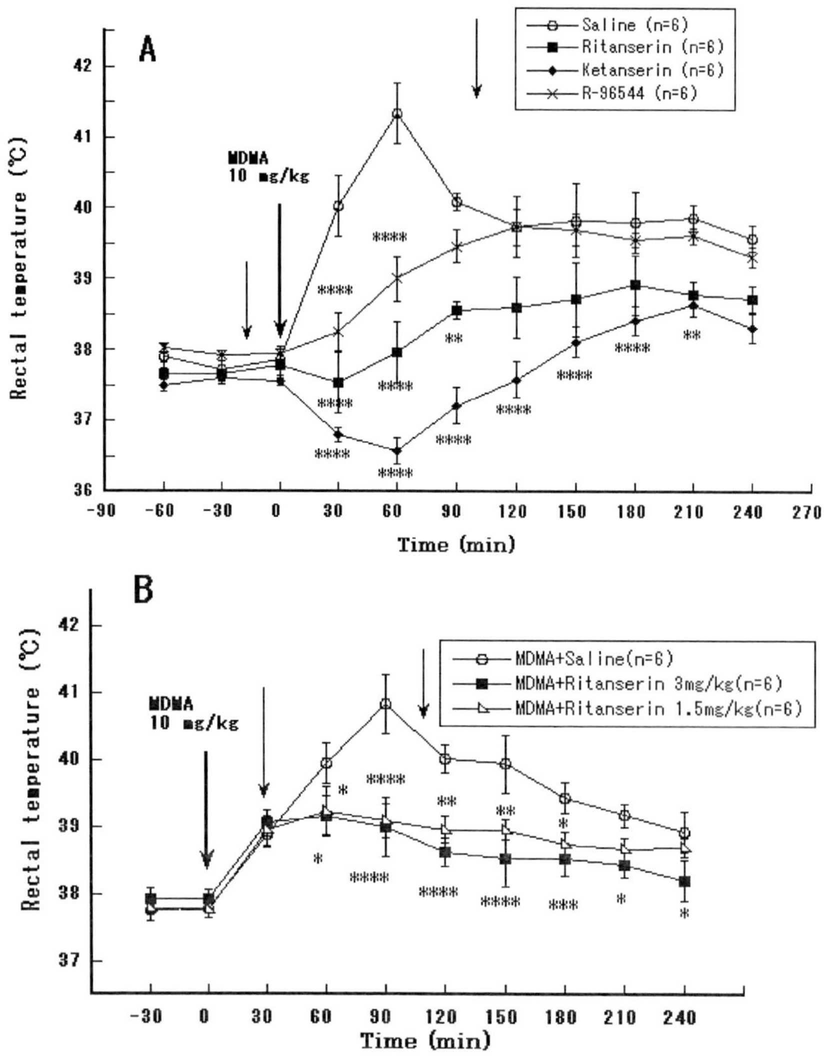

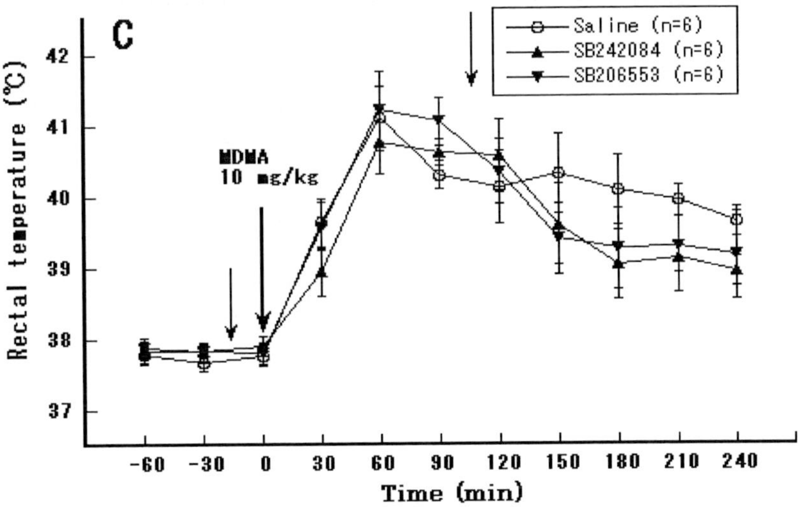

Figure 9. Effects of 5-HT2 antagonists on MDMA-induced hyperthermia. (A) Saline, ritanserin (3 mg/kg), ketanserin (5 mg/kg), and R-96544 (1 mg/kg) were injected i.p. into the rats, and 15 min later, MDMA (10 mg/kg s.c.) was injected. (B) MDMA (10 mg/kg s.c.) was injected, and 30 min later, ritanserin (1.5 mg/kg and 3 mg/kg i.p.) was injected. (C) Saline, SB 242084 (2.5 mg/kg), and SB206553 (3 mg/kg) were injected i.p. into the rats, and 15 min later, MDMA (10 mg/kg s.c.) was injected. Values are presented as means ± SEM. Statistical differences among the groups are indicated as follows: *, $p < 0.05$; **, $p < 0.01$; ***, $p < 0.001$; and ****, $p < 0.0001$ (vs. saline) (Shioda et al., 2008).

Based on the results of our body temperature and microdialysis studies, increases in the 5-HT level and consequent 5-HT2A receptor activation appear to play important roles in MDMA-induced hyperthermia.

Dopamine and MDMA-Induced Hyperthermia

MDMA not only induces acute release of 5-HT in the hypothalamus, it also produces release of DA. Injection of MDMA (10 mg/kg s.c.) induced an increase in the DA level of approximately 10-fold in the hypothalamus (Figure 10A). Our previous study (Shioda et al., 2008) investigated the effects of several DA antagonists on MDMA-induced hyperthermia.

Haloperidol, a D1 and D2 receptor antagonist, and SCH 23390, a selective D1 receptor antagonist, significantly attenuated and reversed MDMA-induced hyperthermia (Figure 11A, B), but the D2 receptor antagonists sulpiride and L-741626 did not influence MDMA-induced hyperthermia (Figure 11C). These results are consistent with those of Mechan et al. (2002), demonstrating that D1 receptor activity is closely related to MDMA-induced hyperthermia.

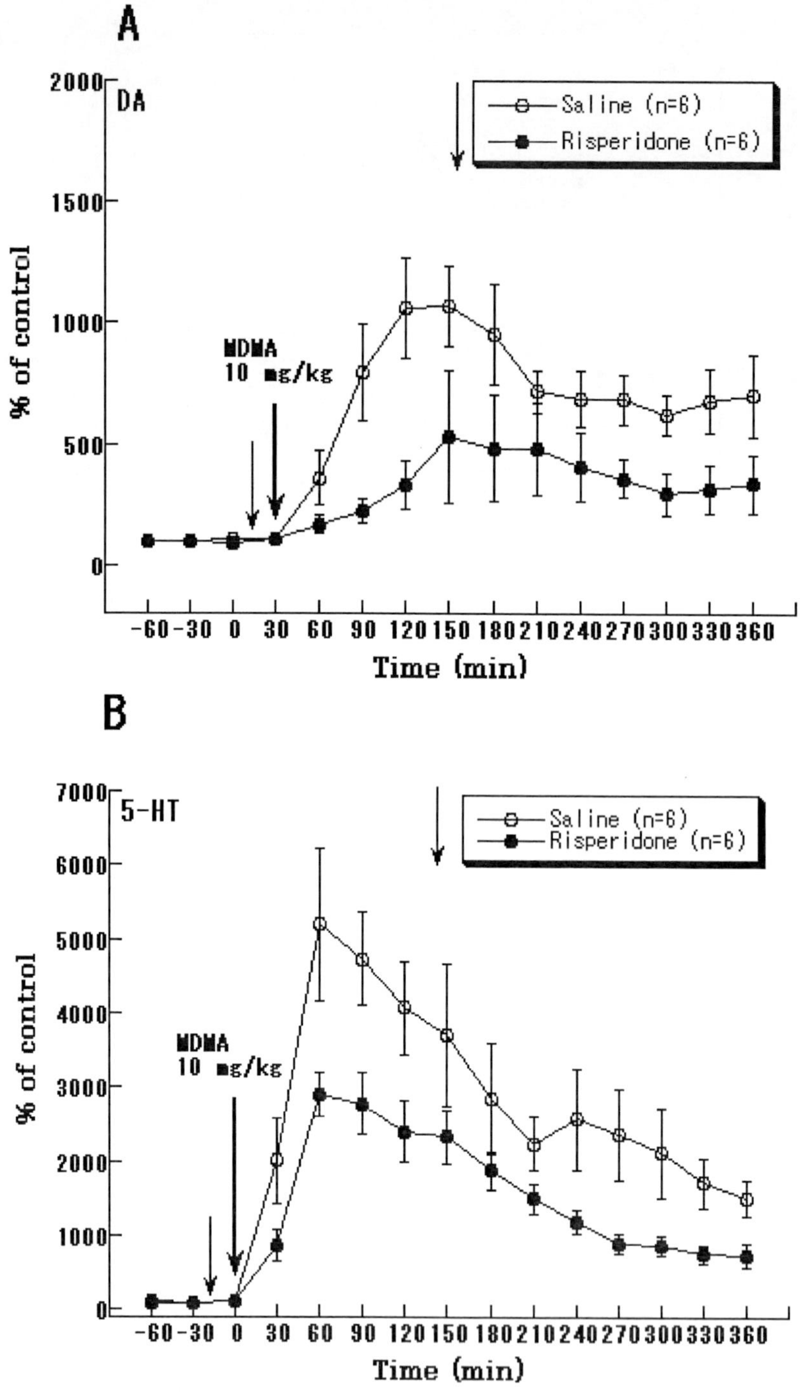

Figure 10. Extracellular levels of DA and 5-HT in the hypothalamus after MDMA administration and the effect of risperidone pretreatment on these levels. Saline and risperidone (0.5 mg/kg i.p.) were injected into the rats, and 15 min later, MDMA (10 mg/kg s.c.) was injected. The average DA and 5-HT concentrations before the MDMA injection (2 measurements performed every 30 min) were used as the baseline values (= 100%). The DA and 5-HT values are presented as mean % of control ± SEM. Statistical differences among the groups are indicated as follows: *, $p < 0.05$; **, $p < 0.01$; ***, $p < 0.001$; and ****, $p < 0.0001$ (vs. saline) (Shioda et al., 2008).

Glutamate, NO, and MDMA-Induced Hyperthermia

Glutamate has been reported to be involved in MDMA-induced hyperthermia, because the glutamate receptor antagonist MK-801 attenuated MDMA-induced hyperthermia in rats (Hewitt et al., 1994). NO may also play a role in MDMA-induced hyperthermia. MDMA results in NO concentration increases (Darvesh et al., 2005) and the NO synthesis inhibitor 7-NI attenuates MDMA-induced hyperthermia (Colado et al., 2001).

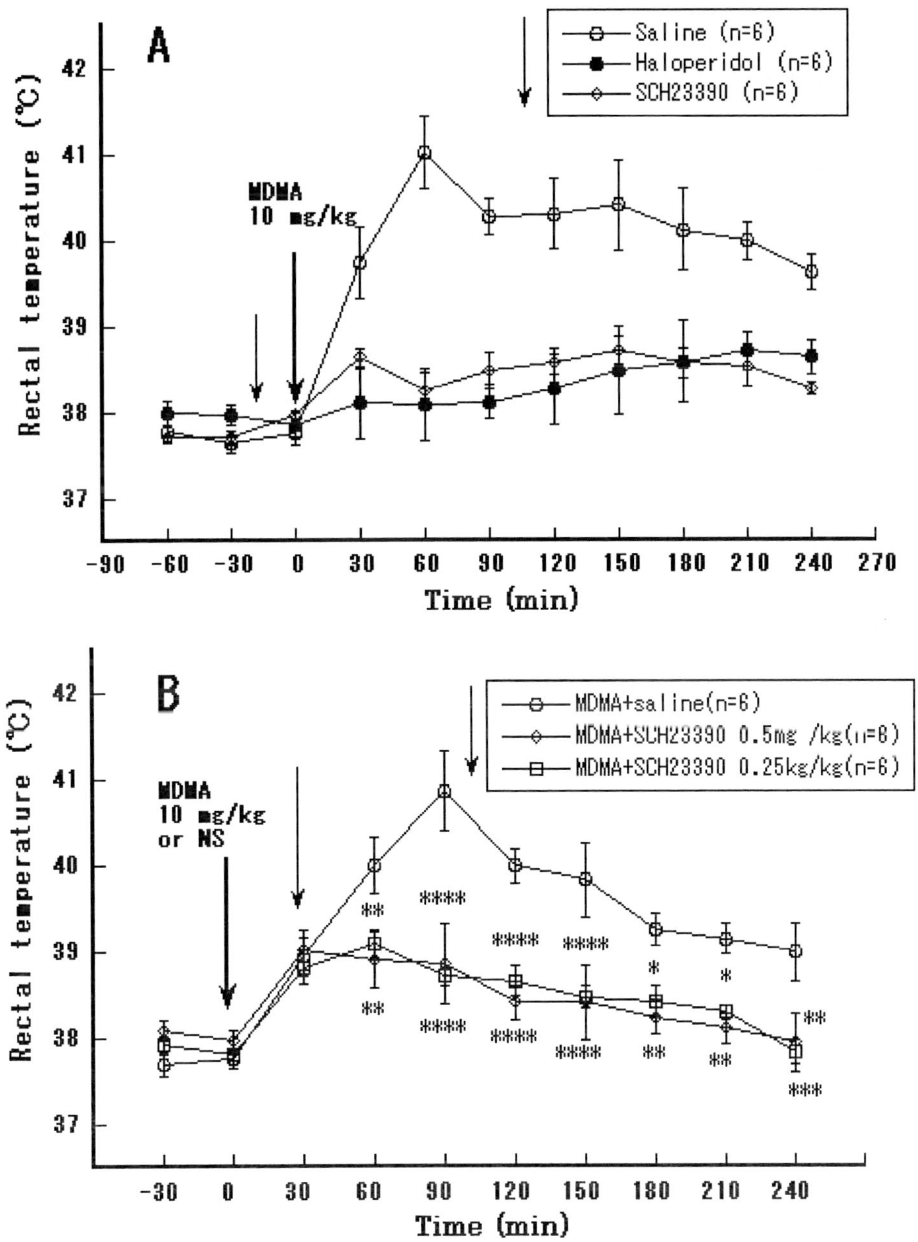

Figure 11. (Continued)

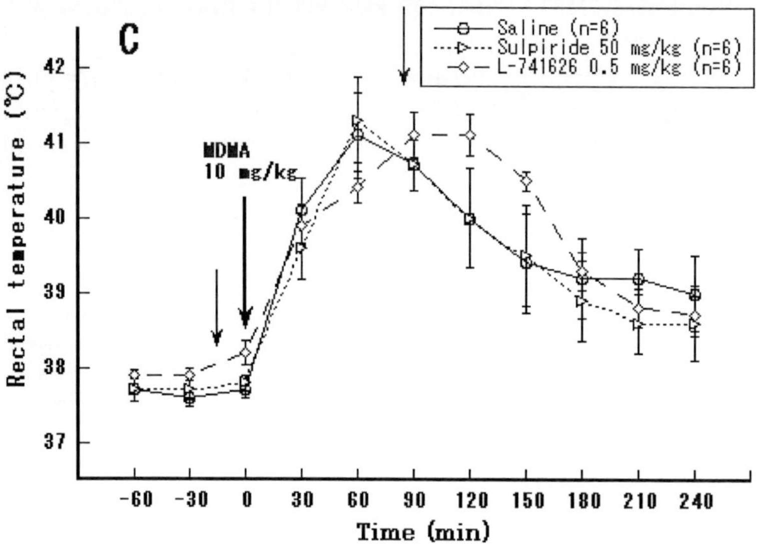

Figure 11. Effects of DA receptor antagonists on MDMA-induced hyperthermia. (A) Saline, haloperidol (0.5 mg/kg), and SCH 23390 (0.5 mg/kg) were injected i.p. into the rats, and 15 min later, MDMA (10 mg/kg s.c.) was injected. (B) MDMA (10 mg/kg s.c.) was injected into the rats, and 30 min later, SCH 23390 (0.25 and 0.5 mg/kg i.p.) was injected. (C) Saline, sulpiride (50 mg/kg), and L-741626 (0.5 mg/kg) were injected i.p. into the rats, and 15 min later, MDMA (10 mg/kg s.c.) was injected. Values are presented as means ± SEM. Statistical differences among the groups are indicated as follows: *, $p < 0.05$; **, $p < 0.01$; ***, $p < 0.001$; and ****, $p < 0.0001$ (vs. saline) (Shioda et al., 2008).

While these studies suggest that glutamate and NO function may be associated with MDMA-induced hyperthermia, it is unclear that the glutamate and NO systems play an important role in MDMA-induced hyperthermia compared with those of the 5-HT and DA systems.

Ambient Temperature and MDMA-Induced Hyperthermia

The effects of ambient temperature on MDMA-induced hyperthermia are well known (Gordon et al., 1991; Green et al., 2005; Shioda et al., 2008); higher ambient temperatures reinforce and lower ambient temperatures attenuate MDMA-induced hyperthermia. When rats were injected with MDMA (10 mg/kg s.c.) at a low ambient temperature (20 °C), their rectal temperatures initially decreased 30 min after MDMA administration and then gradually increased to ~38.5 °C. A low ambient temperature (20 °C) significantly attenuated MDMA-induced hyperthermia compared with a normal ambient temperature (24 °C) (Figure 12).

The detailed mechanism by which ambient temperature influences MDMA-induced hyperthermia is unclear. However, a high ambient temperature (30°C) significantly reinforced MDMA-induced DA and 5-HT releases in the brain compared with a low ambient temperature (20 °C) (O'Shea et al., 2005). Higher ambient temperature is assumed to enhance 5-HT2A receptor responsivity and DAT function (Xie et al., 2000; Zang et al., 2011). Thus, it is suggested that hyperactivity of 5-HT2A receptor and DAT function, and increased release of 5-HT and DA under worm ambient temperature may reinforce MDMA-induced hyperthermia.

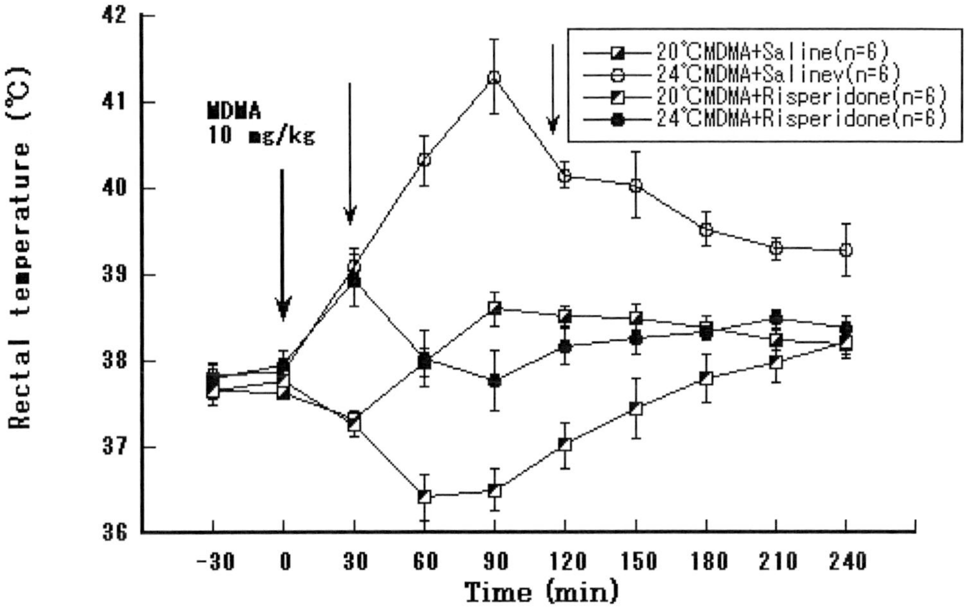

Figure 12. Effects of risperidone on MDMA-induced hyperthermia under low ambient temperature conditions (20°C). MDMA (10 mg/kg s.c.) was injected into the rats, and 30 min later, saline and risperidone (0.5 mg/kg i.p.) were injected at normal and low ambient temperatures (24°C and 20°C, respectively). Statistical differences among the groups were as follows: 60 min, 90 min ($p < 0.0001$), and 120 min ($p < 0.01$) (saline + 10 mg/kg MDMA group, 20°C vs. 0.5 mg/kg risperidone + 10 mg/kg MDMA group, 20°C); 30 min ($p < 0.0001$), 60 min ($p < 0.001$), and 90 min ($p < 0.05$) (0.5 mg/kg risperidone + 10 mg/kg MDMA group, 20°C vs. 0.5 mg/kg risperidone + 10 mg/kg MDMA group, 20°C).

Pharmacological Treatment of MDMA-Induced Hyperthermia

Standard treatment for MDMA-induced hyperthermia is limited to measures such as body cooling and fluid replacement, and there is no established pharmacological therapy. Dantrolene, a specific drug for treatment of malignant hyperthermia and neuroleptic malignant syndrome (Nisijima et al., 2007), has been recommended for treatment of MDMA-induced hyperthermia (Kunitz et al., 2003; Mallick and Bodenham, 1997). Recently, a systematic review of the efficacy of dantrolene for MDMA-induced hyperthermia has been published (Grunau et al., 2010); however, its effectiveness remains controversial (Hall and Henry, 2006; Rusyniak and Sprague; 2005). Therefore, a well-established pharmacotherapy for MDMA-induced hyperthermia is needed. The mechanism of MDMA-induced hyperthermia is assumed to be primarily D1 and 5-HT2A receptor hyperactivity, similar to that of METH-induced hyperthermia. Thus, risperidone may be useful for both METH-induced and MDMA-induced hyperthermia. In our study (Shioda et al., 2008), when rats were injected with MDMA (10 mg/kg s.c.), rectal temperature increased immediately to >41°C.

Risperidone (0.5 mg/kg i.p.) administration 15 min prior to MDMA injection significantly abolished MDMA-induced hyperthermia (Figure 13A).

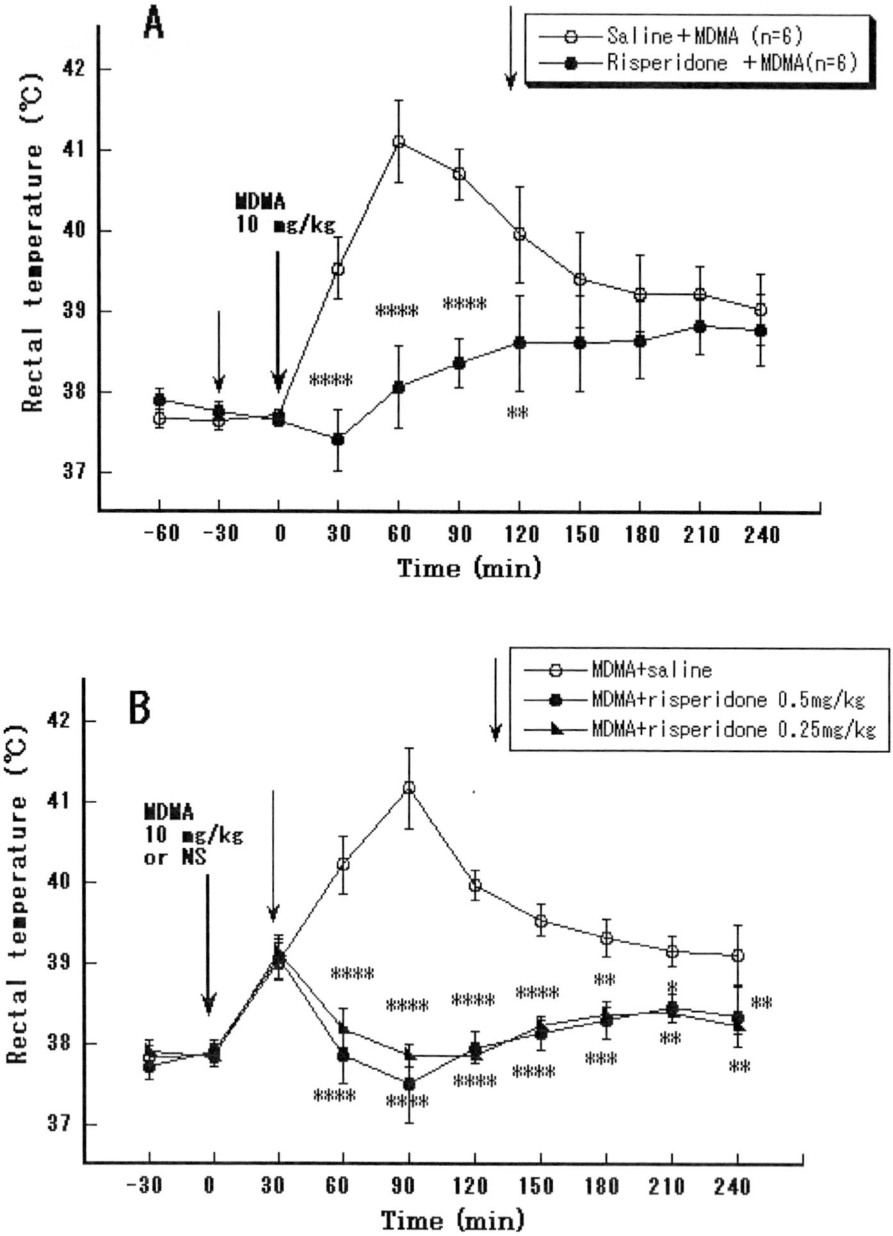

Figure 13. Effects of risperidone on MDMA-induced hyperthermia. (A) Saline and risperidone (0.5 mg/kg i.p.) were injected into the rats, and 15 min later, MDMA (10 mg/kg s.c.) was injected. (B) MDMA (10 mg/kg s.c.) was injected, and 30 min later, risperidone (0.25 and 0.5 mg/kg i.p.) was injected. Values are presented as means ± SEM. Statistical differences among the groups are indicated as follows: *, $p < 0.05$; **, $p < 0.01$; ***, $p < 0.001$; and ****, $p < 0.0001$ (vs. saline).

Furthermore, risperidone (0.25 and 0.5 mg/kg i.p.) administration 30 min after MDMA injection significantly suppressed MDMA-induced hyperthermia (Figure 13B). Meanwhile, when MDMA (10 mg/kg s.c.) was injected, the DA and 5-HT levels in the anterior hypothalamus increased by approximately 10- and 50-fold (Figure 10A,B). Risperidone (0.5 mg/kg i.p.) administration 15 min prior to MDMA injection significantly attenuated DA and 5-HT concentration increases (Figure 10A, B).

Previous studies have reported that 5-HT2A receptor antagonists, including ketanserin and MDL 100907, suppressed MDMA-induced DA release in the striatum (Nash, 1990; Schmidt et al., 1992) and that 5-HT2A receptor activation increased DA synthesis and release in the rat brain (Huang and Nichols, 1993; Pehek et al., 2006).

Benamar et al. (2008) reported that SCH 23390, a selective D1 receptor antagonist, suppressed MDMA-induced DA release in the anterior hypothalamus. Attenuation by risperidone of MDMA-induced DA release is consistent with these previous findings; therefore, the 5-HT2A and D1 receptor antagonistic effect of risperidone is assumed to suppress MDMA-induced DA release.

As noted above, ritanserin, a selective 5-HT2A antagonist, has been reported to suppress METH-induced 5-HT release in the pre-frontal cortex (Ago et al., 2007), and it was suggested that 5-HT2A receptor activation enhanced METH-induced 5-HT release. As the MDMA drug formula and pharmacological action are analogous to those of METH, suppression by risperidone of MDMA-induced 5-HT release may be due to its potent 5-HT2A antagonistic effect.

In addition, because there are large percentages of METH in tablets sold as MDMA (Clemens et al., 2007), risperidone is a potentially useful drug for treatment of MDMA-induced hyperthermia.

As described above, lower ambient temperatures (20°C) attenuated MDMA-induced hyperthermia compared with normal ambient temperature (24°C) (Figure 12). In addition, a combination therapy of risperidone (0.5 mg/kg) administration and low ambient temperature (20°C) completely inhibited MDMA induced hyperthermia (Figure 12). Therefore, risperidone and body cooling combination therapy may have a synergistic effect on human MDMA-induced hyperthermia.

3. Cocaine

Cocaine is the most widely used illegal psychostimulant drug in the USA. Approximately 4.8 million Americans aged 12 or older had abused cocaine at least once in the year prior to being surveyed in 2009 (National Survey on Drug Use and Health, Substance Abuse and Mental Health Administration web site:http://www.samhsa.gov/).

Cocaine abuse is a major health issue in many parts of the world, including the USA, Latin America, and Europe. Adverse effects of the use of cocaine include excited delirium, hypertensive crisis, acute myocardial infarction, tachycardia, and arrhythmia. Cocaine-related mortality can occur when patients present with hyperthermia (Bauwens et al., 1989; Campbell, 1988; Kosten and Kleber, 1988; Loghmanee and Tobak, 1986).

The Pharmacological Mechanisms of Cocaine

The mechanism of cocaine-induced hyperthermia in humans is complicated. Crandall et al. (2002) demonstrated that cocaine impairs heat dissipation due to impairment of sweating and cutaneous vasodilation. In addition, cocaine is believed to increase monoamine extracellular concentrations by competitively inhibiting monoamine transporters. Its inhibitory effect on DAT is particularly intense, leading to marked elevations in DA concentrations (Jones et al., 1995). Previous studies using microdialysis have demonstrated that cocaine increases intracellular monoamine concentrations, including DA, 5-HT, and NA, in the brain (Chen and Reith, 1994; Li et al., 1996; Reith et al., 1997). Reith et al. (1997) found that the degree of elevated monoamine concentrations differs in various regions of the brain. After cocaine (20 mg/kg i.p) was injected, DA, 5-HT, and NA levels in the VTA increased approximately 3.5-fold, 4.5-fold, and 3.5-fold, respectively. In the nucleus accumbens (NACC), cocaine (20 mg i.p.) increased DA, 5-HT, and NA levels approximately 4.5-fold, 4-fold, and 3.5-fold, respectively. Compared with the doses of METH (10 mg/kg) and MDMA (10 mg/kg) used in our studies (Shioda et al., 2008, 2010), a higher dose of cocaine (20 mg/kg) was injected in these studies; however, the relative DA and 5-HT concentration increases were lower.

Differences in the brain regions and injection routes between these studies may have contributed to the observed differences in DA and 5-HT concentration increases, although differences in the mechanism of monoamine concentration increases between cocaine, METH, and MDMA are also assumed to be important. Cocaine elicits multiple symptoms, including hyperthermia, believed to be caused by hyperactivity of the DA system. However, it is difficult to attribute all of the symptoms of cocaine toxicity solely due to DA concentration increases induced by DA re-uptake inhibition.

There is controversy as to whether cocaine also increases DA release; several investigators have reported that cocaine-induced DA concentration increases were due not only to inhibition of DA re-uptake but also to release of DA (Lee et al., 2001; Stamford et al., 1989). Venton et al. (2006) reported that cocaine increased DA release by mobilization of a synapsin-dependent reserve pool. Thus, cocaine is assumed to have an inhibitory effect on DA re-uptake and to enhance DA release. However, cocaine showed multiple symptoms including hyperthermia and it is difficult to explain that all symptoms are only due to DA hyperactivity. Cocaine molecular mechanisms require further research.

Dopamine and Cocaine-Induced Hyperthermia

Cocaine-induced hyperthermia is believed to be due to an increase in extracellular DA concentrations in the brain, because DA is well known as a neurotransmitter that increases body temperature. In addition, Rockhold et al. (1991) demonstrated that the D1 receptor antagonist SCH23390 abolished hyperthermia induced by cocaine, but the D2 receptor antagonist sulpiride increased rectal temperatures elevated by cocaine in rats. Therefore, the D1 receptor plays an important role in cocaine-induced hyperthermia as well as in MDMA- and METH-induced hyperthermia.

Serotonin and Cocaine-Induced Hyperthermia

It has been demonstrated that cocaine induces extracellular 5-HT concentration increases in the brain. However, there has been minimal investigation of 5-HT and 5-HT subtype receptor functions in cocaine-induced hyperthermia. As described earlier, hyperactivity of the 5-HT2A receptors is involved in the effects of psychostimulants, including METH- and MDMA-induced hyperthermia. Therefore, cocaine-induced 5-HT concentration increases may be a contributing factor in cocaine-induced hyperthermia.

Glutamate, NO, and Cocaine-Induced Hyperthermia

Previous studies have reported extracellular glutamate and NO concentration increases in the brain after cocaine injection (Lee et al., 2010; Smith et al., 1995; Suto et al., 2010; Zayara et al., 2011). Cocaine-induced Glu concentration increases were suppressed by D1, 5-HT2A, and 5-HT2C receptor antagonist treatment (Kalivas and Duffy, 1995; Zayara et al., 2011); therefore, D1, 5-HT2A, and 5-HT2C receptor activities are related to cocaine-induced Glu concentration increases.

D1 receptor antagonist treatment was also demonstrated to attenuate cocaine-induced NO concentration increases (Lee et al., 2010); hence cocaine-elevated NO levels were due to cocaine-induced D1 receptor hyperactivity. It has been demonstrated that Glu and NO concentration elevations are related to the effects of psychostimulants, including METH- and MDMA-induced hyperthermia; thus, Glu and NO may also play a role in cocaine-induced hyperthermia.

Ambient Temperature and Cocaine-Induced Hyperthermia

Previous animal investigations have demonstrated that high ambient temperatures reinforce cocaine-induced hyperthermia (Lomax and Daniel, 1990, 1993). These results are similar to those for other psychostimulants, including METH and MDMA. Bohert et al. (2010) reported that there was a significant relationship between the ambient temperature and accidental cocaine overdose fatalities. It is possible that cocaine-induced hyperthermia may be partially due to 5-HT2A receptor activation under the higher ambient temperatures, because higher temperatures reinforce 5-HT2A receptor activity (Zang et al., 2011). However, the exact mechanism of this effect remains unclear.

Pharmacological Treatment of Cocaine-Induced Hyperthermia

Review of existing papers related to cocaine-induced hyperthermia and treatment indicates that there are relatively few systematically performed preclinical studies. On the other hand, there are some clinical reports of cocaine-induced hyperthermia and its treatment. Kosten and Kleber (1988) reported the clinical resemblance between cases of cocaine intoxication and neuroleptic malignant syndrome (NMS). They suggested that dopamine agonists may be effective for cocaine-induced hyperthermia. Daras et al. (1995) and Tranvetyanon et al. (2001) also suggested the resemblance between cocaine-induced hyperthermia and NMS. Considering that cocaine-induced hyperthermia has characteristics similar to NMS, dantrolene may be effective for cocaine-induced hyperthermia. However, there are few papers reporting the efficacy of dantrolene, and Menaker et al. (2008) demonstrated that dantrolene was not effective in a case with cocaine-induced hyperthermia.

Roth et al. (1988) presented a study of 39 patients with acute rhabdomyolysis after cocaine use. Thirteen of the 39 patients (33%) had temperatures above 40.0 °C, and patients with acute renal failure were more likely to die than patients without.

James et al. (1999) compiled 150 cases of cocaine-associated rhabdomyolysis from previous case reports and found a close relationship between cocaine-associated rhabdomyolysis and fatal excited delirium. Thus, among the fatal complications of cocaine intoxication, rhabdomyolysis appears to have received more attention than hyperthermia. However, rhabdomyolysis is also found in METH (Winslow et al., 2007) and MDMA intoxication (Sprague et al., 2004), and it is not clear whether there is a high incidence of rhabdomyolysis in cocaine intoxication. The order of potency for increasing the core temperature is METH > MDMA > cocaine (Jaehne et al., 2007). This may be the reason that hyperthermia has received less attention in cocaine intoxication. In any case, it is apparent that hyperthermia can occur in cocaine toxicity.

It has been reported that cocaine-induced hyperthermia is seen primarily when there are high ambient temperatures (Crandall et al., 2002). Therefore, early application of cooling techniques, including an ice pack to the body, infusion of iced saline using a nasogastric tube, etc., should be performed. In addition, development of pharmacological treatment for cocaine-induced hyperthermia would be valuable.

Risperidone significantly attenuated cocaine-induced DA and 5-HT concentration increases (Broderick et al., 2003). The 5-HT2A receptor antagonists ketanserin and MDL 100907, and a 5-HT2C receptor antagonist SB242084, significantly attenuated cocaine-elevated DA extracellular concentrations in the rat brain (Zayara et al., 2011.

In addition, Broderick et al. (2004) reported that ketanserin significantly attenuated cocaine-induced 5-HT concentration increases. Our studies suggest that pre-administration of risperidone attenuated not only hyperthermia induced by METH or MDMA, but also increases in DA and 5-HT concentrations in rat brains injected with METH or MDMA. Because previous studies did not address changes in body temperature induced by cocaine, and our studies did not include cocaine-induced hyperthermia, we can only speculate regarding effective pharmacotherapies for cocaine-induced hyperthermia.

As mentioned above, considering that 5-HT as well as DA may be related to body temperature increase in cocaine intoxication, 5-HT2A receptor antagonists and D1 receptor antagonists, including risperidone, ketanserin, haloperidol, and others, may attenuate cocaine-induced hyperthermia. Further studies should be performed to confirm this hypothesis.

REFERENCES

Abekawa T, Ohmori T, Honda M, Ito K, Koyama T (2001) Effect of low doses of L-NAME on methamphetamine-induced dopaminergic depletion in the rat striatum. *J. Neural. Transm.* 108. 1219-1230.

Ago Y, Nakamura S, Kajita N, Uda M, Hashimoto H, Baba A, Matsud T (2007). Ritanserin reverses repeated methamphetamine-induced behavioral and neurochemical sensitization in mice. *Synapse* 61. 757-763.

Barr AM, Panenka WJ, MacEwan GW, Thornton AE, Lang DJ, Honer WG, Lecomte T (2006). The need for speed: an update on methamphetamine addiction. *J. Psychiatry Neurosci.* 31. 301-313

Baumann MH, Clark RD, Franken FH, Rutter JJ, Rothman RB (2008). Tolerance to 3,4-methylenedioxymethamphetamine in rats exposed to single high-dose binges. *Neuroscience* 152. 773-84.

Bauwens JE, Boggs JM, Hartwell PS (1989). Fatal hyperthermia associated with cocaine use. *West J. Med.* 150. 210-212.

Benamar K, Geller EB, Adler MW (2008). A new brain area affected by 3,4-methylenedioxymethamphetamine: A microdialysis-biotelemetry study. *Eur. J. Pharmacol.* 596. 84-88.

Bohnert AS, Prescott MR, Vlahov D, Tardiff KJ, Galea S (2010). Ambient temperature and risk of death from accidental drug overdose in New York City, 1990-2006. *Addiction* 105. 1049-1054.

Bowyer JF, Davies DL, Schmued L, Broening HW, Newport GD, Slikker W Jr., Holson RR (1994). Further studies of the role of hyperthermia in methamphetamine neurotoxicity. *J. Pharmacol. Exp.* Ther. 268. 1571-1580

Bowyer JF, Tank AW, Newport GD, Slikker W, Ali SF, Holson RR (1992).The influence of environmental temperature on the transient effects of methamphetamine on dopamine levels and dopamine release in rat striatum. *J. Pharmacol. Exp. Ther.* 260. 817-824

Broderick PA, Olabisi OA, Rahni DN, Zhou Y (2004). Cocaine acts on accumbens monoamines and locomotor behavior via a 5-HT2A/2C receptor mechanism as shown by ketanserin: 24-h follow-up studies. *Prog. Neuropsychopharmacol Biol. Psychiatry.* 28. 547-557.

Broderick PA, Rahni DN, Zhou Y (2003). Acute and subacute effects of risperidone and cocaine on accumbens dopamine and serotonin release using in vivo microvoltammetry on line with open-field behavior. *Prog. Neuropsychopharmacol. Biol. Psychiatry* 27. 1037-1054.

Broening HW, Morford LL, Vorhees CV (2005). Interactions of dopamine D1 and D2 receptor antagonists with D-methamphetamine-induced hyperthermia and striatal dopamine and serotonin reductions. *Synaps*e 56. 84-93.

Bymaster FP, Calligaro DO, Falcone JF, Marsh RD, Moore NA, Tye NC, Seeman P, Wong DT (1996). Radioreceptor binding profile of the atypical antipsychotic olanzapine. *Neuropsychopharmacology* 14. 87-96.

Callahan BT, Ricaurte GA (1998). Effect of 7-nitroindazole on body temperature and methamphetamine-induced dopamine toxicity. *Neuroreport* 9. 2691-2695.

Campbell BG (1988). Cocaine abuse with hyperthermia, seizures and fatal complications. *Med. J. Aust.* 149. 387-389.

Capela JP, Carmo H, Remião F, Bastos ML, Meisel A, Carvalho F (2009). Molecular and cellular mechanisms of ecstasy-induced neurotoxicity: an overview. *Mol. Neurobiol.* 39. 210-271.

Chan P, Chen JH, Lee MH, Deng JF (1994). Fatal and nonfatal methamphetamine intoxication in the intensive care unit. *J. Toxicol. Clin. Toxicol.* 32. 147-155.

Chen J, Wersinger C, Sidhu A (2003). Chronic stimulation of D1 dopamine receptors in human SK-N-MC neuroblastoma cells induces nitric-oxide synthase activation and cytotoxicity. *J. Biol. Chem.* 278. 28089-28100.

Chen NH, Reith ME (1994). Autoregulation and monoamine interactions in the ventral tegmental area in the absence and presence of cocaine: a microdialysis study in freely moving rats. *J. Pharmacol. Exp. Ther.* 271.1597-1610.

Clemens KJ, McGregor IS, Hunt GE, Cornish JL (2007). MDMA, methamphetamine and their combination: possible lessons for party drug users from recent preclinical research. *Drug Alcohol. Rev.* 26. 9-15.

Colado MI, Camarero J, Mechan AO, Sanchez V, Esteban B, Elliott JM, Green AR (2001). A study of the mechanisms involved in the neurotoxic action of 3,4-methylenedioxymethamphetamine (MDMA, 'ecstasy') on dopamine neurones in mouse brain. *Br. J. Pharmacol.* 134. 1711-1723.

Crandall CG, Vongpatanasin W, Victor RG (2002). Mechanism of cocaine-induced hyperthermia in humans. *Ann. Intern Med.* 136. 785-791.

Crespi D, Mennini T, Gobbi M (1997). Carrier-dependent and Ca(2+)-dependent 5-HT and dopamine release induced by (+)-amphetamine, 3,4-methylendioxymethamphetamine, p-chloroamphetamine and (+)-fenfluramine. *Br. J. Pharmacol.* 121. 1735-1743.

Dar KJ, McBrien ME (1996). MDMA induced hyperthermia: report of a fatality and review of current therapy. *Intensive Care Med.* 22. 995-996.

Daras M, Kakkouras L, Tuchman AJ, Koppel BS (1995). Rhabdomyolysis and hyperthermia after cocaine abuse: a variant of the neuroleptic malignant syndrome? *Acta Neurol. Scand* 92. 161-165.

Darke S, Kaye S, McKetin R, Duflou J (2008). Major physical and psychological harms of methamphetamine use. *Drug Alcohol. Rev.* 27. 253-262

Darvesh AS, Yamamoto BK, Gudelsky GA (2005). Evidence for the involvement of nitric oxide in 3,4-methylenedioxymethamphetamine-induced serotonin depletion in the rat brain. *J. Pharmacol. Exp. Ther.* 312. 694-701.

Davidson C, Gow AJ, Lee TH, Ellinwood EH (2001). Methamphetamine neurotoxicity: necrotic and apoptotic mechanisms and relevance to human abuse and treatment. *Brain Res. Brain Res. Rev.* 36. 1-22

Dowsett RP (1996). Deaths attributed to "ecstasy" overdose. *Med. J. Aust.* 164. 700.

Funahashi M, Kohda H, Hori O, Hayashida H, Kimura H (1990). Potentiating effect of morphine upon d-methamphetamine-induced hyperthermia in mice. Effects of naloxone and haloperidol. *Pharmacol. Biochem. Behav.* 36. 345-350.

Gołembiowska K, Konieczny J, Wolfarth S, Ossowska K (2003). Neuroprotective action of MPEP, a selective mGluR5 antagonist, in methamphetamine-induced dopaminergic neurotoxicity is associated with a decrease in dopamine outflow and inhibition of hyperthermia in rats. *Neuropharmacology* 45. 484-492.

Gordon CJ, Watkinson WP, O'Callaghan JP, Miller DB (1991). Effects of 3,4-methylenedioxymethamphetamine on autonomic thermoregulatory responses of the rat. *Pharmacol. Biochem. Behav.* 38. 339-344.

Gough B, Iman SZ, Blough B, Slikker W, Ali S (2002). Comparative effects of substituted amphetamines (PMA, MDMA, and METH) on monoamines in rat caudate: a microdialysis study. *Ann. NY Acad. Sci.* 965. 410-420.

Grahame-Smith DG (1971). Studies in vivo on the relationship between brain tryptophan, brain 5-HT synthesis and hyperactivity in rats treated with a monoamine oxidase inhibitor and L-tryptophan. *J. Neurochem.* 18. 1053-1066.

Green AR, O'Shea E, Saadat KS, Elliott JM, Colado MI (2005). Studies on the effect of MDMA ('ecstasy') on the body temperature of rats housed at different ambient room temperatures.*Br. J. Pharmacol.* 146. 306-312.

Grunau BE, Wiens MO, Greidanus M. (2010). Dantrolene for the treatment of MDMA toxicity. *CJEM.* 12. 435-442

Hall AP, Henry JA (2006). Acute toxic effects of 'Ecstasy' (MDMA) and related compounds: overview of pathophysiology and clinical management. *Br. J. Anaesth.* 96. 678-685.

Han DD, Gu HH (2006). Comparison of the monoamine transporters from human and mouse in their sensitivities to psychostimulant drugs. *BMC Pharmacol.* 6: 6.

He J, Xu H, Yang Y, Zhang X, Li XM (2004). Neuroprotective effects of olanzapine on methamphetamine-induced neurotoxicity are associated with an inhibition of hyperthermia and prevention of Bcl-2 decrease in rats. *Brain Res.* 1018.186-192

He J, Xu H, Yang Y, Zhang X, Li XM (2005). Chronic administration of quetiapine alleviates the anxiety-like behavioural changes induced by a neurotoxic regimen of dl-amphetamine in rats. *Behav. Brain Res.* 160. 178-187

He J, Yang Y, Yu Y, Li X, Li XM (2006). The effects of chronic administration of quetiapine on the methamphetamine-induced recognition memory impairment and dopaminergic terminal deficit in rats.*Behav. Brain Res.* 172. 39-45

Hewitt KE, Green AR (1994). Chlormethiazole, dizocilpine and haloperidol prevent the degeneration of serotonergic nerve terminals induced by administration of MDMA ('Ecstasy') to rats. *Neuropharmacology* 33. 1589-1595.

Homer BD, Solomon TM, Moeller RW, Mascia A, DeRaleau L, Halkitis PN (2008). Methamphetamine abuse and impairment of social functioning: a review of the underlying neurophysiological causes and behavioral implications. *Psychol. Bull.* 134. 301-310

Huang X, Nichols DE (1993). 5-HT2 receptor-mediated potentiation of dopamine synthesis and central serotonergic deficits. *Eur. J. Pharmacol.* 23. 291-296.

Iravani MM, Asari D, Patel J, Wieczorek WJ, Kruk ZL (2000). Direct effects of 3,4-methylenedioxymethamphetamine (MDMA) on serotonin or dopamine release and uptake in the caudate putamen, nucleus accumbens, substantia nigra pars reticulata, and the dorsal raphé nucleus slices. *Synapse* 36. 275-285.

Ishigami A, Kubo S, Gotohda T, Tokunaga I (2003). The application of immunohistochemical findings in the diagnosis in methamphetamine-related death-two forensic autopsy cases. *J. Med. Invest.* 50.112-116

Itzhak Y, Ali SF (1996). The neuronal nitric oxide synthase inhibitor, 7-nitroindazole, protects against methamphetamine-induced neurotoxicity in vivo. *J. Neurochem.* 67. 1770-1773.

Itzhak Y, Martin JL, Ail SF (2000). nNOS inhibitors attenuate methamphetamine-induced dopaminergic neurotoxicity but not hyperthermia in mice. *Neuroreport* 11. 2943-2946.

Jaehne EJ, Salem A, Irvine RJ (2007). Pharmacological and behavioral determinants of cocaine, methamphetamine, 3,4-methylenedioxymethamphetamine, and para-methoxyamphetamine-induced hyperthermia. *Psychopharmacology* (Berl). 194. 41-52.

James RA, Heath M, Chartes W (1999). Cocaine-associated rhabdomyolysis and excited delirium: different stages of the same syndrome. *Am. J. Forensic Med. Pathology*. 20. 120-127.

Jones SR, Garris PA, Wightman RM (1995). Different effects of cocaine and nomifensine on dopamine uptake in the caudate-putamen and nucleus accumbens. *J. Pharmacol. Exp. Ther*. 274. 396-403.

Kalivas PW, Duffy P (1995). D1 receptors modulate glutamate transmission in the ventral tegmental area. *J. Neurosci*. 15. 5379-88.

Kosten TR, Kleber HD (1988). Rapid death during cocaine abuse: a variant of the neuroleptic malignant syndrome? *Am. J. Drug Alcohol Abuse*. 14. 335-346.

Kunitz O, Ince A, Kuhlen R, Rossaint R (2003). Hyperpyrexia and rhabdomyolysis after ecstasy (MDMA) intoxication] *Anaesthesist*. 52. 511-515.

Kojima T, Une I, Yashiki M, Noda J, Sakai K, Yamamoto K (1984). A fatal methamphetamine poisoning associated with hyperpyrexia. *Forensic Sci. Int*. 24. 87-93

Kosten TR, Kleber HD (1988). Rapid death during cocaine abuse: a variant of the neuroleptic malignant syndrome? *Am. J. Drug Alcohol Abuse*. 14. 335-346.

Kuczenski R, Segal DS, Cho AK, Melega W (1995). Hippocampus norepinephrine, caudate dopamine and serotonin, and behavioral responses to the stereoisomers of amphetamine and methamphetamine. *J. Neurosci*. 15. 1308-1317.

Lee TH, Balu R, Davidson C, Ellinwood EH (2001). Differential time-course profiles of dopamine release and uptake changes induced by three dopamine uptake inhibitors. *Synapse*. 41. 301-310.

Lee DK, Koh WC, Shim YB, Shim I, Choe ES (2010). Repeated cocaine administration increases nitric oxide efflux in the rat dorsal striatum. *Psychopharmacology* (Berl). 208. 245-256.

Li MY, Yan QS, Coffey LL, Reith ME (1996). Extracellular dopamine, norepinephrine, and serotonin in the nucleus accumbens of freely moving rats during intracerebral dialysis with cocaine and other monoamine uptake blockers. *J. Neurochem*. 66. 559-568.

Loghmanee F, Tobak M (1986). Fatal malignant hyperthermia associated with recreational cocaine and ethanol abuse. *Am. J. Forensic Med. Pathol*. 7. 246-248.

Lomax P, Daniel KA (1990). Cocaine and body temperature in the rat: effects of ambient temperature. *Pharmacology* 40. 103-109.

Lomax P, Daniel KA (1993). Cocaine and body temperature: effect of exercise at high ambient temperature. *Pharmacology*. 46.164-172.

Mallick A, Bodenham AR (1997). MDMA induced hyperthermia: a survivor with an initial body temperature of 42.9°C. *J. Accid. Emerg. Med*. 14. 336-338.

Makisumi T, Yoshida K, Watanabe T, Tan N, Murakami N, Morimoto A (1998).Sympatho-adrenal involvement in methamphetamine-induced hyperthermia through skeletal muscle hypermetabolism. *Eur. J. Pharmaco*l. 363.107-112.

Mazzola-Pomietto P, Aulakh CS, Wozniak KM, Hill JL, Murphy DL (1995). Evidence that 1-(2,5-dimethoxy-4-iodophnyl)-2-aminopropane (DOI)-induced hyperthermia in rats is mediated by stimulation of 5-HT2A receptors. *Psychopharmacol.* 117. 193-199.

Mechan AO, Esteban B, O'Shea E, Elliott JM, Colado MI, Green AR (2002). The pharmacology of the acute hyperthermic response that follows administration of 3,4-methylenedioxymethamphetamine (MDMA, `ecstasy') to rats. *Br. J. Pharmacol.* 135. 170-180.

Menaker J, Farcy DA, Boswell SA, Stein DM, Dutton RP, Hess JR, Scalea TM (2008). Cocaine-induced agitated delirium with associated hyperthermia: a case report. *J. Emerg. Med.* doi:10.1016/j.jemermed.2008.02.055

Mueller PD, Korey WS (1998). Death by "ecstasy": the serotonin syndrome? Ann Emerg Med. 32. 377-380.

Myles BJ, Jarrett LA, Broom SL, Speaker HA, Sabol KE (2008). The effects of methamphetamine on core body temperature in the rat--part 1: chronic treatment and ambient temperature. *Psychopharmacology* (Berl). 198. 301-311.

Nash JF (1990). Ketanserin pretreatment attenuates MDMA-induced dopamine release in the striatum as measured by in vivo microdialysis. *Life Sci.* 47. 2401-2408.

Nash JF, Yamamoto BK (1992). Methamphetamine neurotoxicity and striatal glutamate release: comparison to 3,4-methylenedioxymethamphetamine. *Brain Res.* 581. 237-243.

Nicola SM, Kombian SB, Malenka RC (1996). Psychostimulants depress excitatory synaptic transmission in the nucleus accumbens via presynaptic D1-like dopamine receptors. *J. Neurosci.* 16. 1591-1604

Nisijima K, Shioda K, Iwamura T (2007). Neuroleptic malignant syndrome and serotonin syndrome. *Progress in Brain Res.* 41. 81-104.

O'Connor B. (1994). Hazards associated with the recreational drug 'ecstasy'. *Br. J. Hosp. Med.* 52: 507. 510-514.

O'Dell SJ, Weihmuller FB, Marshall JF (1993). Methamphetamine-induced dopamine overflow and injury to striatal dopamine terminals: attenuation by dopamine D1 or D2 antagonists. *J. Neurochem.* 60. 1792-1799.

Ohmori T, Abekawa T, Koyama T (1996). The role of glutamate in behavioral and neurotoxic effects of methamphetamine. *Neurochem. Int.* 29. 301-307.

O'Shea E, Escobedo I, Orio L, Sanchez V, Navarro M, Green AR, Colado MI (2005). Elevation of ambient room temperature has differential effects on MDMA-induced 5-HT and dopamine release in striatum and nucleus accumbens of rats. *Neuropsychopharmacology* 30. 1312-1323.

Park DJ, West AR (2009). Regulation of striatal nitric oxide synthesis by local dopamine and glutamate interactions. *J. Neurochem.* 111. 1457-1465.

Parrott AC (2002). Recreational Ecstasy/MDMA, the serotonin syndrome, and serotonergic neurotoxicity. *Pharmacology Biochem. Behavior.71.* 837-844.

Pehek EA, Nocjar C, Roth BL, Byrd TA, Mabrouk OS (2006). Evidence for the preferential involvement of 5-HT2A serotonin receptors in stress- and drug-induced dopamine release in the rat medial prefrontal cortex. *Neuropsychopharmacology*. 31.265-277.

Porras G, Di Matteo V, Fracasso C, Lucas G, De Deurwaerdère P, Caccia S, Esposito E, Spampinato U (2002). 5-HT2A and 5-HT2C/2B receptor subtypes modulate dopamine release induced in vivo by amphetamine and morphine in both the rat nucleus accumbens and striatum. *Neuropsychopharmacology* 26. 311-324.

Reith ME, Li MY, Yan QS (1997). Extracellular dopamine, norepinephrine, and serotonin in the ventral tegmental area and nucleus accumbens of freely moving rats during intracerebral dialysis following systemic administration of cocaine and other uptake blockers. *Psychopharmacology* (Berl). 134. 309-317.

Rockhold RW, Carver ES, Ishizuka Y, Hoskins B, Ho IK (1991). Dopamine receptors mediate cocaine-induced temperature responses in spontaneously hypertensive and Wistar-Kyoto rats. *Pharmacol. Biochem. Behav.* 40. 157-162.

Roth D, Alarcón FJ, Fernandez JA, Preston RA, Bourgoignie JJ (1988). Acute rhabdomyolysis associated with cocaine intoxication. *N. Engl. J. Med.* 319. 673-677.

Rusyniak DE, Sprague JE (2005). Toxin-induced hyperthermic syndromes. *Med. Clin. N Am.* 89. 1277-1296.

Sammut S, Dec A, Mitchell D, Linardakis J, Ortiguela M, West AR. (2006) Phasic dopaminergic transmission increases NO efflux in the rat dorsal striatum via a neuronal NOS and a dopamine D(1/5) receptor-dependent mechanism. *Neuropsychopharmacology*. 31. 493-505

Sanchez V, Zeini M, Camarero J, O'Shea E, Bosca L, Green AR, Colado MI (2003). The nNOS inhibitor, AR-R17477AR, prevents the loss of NF68 immunoreactivity induced by methamphetamine in the mouse striatum. *J. Neurochem.* 85. 515-524.

Schep LJ, Slaughter RJ, Beasley DM (2010). The clinical toxicology of metamphetamine. *Clin. Toxicol (Phila).* 48.675-694.

Schifano F, Corkery J, Naidoo V, Oyefeso A, Ghodse H (2010). Overview of amphetamine-type stimulant mortality data--UK, 1997-2007. *Neuropsychobiology.* 61.122-130.

Scott JC, Woods SP, Matt GE, Meyer RA, Heaton RK, Atkinson JH, Grant I (2007). Neurocognitive effects of methamphetamine: a critical review and meta-analysis. N*europsychol. Rev.* 17. 275-297

Schmidt CJ, Fadayei GM, Sullivan CK, Taylor VL (1992). 5-HT2 receptors exert a state-dependent regulation of dopaminergic function: studies with MDL 100,907 and the amphetamine analogue, 3,4-methylenedioxymethamphetamine. *Eur. J. Pharmacol.* 223. 65-74.

Seger D (2010). Cocaine, metamfetamine, and MDMA abuse: the role and clinical importance of neuroadaptation. *Clin. Toxicol (Phila).* 48.695-708.

Sheng P, Cerrut,i C, Ali S, Cadet J (1996). Nitric oxide is a mediator of methamphetamine (METH)-induced neurotoxicity. In vitro evidence from primary cultures of mesencephalic cells. *Ann. N Y Acad. Sci.* 801.174-186.

Shimada A, Yamaguchi K, Yanagita T (1996). Neurochemical analysis of the psychotoxicity of methamphetamine and cocaine by microdialysis in the rat brain. *Ann. N Y Acad. Sci.* 801.361-370.

Shioda K, Nisijima K, Yoshino T, Kato S (2010). Effect of risperidone on acute methamphetamine-induced hyperthermia in rats. *Drug Alcohol Depend.* 111. 241-249.

Shioda K, Nisijima K, Yoshino T, Kuboshima K, Iwamura T, Yui K, Kato S (2008). Risperidone attenuates and reverses hyperthermia induced by 3,4-methylenedioxymethamphetamine (MDMA) in rats. *Neurotoxicology.* 29. 1030-1036.

Smith JA, Mo Q, Guo H, Kunko PM, Robinson SE (1995). Cocaine increases extraneuronal levels of aspartate and glutamate in the nucleus accumbens. *Brain Res.* 683. 264-269.

Sprague JE, Brutcher RE, Mills ED, Caden D, Rusyniak DE (2004). Attenuation of 3,4-methylenedioxymethamphetamine (MDMA, ecstasy)-induced rhabdomyolysis with α1- plus β3-adrenoceptor antagonists. *Br. J. Pharmacol.* 142. 667-670.

Stamford JA, Kruk ZL, Millar J (1989). Dissociation of the actions of uptake blockers upon dopamine overflow and uptake in the rat nucleus accumbens: in vivo voltammetric data. *Neuropharmacology.* 28. 1383-1388.

Stephans SE, Yamamoto BK (1994). Methamphetamine-induced neurotoxicity: roles for glutamate and dopamine efflux. *Synapse.* 17. 203-209.

Suto N, Ecke LE, You ZB, Wise RA (2010).Extracellular fluctuations of dopamine and glutamate in the nucleus accumbens core and shell associated with lever-pressing during cocaine self-administration, extinction, and yoked cocaine administration. *Psychopharmacology* (Berl). 211. 267-75.

Tanvetyanon T, Dissin J (2001). Hyperthermia and chronic pancerebellar syndrome after cocaine abuse. *Arch. Intern. Med.* 161. 608-610.

Vanattou-Saïfoudine N, McNamara R, Harkin A (2010). Mechanisms mediating the ability of caffeine to influence MDMA ('Ecstasy')-induced hyperthermia in rats. *Br. J. Pharmacol.* 160. 860-877.

Venton BJ, Seipel AT, Phillips PE, Wetsel WC, Gitler D, Greengard P, Augustine GJ, Wightman RM (2006). Cocaine increases dopamine release by mobilization of a synapsin-dependent reserve pool. *J. Neurosci.* 26. 3206-3209.

Winslow BT, Voorhees KI, Pehl KA (2007). Methamphetamine abuse. *Am. Fam. Physicians.* 76. 1169-1174.

Wood MD, Scott C, Clarke K, Cato KJ, Patel N, Heath J, Worby A, Gordon L, Campbell L, Riley G, Davies CH, Gribble A, Jones DN (2006). Pharmacological profile of antipsychotics at monoamine receptors: atypicality beyond 5-HT2A receptor blockade. *CNS Neurol. Disord Drug Targets.* 5. 445-452.

Xie T, McCann UD, Kim S, Yuan J, Ricaurte GA (2000). Effect of temperature on dopamine transporter function and intracellular accumulation of methamphetamine: implications for methamphetamine-induced dopaminergic neurotoxicity. *J. Neurosci.* 20. 7838-7845.

Yamada J, Sugimoto Y, Wakita H, Horisaka K (1998). The involvement of serotonergic and dopaminergic systems in hypothermia induced in mice by intracerebroventricular injection of serotonin. *Jpn. J. Pharmacol.* 48. 145-148.

Yamawaki S, Laim H, Horita A (1983). Dopaminergic and serotonergic mechanisms of thermoregulation: mediation of thermal effects of apomorphine and dopamine. *J. Pharmacol. Exp. Ther.* 227. 383-388.

Zayara AE, McIver G, Valdivia PN, Lominac KD, McCreary AC, Szumlinski KK (2011). Blockade of nucleus accumbens 5-HT2A and 5-HT2C receptors prevents the expression of cocaine-induced behavioral and neurochemical sensitization in rats. *Psychopharmacology* (Berl). 213. 321-335.

Zhang G, Tao R (2011). Enhanced responsivity of 5-HT(2A) receptors at warm ambient temperatures is responsible for the augmentation of the 1-(2,5-dimethoxy-4-iodophenyl)-2-aminopropane (DOI)-induced hyperthermia. *Neurosci. Lett.* 490. 68-71.

In: Hyperthermia: Recognition, Prevention and Treatment ISBN 978-1-61942-275-9
Editor: Xing-Hua Gao and Hong-Duo Chen, 199-208

Chapter 6

PLASMA HSP72 AS A DIAGNOSTIC AID IN CASES OF HEATSTROKE

Patricia A. Ruell and Martin W. Thompson
Faculty of Health Sciences, The University of Sydney,
Sydney, NSW, Australia

ABSTRACT

Heat illnesses range from heat fatigue, heat rashes, disorders resulting from water or salt depletion, including heat exhaustion and heat cramps, to the most serious disorder, heatstroke. Heatstroke is a medical emergency and the resulting physiological changes can be profound and life-threatening. The mortality rate can be as high as 50%. Classic heatstroke tends to affect the very young or old, while exertional heatstroke mainly affects young, fit individuals. Many of the physiological changes found in exertional heatstroke are difficult to differentiate from those associated with intense physical exertion.

The diagnosis of heatstroke is based on an elevated body core temperature (>41°C) and signs of central nervous system disturbance such as disorientation and combative behavior. The presence or absence of sweating is no longer considered diagnostic. It would be useful if there were a laboratory test to confirm a diagnosis of heatstroke, as body core temperature has often returned to normal by the time the patient has been transported to hospital. Heat shock protein 72 (Hsp72) is markedly elevated by heat and/or other stressors to the cell. Hsp72 is also found in the plasma and is elevated immediately after exercise; it is thought to act as a chemokine in the plasma, initiating an inflammatory response.

Plasma Hsp72 is a potentially useful marker of heatstroke, and several studies are discussed that have measured plasma Hsp72 in cases of heatstroke. Further research is needed to examine the mechanism of the increased plasma concentrations of Hsp72 in heatstroke, as passive hyperthermia alone does not appear to lead to the same increase in plasma Hsp72 levels.

INTRODUCTION

Heat Illnesses

The heat illnesses range from relatively benign conditions such as heat cramps, to the most serious disorder, heatstroke. The two types of heatstroke, classic and exertional, occur in quite different populations.

Classic heatstroke affects mainly the very young and old with the disorder often complicated by a pre-existing medical condition. Elderly people without access to air conditioning and those who are more dependent are at greater risk of heat-related mortality during a heat wave (Davido et al., 2006; Semenza et al., 1996). Exertional heatstroke affects mainly physically fit young individuals and is the third leading cause of death in athletes in the United States (Moreau and Deeter, 2005).

Some occupations are more at risk with military personnel and those working in hot environments such as firefighters more likely to suffer heatstroke. Even when treated, the mortality rate ranges from 10 to 50% and a significant number suffer long term neurological damage (7-14%) (Bouchama, 1995). While earlier publications considered absence of sweating to be diagnostic for the disorder (heatstroke) in conjunction with central nervous system disturbance and body temperature above 40.6°C (Leithead and Lind, 1964), it is now considered that patients with lower body temperatures can be at risk, and cessation of sweating is not necessarily diagnostic.

Diagnosis of Heat Illnesses

Heatstroke is a medical emergency with initial clinical intervention usually taken without reference to laboratory tests. Physiological changes can be profound and life-threatening. In classical heatstroke, patients have respiratory alkalosis, while those with exertional heatstroke have both respiratory alkalosis and lactic acidosis (Bouchama and Knochel, 2002). In exertional heatstroke, many of the physiological changes are difficult to differentiate from those associated with intense physical exertion. While hypoglycemia has been found in some collapsed runners (Sutton et al., 1972), often plasma glucose is elevated after completing a race in both healthy runners and those collapsing with heat illness (Richards et al., 1979). The heat illnesses range from relatively benign disorders to heatstroke, which may be fatal, and the physiological changes will differ depending on the severity of the disorder.

Some attempts have been made to utilize laboratory data as an aid in diagnosis, for example elevated growth hormone appears to be prognostic in cases of classical heatstroke (Alzeer et al., 1996).

Most investigations have been exploratory in nature, attempting to delineate the mechanisms leading to heatstroke. A variety of laboratory tests have identified abnormalities in the blood and plasma of patients with heat illness, including alterations in plasma phosphate (Knochel and Caskey, 1977), endotoxin, cytokines (Bouchama et al., 1991), and lymphocyte distribution (DuBose et al., 2003).

The Heat Shock Proteins

The heat shock proteins are a family of proteins that are classified according to their molecular weight (Kiang and Tsokos, 1998). The most often studied protein is heat shock protein 72 of molecular weight 72 kDa.

The concentration of Hsp72 is rapidly upregulated in the cell in response to heat or other stressful conditions, and the increased concentration of Hsp72 has a protective function in the cell. Hsp72 acts by refolding damaged proteins, and transporting other proteins either into other organelles, or to the lysosymes for degradation. While Hsp72 was for many years considered to be solely an intracellular protein, it is now known to be found in the extracellular space, in the serum, saliva and urine. Serum Hsp72 is increased by a number of medical conditions, including cardiovascular disease and some cancers. In healthy individuals levels of serum Hsp72 are increased immediately after exercise (Walsh et al., 2001). While intracellular Hsp72 has a protective function in the cell, the function of Hsp72 in serum is less well established. Hsp72 is thought to be a danger signal in the blood, activating a cytokine cascade when the level is increased. Extracellular Hsp72 binds to receptors on leucocytes including monocytes, and macrophages. After binding of Hsp72 to the cell membrane of human monocytes, a rapid intracellular calcium flux is initiated, leading to upregulation of the pro-inflammatory cytokines TNF-α, IL-1β and IL-6 (Asea et al., 2000).

Induction of Plasma Hsp72 by Passive Heating and Exercise

Models of heatstroke may have elements of passive heating and/or exercise-heat exposure protocols depending on whether the heatstroke to be investigated is classic or exertional. Some human and animal experiments have examined plasma Hsp72 in the context of passive heating and/or exercise and we will briefly examine the plasma Hsp72 response in these models before looking at plasma Hsp72 and heatstroke.

Passive Heating

Passive heating elevates body temperatures to levels that are found in some cases of classic and exertional heatstroke. In one experiment, female rats were passively heated for 60 minutes with a final core temperature of less than 40°C (Ogura et al., 2008). Plasma Hsp72 was not significantly elevated in this experiment. A previous study in mice also did not find an increase in serum Hsp72 following 6 hours exposure to a rectal temperature (T_{rec}) of 39.5-40.0°C (Ostberg et al., 2002). In a human model, young men were passively heated such that their core temperature was elevated by 2.3°C to around 38.5°C (Whitham et al., 2007). Plasma Hsp72 concentrations were not higher than in a control trial where T_{rec} was not elevated. Our laboratory investigated plasma Hsp72 in young men in a passive heating experiment where core temperature was elevated to 39.7°C (Ruell et al., 2009). Plasma Hsp72 was not elevated immediately after passive heating at this higher temperature (Figure 1). In some subjects, plasma Hsp72 was measured 2.5 hours after exercise, and no elevation was detected at that time point.

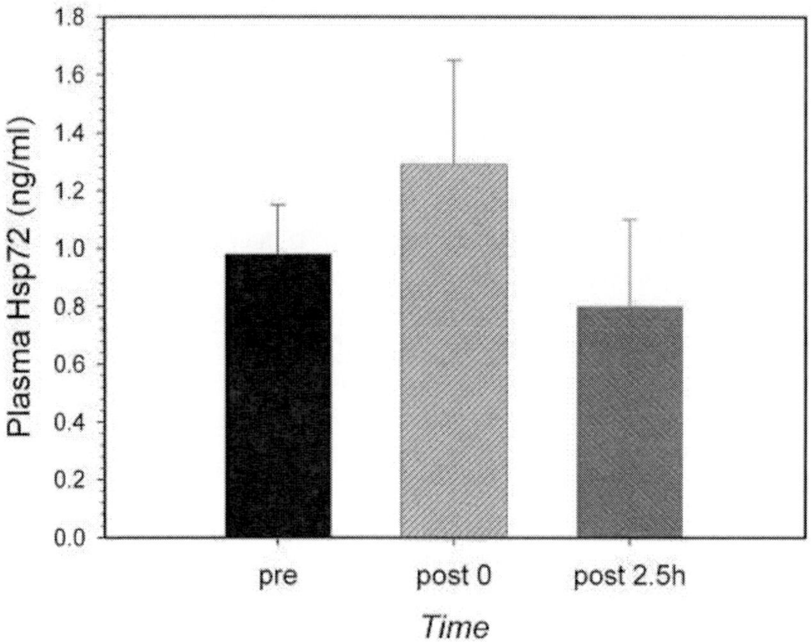

Figure 1. There were no significant changes in plasma Hsp72 concentration following passive heating. Eleven young men were passively heated in a bath for approximately 70 minutes (post 2.5h, n=4) (Ruell et al., 2009).

Exercise

There are now a large number of studies showing that plasma Hsp72 is increased following intense exercise. While both plasma and serum Hsp72 have been measured, it has been shown that values measured in plasma are higher than those in serum (Whitham and Fortes, 2006). Many studies now report using a similar method for analysis of plasma Hsp72 (Elisa) making comparison between studies easier. Plasma Hsp72 is affected both by exercise intensity and duration, with values reaching 4.2 ng/ml after 60 min exercise (75% $\dot{V}O_{2max}$) and 16.3 ng/ml following marathon running (Yamada et al., 2008). Immediately after a 14 km race, plasma Hsp72 was 17.0 ± 2.6 ng/ml (n=7) in runners who finished the race with no health problems (Ruell et al., 2006).

Whilst the T_{rec} in the study of control runners following the 14 km race and young men passively heated (Ruell et al., 2009) were quite similar (39.8 ± 0.2 °C and 39.7 ± 0.1 °C respectively), plasma Hsp72 was not elevated by passive heating. The duration of passive heating was longer than race time (73 vs 60 min), indicating that the duration of heating was not a factor in the different plasma Hsp72 response. Plasma Hsp72 reached a peak immediately at the end of exercise in a previous study (Walsh et al., 2001), and it was assumed that plasma Hsp72 would be highest at the end of heating. However, in case the response was for some reason delayed following passive heating, plasma from four subjects was tested 2.5 hours after heating, and again no significant increase was found. Rectal temperature was only measured in the 14 km race at the end of exercise, and it is possible that higher temperatures were achieved before this time point. To clarify this point future studies are required, monitoring T_{rec} during the entire race using a data logger.

During exercise there may be regional differences in tissue temperature that could explain the different response of plasma Hsp72 to passive and exertional hyperthermia. Rowell et al., (1968) found that at exhaustion following very intense exercise, liver temperature sometimes exceeded T_{rec} by as much as 1.5°C, while during passive heating liver temperature was usually slightly less than T_{rec}. This suggests that there is a relatively greater vasoconstrictive drive to the hepatosplanchnic tissues during intense exercise compared with passive heat exposure. A higher liver temperature during exercise might be the stimulus for increased levels of plasma Hsp72, because the hepatosplanchnic tissues appear to be a source of Hsp72 released into the circulation during exercise (Febbraio et al., 2002).

Plasma Hsp72 and Heat Stroke

There are a growing number of research studies using the presence or absence of plasma Hsp72 as a prognostic marker for particular diseases. Some studies have also determined levels of plasma Hsp72 in patients with heatstroke. A report in humans measured levels of serum Hsp72 and antibodies to Hsp72 in patients with classic heatstroke (Wang et al., 2001). It was found that serum Hsp72 was significantly lower than in healthy subjects exposed to the same hot environment, while antibody levels were higher in patients with severe heatstroke. In another group of patients with exertional heat illness, antibodies to heat shock proteins were significantly higher than in control subjects (Wu et al., 2001). The patients with the more severe symptoms had the highest antibody titers against Hsp72.

In a more recent study, plasma Hsp72 was measured following a 14 km running race and the results in the control runners have been discussed in the previous section (Ruell et al., 2006). An elevated level of plasma Hsp72 was found in all groups of subjects compared with normal resting values. A higher level of plasma Hsp72 was observed in those patients with neurological symptoms and those who collapsed (n=13), compared with the post-exercise values in the controls. Three runners in this group had plasma Hsp72 levels over 50ng/ml and all exhibited signs of confusion or combative behaviour. One runner (plasma Hsp72 of 51.8 ng/ml) had a T_{rec} of 42°C and showed symptoms of nausea and vomiting.

A study of patients with classic heat stroke admitted to hospital during the 2003 heatwave in Paris measured a number of plasma constituents in 18 patients with heatstroke (Huisse et al., 2008). The median Hsp72 level was 287 ng/ml with a range from 10-1371 ng/ml. The patients were older (median 68 years) with a high maximum core temperature (41.4°C) and a high mortality rate (14/18 died). High levels of plasma IL-6 were found in the Paris study that correlate with the severity of illness and this is consistent with other studies (Leon and Helwig, 2010).

In a baboon model of passive heatstroke, plasma Hsp72 was released into the circulation with levels much higher in the severe compared with the moderate heatstroke group (Dehbi et al., 2010). The mortality rate amongst the baboons in the severe heatstroke group was 67%. The levels of plasma Hsp72 were higher in the non-survivors, and were correlated with markers of tissue damage. Plasma Hsp72 levels as high as 169 ng/ml were measured in the severe heatstroke group. Following heatstroke, Hsp72 levels were approximately 50 ng/ml one hour post and increased to a maximum level 18 hours post heat exposure after which time Hsp72 levels declined.

The results from the aforementioned four studies are summarized in Table 1. The analysis of serum Hsp72 in the first study used a semi-quantitative method (Wang et al., 2001) so values of plasma Hsp72 are not given. The later publications show high plasma/serum Hsp72 values in patients with heatstroke, while in the 2001 study values were lower than in high temperature controls and it is not clear why there is a different trend. The methodology used for analysis of serum Hsp72 in the 2001 report was a western blot method which is quite different to the Elisa method used in the later three studies. It is also possible that the patients in the first study were not as ill, with body temperature given as >38°C, while in the other studies temperatures as high as 42°C or greater were recorded.

Studies of Plasma Hsp72 and Trauma

It has been suggested that elevated serum Hsp72 is a generalized marker of stressed cells or tissues, and is an indicator of exposure of the organism to near lethal stress (Hecker and McGarvey, 2011). Several studies have observed high levels of plasma Hsp72 following trauma. Pittet et al (2002) showed that trauma patients (37% with a head injury) had serum Hsp72 levels as high as 150 ng/ml.

Da Rocha et al (2005) investigated males with severe traumatic brain injury, and found serum Hsp72 as high as 143 ng/ml soon after hospital admission. The 2002 study demonstrated that a high serum Hsp72 predicted a more favourable outcome for the patient, in contrast with the later study where those who died had a higher serum Hsp72 concentration on entry to the study. The two studies had patients with different ages and proportion of head injuries that might explain the difference between the results.

Table 1. Several studies have measured both body temperature and plasma Hsp72 concentration in patients with heatstroke. Markedly elevated levels of plasma Hsp72 were noted in several of these studies

Population	Number studied	Classic/ EHS	Plasma Hsp72 (ng/ml)	T_{rec} (°C)	Ref.
Older men and women	58	Classic - severe		>38	(Wang et al., 2001)
Young males	13	EHS	38 (21-57) (Immediately post race)	41 (39-42)	(Ruell et al., 2006)
Older men and women	18	Classic	287 (10-1371)	41.4 (39.1-42.4)	(Huisse et al., 2008)
Young baboons (Non-survivors)	6	Classic	145 (121-169) (3 hours post heating)	43.5 (43.3-43.8)	(Dehbi et al., 2010)

In another study, plasma Hsp72 was as high as 49.5 ng/ml in patients following liver resection who had post operative organ dysfunction (Kimura et al., 2004).

Origin of Plasma Hsp72

The presence of Hsp72 in serum is a relatively recent finding, and its function remains to be elucidated. Because Hsp72 is an intracellular protein, it is not known how the protein enters the serum nor which tissues release Hsp72 into the circulation. A role for extracellular Hsp72 has been proposed in initiating a cytokine response (Asea et al., 2002). The protein could be released by damaged cells and thus represent a danger signal in the cell. As such Hsp72 appears to act as a chemokine in the serum, initiating an inflammatory response (Dybdahl et al., 2002). In the case of heatstroke, it is possible that high levels of plasma Hsp72 initiate a pathological sequence, leading in some cases to multi-organ failure and death. The Hsp72 must be released from a cell or tissue, and this could be following cell necrosis, or active release. It has been shown in vitro that Hsp72 is released from peripheral blood mononuclear cells possibly via lipid rafts (Hunter-Lavin et al., 2004). Other studies suggest that Hsp72 is secreted in exosomes, small membrane-bound particles (30-100nm) that are released by several cell types such as macrophages and lymphocytes and are found in body fluids including serum and urine (Zhang and Grizzle, 2011). Exosomes contain a wide variety of molecules ranging from peptides to RNA molecules and lipids and they may be involved in signaling at sites distant to their formation. It is not clear whether stress increases the exosome secretory rate or increases the Hsp72 content of each exosome (Whitham and Fortes, 2008).

Other Heat Shock Proteins and Heatstroke

While Hsp72 has been the heat shock protein most commonly studied, more recent research indicates that other heat shock proteins may also have a role. Plasma Hsp60 was elevated in heatstroke patients (83 ng/ml) compared with healthy controls (3 ng/ml) (Huisse et al., 2008). This is in contrast to the baboon heatstroke model where plasma Hsp60 was not detected. In a group of young men with exertional heat illness, the antibody titre against Hsp60 was not higher than a control group, while the antibody titre against Hsp90 was significantly higher than a control group (Wu et al., 2001).

Conclusion

The findings in several studies of elevated plasma Hsp72 levels in hospitalized patients with classic heatstroke, runners with early signs of exertional heatstroke (high temperature with signs of neurological dysfunction) and a baboon heatstroke model suggest that a high plasma Hsp72 concentration is associated with serious illness and a poor prognosis. When simpler, less expensive methods of Hsp72 become available, this may form the basis for a useful test for distinguishing those patients most at risk of developing heatstroke. Plasma

Hsp72 appears to be a useful marker for diagnosis of heatstroke and further studies investigating the association between heatstroke and plasma Hsp72 are warranted.

References

Alzeer, A., Al Arifi, A., El-Hazmi, M., Warsy, A.S., and Nylen, E.S. (1996). Thermal regulatory dysfunction of growth hormone in classical heat stroke? *European Journal of Endocrinology*, 134, 727-730.

Asea, A., Kraeft, S.-K., Kurt-Jones, E., Stenvenson, M., Chen, L., Finberg, R., Koo, G., and Calderwood, S. (2000). HSP70 stimulates cytokine production through a CD14-dependant pathway, demonstrating its dual role as a chaperone and cytokine. *Nature Medicine*, 6, 435-442.

Asea, A., Rehli, M., Kabingu, E., Boch, J.A., Bare, O., Auron, P.E., Stevenson, M.A., and Calderwood, S.K. (2002). Novel signal transduction pathway utilized by extracellular HSP70. *The Journal of Biological Chemistry*, 277, 15028-15034.

Bouchama, A. (1995). Heatstroke: a new look at an ancient disease. *Intensive Care Medicine*, 21, 623-625.

Bouchama, A., and Knochel, J.P. (2002). Medical Progress: Heat Stroke. *New England Journal of Medicine*, 346, 1978-1988.

Bouchama, A., Parhar, R., ElYazigi, A., Sheth, K., and Al-Sedairy, S. (1991). Endotoxemia and release of tumour necrosis factor and interleukin 1a in acute heatstroke. *Journal of Applied Physiology*, 70, 2640-2644.

Da Rocha, A.B., Zanoni, C., De Freitas, G.R., Andre, C., Himelfarb, S., Schneider, R.F., Grivicich, I., Borges, L., Schwartsmann, G., Kaufmann, M., et al. (2005). Serum Hsp70 as an Early Predictor of Fatal Outcome after Severe Traumatic Brain Injury in Males. *Journal of Neurotrauma*, 22, 966-977.

Davido, A., Patzak, A., Dart, T., Sadier, M.P., Meraud, P., Masmoudi, R., Sembach, N., and Cao, T.H. (2006). Risk factors for heat related death during the August 2003 heat wave in Paris, France, in patients evaluated at the emergency department of Hôpital Européen Georges Pompidou. *Emergency Medical Journal*, 23, 515-518.

Dehbi, M., Baturcam, E., Eldali, A., Ahmed, M., Kwaasi, A., Chishti, M.A., and Bouchama, A. (2010). Hsp-72, a candidate prognostic indicator of heatstroke. *Cell Stress and Chaperones*, 15, 593-603.

DuBose, D.A., Wenger, C.B., Flinn, S.D., Judy, T.A., Dubovtsev, A.I., and Morehouse, D.H. (2003). Distribution and mitogen response of peripheral blood lymphocytes after exertional heat injury. *Journal of Applied Physiology*, 95, 2381-2389.

Dybdahl, B., Wahba, A., Lien, E., Trude, H., Waage, A., Qureshi, N., Sellevold, O., Espevik, T., and Sundan, A. (2002). Inflammatory response after open heart surgery: release of heat shock protein 70 and signalling through toll-like receptor 4. *Circulation,* 105, 685-690.

Febbraio, M.A., Ott, P., Nielsen, H.B., Steensberg, A., Keller, C., Krustrup, P., Secher, N.H., and Pedersen, B.K. (2002). Exercise induces hepatosplanchnic release of heat shock protein 72 in humans. *Journal of Physiology,* 544.3, 957-962.

Hecker, J.G., and McGarvey, M. (2011). Heat shock proteins as biomarkers for the rapid detection of brain and spinal cord ischemia: a review and comparison to other methods of detection in thoracic aneurysm repair. *Cell Stress and Chaperones*, 16, 119-131.

Huisse, M., Pease, S., Hurtado-Nedelec, M., Arnaud, B., Malaquin, C., Wolff, M., Gougerot-Pocidalo, M., Kermarrec, N., and Chollet-Martin, S. (2008). Leukocyte activation: The link between inflammation and coagulation during heatstroke. A study of patients during the 2003 heat wave in Paris. *Critical Care Medicine*, 36, 2288-2295.

Hunter-Lavin, C., Davies, E.L., Bacelar, M.F.V.G., Marshall, M.J., Andrew, S.M., and Williams, J.H.H. (2004). Hsp70 release from peripheral blood mononuclear cells. *Biochemical and Biophysical Research Communications*, 324, 511-517.

Kiang, J.G., and Tsokos, G.C. (1998). Heat shock protein 70 kDa: molecular biology, biochemistry, and physiology. *Pharmacology and Therapeutics*, 80, 183-201.

Kimura, M., Itoh, H., Ambiru, S., Shimizu, H., Togawa, A., Yoshidome, H., Ohtsuka, M., Shimamura, F., Kato, A., Nukui, Y., et al. (2004). Circulating heat-shock protein 70 is associated with postoperative infection and organ dysfunction after liver resection. *American Journal of Surgery*, 187, 777-784.

Knochel, J.P., and Caskey, J.H. (1977). The mechanism of hypophosphatemia in acute heat stroke. *JAMA,* 238, 425-426.

Leithead, C., S., and Lind, A.R. (1964). *Heat stress and heat disorders.* (London, Cassell).

Leon, L.R., and Helwig, B.G. (2010). Heat stroke: role of the systemic inflammatory response. *Journal of Applied Physiology*, 109, 1980-1988.

Moreau, T.P., and Deeter, M. (2005). Heatstroke-predictable, preventable, treatable. *JAAPA,* 18, 30-35.

Ogura, Y., Naito, H., Akin, S., Ichinoseki-Sekine, N., Kurosaka, M., Kakigi, R., Sugiura, T., Powers, S.K., Katamoto, s., and Demirel, H.A. (2008). Elevation of body temperature is an essential factor for exercise-increased extracellular heat shock protein 72 level in rat plasma. *Am. J. Physiol. Regul. Integr. Comp. Physiol*, 294, R1699-R1607.

Ostberg, J.R., Kaplan, K.C., and Repasky, E.A. (2002). Induction of stress proteins in a panel of mouse tissues by fever-range whole body hyperthermia. *International Journal of Hyperthermia*, 18, 552-562.

Pittet, J.-F., Lee, H., Morabito, D., Howard, M.B., Welch, W.J., and Mackersie, R.C. (2002). Serum levels of Hsp72 measured early after trauma correlate with survival. *Journal of Trauma*, 52, 611-617.

Richards, D., Richards, R., Schofield, P.J., and Ross, V. (1979). Biochemical and haematological changes in Sydney's the Sun City-to-Surf fun runners. *Medical Journal of Australia*, 2, 449-453.

Rowell, L.R., brengelmann, G.L., Blackmon, J.R., Twiss, R.D., and Kusumi, F. (1968). Splanchnic blood flow and metabolism in heat-stressed man. *Journal of Applied Physiology*, 24, 475-484.

Ruell, P., Thompson, M., Chapman, P., and Hoffman, K. (2009). Passive heating at high core temperature does not increase plasma HSP72. *Paper presented at: 14th Annual Congress ECSS* (Oslo, Norway).

Ruell, P.A., Thompson, M.W., Hoffman, K.M., Brotherhood, J.R., and Richards, D.A. (2006). Plasma Hsp72 is higher in runners with more serious symptoms of exertional heat illness. *European Journal of Applied Physiology*, 97, 732-736.

Semenza, J.C., Rubin, C.H., Falter, K.H., Selanikio, J.D., Flanders, W.D., Howe, H.L., and Wilhelm, J.L. (1996). Heat-related deaths during the July 1995 heat wave in Chicago. *The New England Journal of Medicine*, 335, 84-90.

Sutton, J., Coleman, M.J., Millar, A.P., Lazarus, L., and Russo, P. (1972). The medical problems of mass participation in athletic competition. The "City-to-Surf" race. *Medical Journal of Australia*, 2, 127-133.

Walsh, R.C., Koukoulas, I., Garnham, A., Moseley, P.L., Hargreaves, M., and Febbraio, M.A. (2001). Exercise increases serum Hsp72 in humans. *Cell Stress and Chaperones*, 6, 386-393.

Wang, Z.-Z., Wang, C.-L., Wu, T.-C., Pan, H.-N., Wang, S.-K., and Jiang, J.-D. (2001). Autoantibody response to heat shock protein 70 in patients with heatstroke. *American Journal of Medicine*, 111, 654-657.

Whitham, M., and Fortes, M.B. (2006). Effect of blood handling on extracellular Hsp72 concentration after high intensity exercise in humans. *Cell Stress and Chaperones*, 11, 304-308.

Whitham, M., and Fortes, M.B. (2008). Heat shock protein 72: release and biological significance during exercise. *Frontiers in Bioscience*, 13, 1328-1339.

Whitham, M., Laing, S.J., Jackson, A., Maassen, N., and Walsh, N.P. (2007). Effect of exercise with and without a thermal clamp on the plasma heat shock protein 72 response. *Journal of Applied Physiology*, 103, 1251-1256.

Wu, T., Chen, S., Xiao, C., Wang, C.-Y., Pan, Q., Wang, Z., Xie, M., Mao, Z., Wu, Y.Y., and Tanguay, R.M. (2001). Presence of antibody against hsp71 in patients with acute heat-induced illness. *Cell Stress and Chaperones*, 6, 113-120.

Yamada, P., Amorim, F., Moseley, P., and Schneider, S. (2008). Heat Shock Protein 72 Response to Exercise in Humans. *Sports Medicine*, 38, 715-733.

Zhang, H.-G., and Grizzle, W.E. (2011). Exosomes and cancer: a newly described pathway of immune suppression. *Clinical Cancer Research*, 17, 969-964.

In: Hyperthermia: Recognition, Prevention and Treatment ISBN 978-1-61942-275-9
Editor: Xing-Hua Gao and Hong-Duo Chen, 209-214

Chapter 7

HYPERTHERMIA: PATHOLOGICAL FINDINGS AND RECOGNITION AT FORENSIC AUTOPSIES

Yasuo Bunai, Kayoko Akaza and Atsushi Nagai
Department of Legal Medicine, Graduate School of Medicine,
Gifu University, Gifu, Japan

ABSTRACT

Heatstroke is a thermal insult induced by an imbalance between accumulation of heat in the body and heat loss from the body. The insult leads to central nervous system disturbances ranging from confusion to deep coma. A multisystem insult follows and leads to a fatal outcome. Heatstroke is associated with certain clinical complications. However, the postmortem findings of heatstroke, including the histopathological findings, are nonspecific, and some fatal heatstroke cases are undoubtedly misdiagnosed as ischemic heart disease at autopsy. If rectal temperature is not measured on admission or there is a long interval between collapse and admission, clinicians are not reminded of the possibility of heatstroke, and pathologists do not receive important information. On the other hand, right-sided cardiac dilation, subendocardial hemorrhage in the left ventricle, severe pulmonary congestion, and venous distention, particularly in the cerebrum, are typical pathological findings associated with heatstroke and remind pathologists of the possibility of the diagnosis. Moreover, the early occurrence and rapid development of postmortem changes brought on by high body temperature at the time of death are also typical findings after death in a state of hyperthermia.

INTRODUCTION

A hot environment or working and exercising in a hot environment would result in hyperthermia, and they often result in heatstroke. Heatstroke is characterized by an elevated core body temperature and central nervous system dysfunction, and becomes complicated by multi-organ failure, endotoxinemia, and increased peripheral blood cytokine levels [1]. The hyperthermic patient (rectal temperature 41^0C to 43^0C) is in grave danger of death [2]. The mortality rate is particularly high, and will be increased by global warming [2,3]. Heatstroke can occur at any age, but is more common in susceptible persons, i.e., alcoholics, the elderly, including elderly persons in institutional care, with cardiovascular diseases or cognitive impairment, and taking diuretics or multiple medications, and psychiatric patients taking antipsychotic agents [4]. In forensic practice, heatstroke victims dead during excited delirium after taking pharmacologic agents, such as cocaine and amphetamine-like agents, are also subjects [5-7]. Infants and children are also at risk and they would be the neglected victims in forensic autopsy, especially if left in closed cars in direct sunlight.

Hyperthermia and Its Complications

Hyperthermia is characterized by a normothermic setting of the thermoregulatory center in conjunction with an uncontrolled increase in body temperature that exceeds the body's ability to dissipate heat. Exogenous heat exposure and endogenous heat production are mechanisms that lead to dangerously high body temperatures.

Since the physiologic disturbance and management of heat illness are different, it must be diagnosed correctly [2,3]. Heat cramps are associated with acute salt loss and are treated by electrolyte replacement. Heat exhaustion and heat prostration, which are manifested by progressive lassitude and inability to work, are followed by severe headache, vomiting, tachycardia, and hypotension. Water depletion and hypovolemia are accompanied by hemoconcentration, hypernatremia, and hyperchloremia, and the treatment is rehydration and rest. Heatstroke (sunstroke) is an acute medical emergency and a life-threatening illness characterized by hyperthermia and central nervous system symptoms, and its clinical manifestations are a body temperature above 40°C, delirium, and convulsions or coma. According to whether heatstroke is associated with continuous or strenuous physical labor, it is divided into a classic type and an exertional type [2-4]. The classic type epidemically occurs in a high-temperature environment and occasionally occurs in very young or elderly individuals with poor ability to adapt to heat stress. Exertional heatstroke (EHS) is a state of extreme hyperthermia that occurs when excess heat generated as a result of physical exercise exceeds the body's ability to dissipate it. In contrast to the heatstroke that occurs in the elderly during heat waves [2-4,8], EHS occurs sporadically in young and very physically active populations, even in comfortable climates, and its onset is sudden. EHS has been reported in military recruits, marathon runners, and other athletes, and in miners and other industrial workers performing strenuous jobs in a hot environment [2-4,9-11]. EHS is characterized by the triad of hyperthermia, altered cerebral function, and hypohidrosis [2,3,11]. The rectal temperature of EHS patients ranges from 40.0°C to 43.0°C, but may be decreased by the time required to transport the patient from the scene to the hospital [9]. Patients with a rectal temperature of 41°C to 43°C are considered in grave danger of death. In most cases, a single predisposing factor, e.g., low physical fitness level, lack of acclimation, overweight, dehydration, scarred skin, alcohol, or drugs, reduces the ability of a physically active individual to sustain heat, and in combination with other factors which can increase the risk of EHS.

The most serious complication of heatstroke is multi-organ dysfunction, including cerebropathy, rhabdomyolysis, acute renal failure, acute respiratory distress syndrome, myocardial injury, hepatic injury, pancreatic injury, hematopenia, hemorrhagic diseases, especially diffuse intravascular coagulation (DIC) [2,3]. Heatstroke is often fatal despite adequate therapeutic measures for hyperthermia, and survivors have permanent nervous dysfunction. Heatstroke mortality of 10% to 50% has been reported [3,4]. According to the United States Centers for Disease Control and Prevention, 4780 deaths were attributed to heat injury in the United States during the period from 1979 to 2002. The elderly, the bedridden, patients taking anticholinergic drugs, antiparkinsonian drugs, or diuretics, and persons confined to a poorly ventilated environment without air conditioning are most susceptible to death from heat injury. Moreover, the results of a survey indicated that weekly mortality from heatstroke in the army might be 1.5 – 5.0/100,000 [12].

Malignant hyperthermia occurs in individuals with an inherited sarcoplasmic reticulum abnormality that causes a rapid increase in intracellular calcium levels after inhalation of halothane and other anesthetics or with succinylcholine. Hyperthermia, increased muscle metabolism, muscle rigidity, rhabdomyolysis, acidosis, cardiovascular instability developed rapidly. The neuroleptic malignant syndrome, serotonin syndrome, anticholinergic poisoning, sympathomimetic poisoning, and excited delirium also cause hyperthermia and are sometimes fatal [7,12].

Pathogenesis of Heatstroke

Heat stress induces heat acclimation, an acute phase response, and a heat stress response involved with the production of heat stress protein. If heat stress exceeds the ability of the body to compensate, central venous pressure decreases, and a significant increase in core body temperature occurs. They lead to thermoregulatory and circulatory failure and to an increase in inflammatory cytokines (TNF-α, IL-1 and IL-6) that in turn leads to endotoxinemia and worsening of the circulatory failure. These changes activate the endothelium, which releases vasoactive factors, such as nitrogen oxide and endothelins. The interactions between cytokines, endothelial activate factors, and intracellular ion imbalances lead to a malignant cycle and cause high body temperature, central nervous system abnormalities, dysfunction of hemagglutination and organs [1,3,4,13]. In addition, aging, lack of heat acclimation, and genetic polymorphisms may lead to low expression of heat shock proteins, weakening of the heat-stress response, and severer oxidative injury during heat exposure.

The effects of heat stress on cellular function have been reviewed elsewhere [1,4,13]. Briefly, they include 1) inhibition of DNA synthesis, and transcription, and of RNA splicing and translation; 2) inhibition of cell cycle processes; 3) protein denaturation and misaggregation; 4) increased protein degradation via proteasomal and lysosomal pathways; 5) disruption of cytoskeletal components; 6) metabolic alterations that decreases intracellular ATP production; 7) alterations of membrane permeability that increase intracellular Na+, H+, and Ca+. In mammalian cells, heat shock causes changes in gene expression and in the activity of expressed proteins. This stress response includes an increase in thermotolerance that enables cells to survive subsequent, severer heat stress that is temporarily associated with increased expression of heat shock proteins. During increasingly severe heat exposure, heat shock sets off cells from survive to apoptosis and, in the extreme, to cellular necrosis.

Additionally, cellular damage and necrosis might be in part due to reduction of blood flow and systemic factor.

COMPLICATIONS AND PATHOLOGICAL FINDINGS IN HYPERTHERMIA

Central nervous system: Hyperthermia is always associated with central nervous system disturbances, and their severity is usually directly related to the duration of the hyperthermia and to circulatory failure [2,9,14,15]. Neurological manifestations of hyperthermia include slurred speech, delirium, stupor, lethargy, coma, and seizures. In the majority of cases of heatstroke, the initial clinical manifestation is loss of consciousness. Coma that persists after a return to normothermia is a poor prognostic sign, whereas coma that persists for 24 h and is accompanied by seizures may be followed by a recovery without evident mental or neurological disturbances [2]. Seizures occur in approximately 60% of heatstroke patients and are more common at body temperatures above 41^0C. Cerebral impairment may persist after recovery.

The pathological picture of heatstroke is characterized by brain edema or congestion [2,16]. If the patient dies soon, only minimal pathological changes are observed. If the patient survives for a few days, examination of the brain reveals diffuse swelling and petechial hemorrhages that are widely distributed but particularly common around the third and fourth ventricles. Subarachnoid and subdural hemorrhages have also been described. Neurocyte swelling and shrinkage are observed throughout the brain, but especially in the cerebellum, where the Purkinje cells are often lost. The upper layer of the cerebral cortex is more affected than the lower layer. The striatum, thalamus, and pallidum show changes that are similar to the changes in the cortex, but are less severe [16,17].

Cardiovascular system: Disturbances of the cardiovascular system, including sinus tachycardia, increased cardiac output, and cardiac failure, are also observed. The increase in cardiac output occurs in response to the increase in metabolic rate and oxygen demand, and to the low peripheral resistance due to cutaneous vasodilation [2,9]. Heart failure and insufficient cardiac output are caused by direct myocardial injury and an increase in pulmonary resistance and not by the combined effect of dehydration and blood pooling as a result of vasodilation [2,9]. Pathological changes in the heart consist of subendocardial hemorrhages in the left ventricle and focal necrosis of cardiac muscle fibers [2,9,16]. Myocardial infarction due to hypoperfusion may be observed in patients with underlying coronary diseases.

Respiratory system: Hyperventilation is present in most cases of heatstroke, and, if prolonged, may lead to respiratory alkalosis and tetany. Adult respiratory distress syndrome has also been identified in heatstroke victims and has been linked to the development of disseminated intravascular coagulation (DIC). Pulmonary aspiration is a common complication of coma and convulsions [9]. Pathologically, severe, acute pulmonary edema in the absence of underlying heart disease has been observed in heatstroke victims. Intra-alveolar hemorrhages and pulmonary infarction are also found in autopsy cases [9].

Renal system: Acute renal failure is an uncommon complication of classic heatstroke but a common complication of exertional heatstroke. Hypovolemia caused by shunting of blood to the periphery and profuse sweating precipitates tubular necrosis. Direct thermal injury may lead to renal damage. DIC contributes to renal failure. Myoglobinuria secondary to rhabdomyolysis damages renal tubules further [2,3,9]. Myoglobinuria can be revealed by immunohistochemically detected myoglobin casts at autopsy.

Hepatobiliary system: Liver damage and dysfunction develop in most heatstroke patients. Serum bilirubin and transaminase levels are found to be elevated 12-24 h after the onset of heatstroke and to peak within 18-36 h. Hepatic central lobular necrosis and extensive cholestasis has been observed in autopsy cases [18].

Hemostasis: A hemorrhagic diathesis is frequently observed in heatstroke patients and is manifested by purpura, conjunctival petechiae, bloody diarrhea, hemoptysis, hematuria, and thrombocytopenia. Its etiology is multifactorial and may be related to thermal injury to the vascular endothelium inducing increased platelet aggregation and to direct thermal deactivation of plasma proteins that depletes clotting factors. DIC is a rare complication of classic heatstroke, but is a common complication of exertional heatstroke and is associated with high mortality. DIC occurs secondary to a combination of decreased clotting factor production and activity and increased consumption of intravascular clotting factors [2,3,9,19].

Heatstroke victims also reveal petechial hemorrhages in the thymus, lung, and heart at autopsy. If there is severe dehydration, the blood is viscous, and the hematocrit is high at autopsy.

Practical Problems in Determining Whether the Cause of Death Was Hyperthermia at Autopsy

Some clinical and pathological manifestations, including in the hepatobiliary system and hemostasis, peak 12 hours after the onset of heatstroke, and thus may not be observed in patients who die shortly after its onset [9,16]. The postmortem findings in heatstroke, including the histopathological findings, are nonspecific, and some fatal heatstroke cases are undoubtedly misdiagnosed as ischemic heart disease [10]. If rectal temperature is not measured on admission or there is a long interval between collapse and admission, clinicians are not reminded of the possibility of heatstroke, and pathologists do not receive important information. On the other hand, right-sided cardiac dilation, subendocardial hemorrhage in the left ventricle, severe pulmonary congestion, and venous distention, particularly in the cerebrum, are typical pathological findings that are correlated with heatstroke and observed in heatstroke victims who die shortly after the onset [9]. These changes are also observed in deaths from other causes, but they are closely correlated with the circulatory failure caused by the hyperthermia associated with heatstroke, and remind pathologists of the possibility of the diagnosis. Moreover, postmortem changes develop rapidly after death at high environmental temperatures or when body temperature is high at the time of death. Thus, the early occurrence and rapid development of postmortem changes are the first clue at autopsy that reminds pathologists of the possibility of death associated from hyperthermia [20]. The diagnosis of hyperthermia at autopsy should be made on the basis of the autopsy findings, reconstruction of the scene of death, and investigation of environmental factors at death, as well as clinical reports, if available.

References

[1] Yan YE, Zhao YQ, Wang H, Fan M. Pathophysiological factors underling heat stroke. *Med. Hypotheses* 2006; 67(3): 609-617.

[2] Clowes GHA, O'Donnell TF. Heat stroke. *N. Engl. J. Med.* 1974; 291(11): 564-567.

[3] Bouchama A, Knochel JP. Heatstroke. *N. Engl. J. Med.* 2002; 346 (25): 1978-1988.

[4] Yaqub B, Deeb AL. Heat strokes: Aetiopathogenesis, neurological characteristics, treatment and outcome. *J. Neurol. Sci.* 1998; 156(2): 144-151.

[5] Wetli AV, Mash D, Karch SB. Cocaine-associated agitated delirium and the neuroleptic malignant syndrome. *Am. J. Emerg. Med.* 1996; 14(4): 425-428.

[6] Staley JK, Rothman RB, Rice KC, Partilla J, Mash DC. Kappa2 opioid receptors in limbic areas of human brain are up regulated by cocaine in fatal overdose victims. *J. Neurosci.* 1997; 17(21): 5228-5233.

[7] Bunai Y, Akaza K, Nakamura I, Yamanouchi H, Fukunaga T, Kubo S, Ohya I. Pulmonary changes associated with methamphetamine poisoning. *Acta Crim. Japon* 2002; 68(5): 141-146.

[8] Austin MG, Berry JW. Observations on one hundred cases of heatstroke. *JAMA* 1956; 161(16): 1525-1529.

[9] Shapiro Y, Seidman DS. Field and clinical observations of exertional heat stroke. *Med. Sci. Sports Exerc.* 1990; 22(1): 6-14.

[10] Poter AM. Heat illness and soldiers. *Mil. Med.* 1993; 158(9): 606-609.

[11] Gardner JW, Kark JA, Karnei K, Sanborn JS, Gastaldo E, Burr P, Wenger B. Risk factors predicting exertional heat illness in male Marine Corps recruits. *Med. Sci. Sports Exer* 1996; 28(8): 939-944.

[12] Epstein Y, Hadad E, Shapiro Y. Pathological factors underlying hyperthermia. *J. Thermal. Biol.* 2004; 29(7,8): 487-494.

[13] Sonna LA, Fujita J, Gaffin SL, Lilly CM. Effects of heat and cold stress on mammalian gene expression. *J. Appl. Physiol.* 2002; 92(4): 1725-1742.

[14] O'Donnell TF, Clowers GHA. The circulatory abnormalities of heat stroke. *N. Engl. J. Med.* 1972; 287(15): 734-737.

[15] Romero JJ, Clement PF, Belden C. Neuropsychological sequelae of heat stroke. report of three cases and discussion. *Mil. Med.* 2000; 165(6): 500-503.

[16] Gore I, Issacson NH. The pathology of hyperpyrexia. observations at autopsy in 17 cases of fever therapy. *Am. J. Pathol.* 1946; 25(6): 1029-1059.

[17] Davis RL Robertson DM. *Textbook of Neuropathology.* Baltimore, London, Sydney: Williams and Wilkins; 1985.

[18] Kubo S, Orihara Y, Kitamura O, Ikematsu K, Tsuda R, Nakasono I. An autopsy case of neuroleptic malignant syndrome (NMS) and its immunehistochemical findings of muscle-associated proteins and mitochondria. *Forensic Sci. Int* 2001; 115(1-2): 155-158.

[19] O'Donnell TF. Acute heat stroke. Epidemic, biochemical, renal, and coagulation studies. *JAMA* 1975; 234(8): 824-828.

[20] Bunai y, Akaza K, Jiang WX, Nagai A. Fatal hyperthermia associated excited delirium during arrest. *Leg. Med.* 2008; 10(6): 306-309.

In: Hyperthermia: Recognition, Prevention and Treatment ISBN 978-1-61942-275-9
Editor: Xing-Hua Gao and Hong-Duo Chen, 215-221

Chapter 8

HYPERTHERMIA: THE REVEALING SIGN IN PYODERMA GANGRENOSUM

Benoit Ayestaray
Department of Plastic, Reconstructive and Aesthetic Surgery.
Nîmes University Hospital, pl Professor Robert Debré,
Nîmes, France

ABSTRACT

Pyoderma gangrenosum is the most common neutrophilic dermatosis. It can appear under different forms : ulcerative, pustular, bullous, vegetative or necrotizing. Hyperthermia is usually the first revealing sign, in all these variants, before the skin inflammation occurs. Pyoderma gangrenosum may be mistaken for a skin infection. Despite antibiotics associated with a surgical excision, hyperthermia and skin inflammation increase. The only effective treatments are corticoids and immunomodulators.

INTRODUCTION

Pyoderma gangrenosum is a neutrophilic dermatosis, of unknown aetiology, described by Brocq in 1908 [1,2] and Brunsting in 1930 [3]. The family of neutrophilic dermatosis is characterised histologically by a neutrophilic and lymphocytic infiltrate, without vasculitis or infectious cause. The physiopathology implies an anormal T oligoclonal cellular expansion [4], responding to an antigenic stimulus, with synthesis of proinflammatory cytokines such as interleukin-8 and tumor necrosis factor-α [5]. No infectious agent has been found in the genesis of pyoderma gangrenosum.

Different forms of pyoderma gangrenosum are described according to their clinical signs and the associated diseases [6-21]:

- Ulcerative Pyoderma Gangrenosum begins by small papules evolving to deep worn-edged ulcers, and often occurs after a minor trauma (pathergy development).
- Pustular Pyoderma Gangrenosum is a superficial variant, associated with inflammatory bowel disease.
- Bullous Pyoderma Gangrenosum is another superficial variant, associated with malignant haemopathies.
- Vegetative Pyoderma Gangrenosum is the least aggressive variant, responding to local treatment.
- Necrotizing Pyoderma Gangrenosum is very rare, and characterised by a spreading necrosis on an inflammatory skin.

In all these variants, the first sign to occur is hyperthermia, without any skin inflammation.

CLINIC

Hyperthermia is usually superior to 39°C , without variation. After 24 to 48 hours of evolution, an erythematous skin lesion appears. Thus, it is not rare that a beginning pyoderma gangrenosum can be mistaken for a skin infection, in particularly a cellulitis.

In this case, antibiotics are ineffective. Hyperthermia increases despite a high dose of antibiotics. If the variant is necrotizing, the disease can be mistaken for a necrotizing fasciitis. In this case, surgical excision is not efficient. Hyperthermia will get worst after surgery, because of a pathergy phenomenon (soft tissue inflammation worsened by surgical trauma) [22-25].

BIOLOGY

Biologically, inflammatory syndrome is high : leucocytosis >20000/mm3, CRP > 200 mg/l. Bacteriological samples remain negative. Histological examination finds a neutrophilic infiltrate, frequently associated with a hypodermitis and pyodermitis, without any vasculitis.

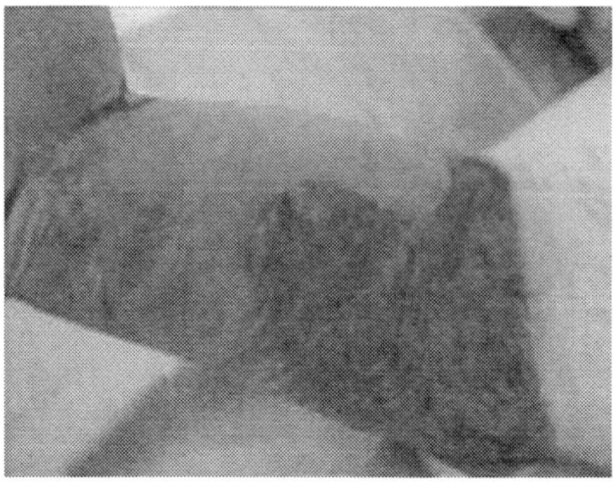

Figure 1. Necrotizing variant of Pyoderma Gangrenosum. A necrotic area is spreading on the inflammatory skin (right arm).

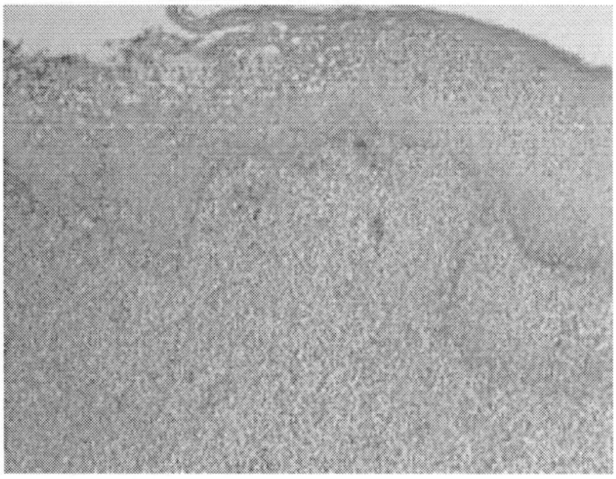

Figure 2. dermic infiltrate of undebased polynuclears ; collection of epidermic abcess coalescing in hypopion (HE coloration, x 100 magnification).

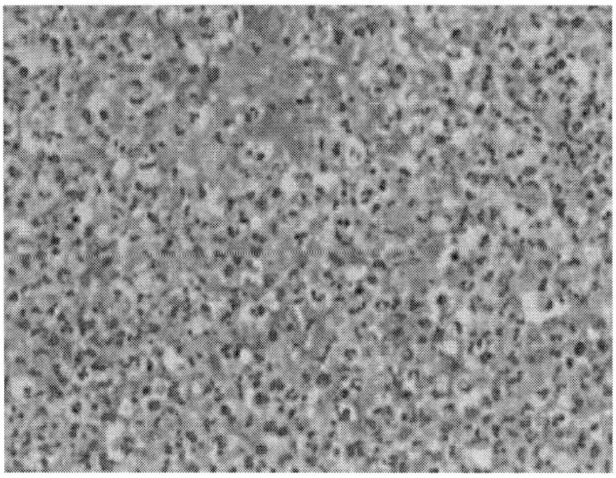

Figure 3. infiltrate of neutrophilic polynuclears (HE coloration, x 200 magnification).

Differential Diagnosis

Pyoderma gangrenosum can hardly be differentiated clinically and biologically from a dermo-hypodermal infection. In these two different diseases, symptomatology is actually dominated by inflammatory aspect of the skin, pain and hyperthermia [26-29]. If hyperthermia increases over 41°C, a myelogram should be performed to look for a lymphoid proliferation. Rarely, a leukemia can overcome [30,31].

Treatment

The only treatment, which can be effective on the hyperthermia secondary to Pyoderma Gangrenosum, are corticoids, immunomodulators [32-49], immunoglobulins [50,51] and other non invasive treatments [52-59]. Usually, a treatment by corticoids per os, started at the dose of 1mg/kg/24h, is sufficient. A clear improvement of hyperthermia occurs after 48 h. If corticoids are not recommended or tolerated, anti-TNF alpha (Tumor Necrosis Factor alpha inhibitors) or intravenous immunoglobulins, must be used.

Conclusion

Hyperthermia is the first revealing sign in Pyoderma Gangrenosum. Prudence is required in view of hyperthermia associated with inflammatory skin lesions, worsening under antibiotics. Corticoids and immunomodulators are the only effective treatment on hyperthermia secondary to Pyodermia Gangrenosum.

References

[1] Farhi D, Wallach D, Avril MF. Pyoderma gangrenosum is 100 years old. From Louis Brocq to biotherapies. *Rev Prat.* 2008 Feb 29;58(4):457-61.

[2] Farhi D. The clinical and histopathological description of geometric phagedenism (pyoderma gangrenosum) by Louis Brocq one century ago. *Arch. Dermatol.* 2008 Jun;144(6):755.

[3] Brunsting HA, Goeckerman WH, O'Leary PA. Pyoderma gangrenosum: clinical and experimental observations in five cases occurring in adults. *Arch. Dermatol.* 1930 22, pp. 665-680.

[4] Brooklyn TN, Williams AM, Dunnill MG, Probert CS. T-cell receptor repertoire in pyoderma gangrenosum: evidence for clonal expansions and trafficking. *Br. J. Dermatol.* 2007 Nov;157(5):960-6.

[5] Oka M. Pyoderma gangrenosum and interleukin 8.*Br. J. Dermatol.* 2007 Dec;157(6):1279-81.

[6] Wustrack KO, Zarem HA. Pyoderma gangrenosum: recognition and management. *Plast Reconstr Surg.* 1978 Sep;62(3):423-8.

[7] Powell FC, Su D, Perry HO. Pyoderma gangrenosum : classification and management. *J. Am. Ac. Dermatol.* 1996 34(3) : 395-409.

[8] Ahmadi S, Powell FC. Pyoderma gangrenosum: uncommon presentations. *Clin. Dermatol.* 2005 Nov-Dec;23(6):612-20.

[9] [9] Brooklyn T, Dunnill G, Probert C. Diagnosis and treatment of pyoderma gangrenosum.*BMJ.* 2006 Jul 22;333(7560):181-4.

[10] Newell EL, Chaudhry SI, Black MM. Acute bullous hemorrhagic pyoderma gangrenosum: pitfalls of diagnostic delay. A case report. *J. Bone Joint Surg Am.* 2008 Jan;90(1):174-7.

[11] Chia MW, Teo L, Tay YK, Poh WT. Pustular pyoderma gangrenosum: an uncommon variant which is easily misdiagnosed. *Dermatol Online J.* 2008 Feb 28;14(2):21.

[12] Spear M. Pyoderma gangrenosum: an overview. *Plast Surg. Nurs.* 2008 Jul-Sep;28(3):154-7.

[13] Carron PN, Yerly S, Ksontini R, et al. Pyoderma gangrenosum: diagnostic and therapeutic challenge. *Rev. Med. Suisse.* 2008 Sep 10;4(170):1938-40, 1942-3.

[14] Cohen PR. Neutrophilic dermatoses: a review of current treatment options. *Am. J. Clin. Dermatol.* 2009;10(5):301-12.

[15] Berbert AL, Mantese SA, Rocha A, et al. Superficial granulomatous pyoderma: report of a case of an uncommon variant of pyoderma gangrenosum. *An. Bras. Dermatol.* 2009 Jul;84(3):285-8.

[16] Mahajan AL, Ajmal N, Barry J, et al. Could your case of necrotising fascitis be Pyoderma gangrenosum? *Br. J. Plast Surg.* 2005 Apr;58(3):409-12.

[17] Heermann R, Kiehl P, Issing PR, Lenarz T. Pyoderma gangraenosum. Case report and comparison with necrotizing fasciitis. HNO. 2002 Mar;50(3):244-7.

[18] Meier F, Berner D, Scherwitz C, et al. An unusual case of pyoderma gangrenosum with necrotizing granulomatous dermatitis. *J. Dtsch Dermatol. Ges.* 2003 Apr;1(4):302-5.

[19] Berry MG, Tavakkolizadeh A, Sommerlad BC. Necrotizing ulceration after breast reduction. *J. R Soc. Med.* 2003 Apr;96(4):186-7.

[20] Barr KL, Chhatwal HK, Wesson SK, et al. Pyoderma gangrenosum masquerading as necrotizing fasciitis. *Am. J. Otolaryngol.* 2009 Jul-Aug;30(4):273-6.

[21] Ayestaray B, Dudrap E, Chartaux E, Verdier E, Joly P. Necrotizing pyoderma gangrenosum : an unusual differential diagnosis of necrotizing fasciitis. *J. Plast Reconstr. Aesthet Surg.* 2010 Aug;63(8):e655-8.

[22] Fulbright RK, Wolf JE, Tschen JA. Pyoderma gangrenosum at surgery sites. *J. Dermatol. Surg Oncol.* 1985 Sep;11(9):883-6.

[23] Ouazzani A, Berthe JV, de Fontaine S. Post-surgical pyoderma gangrenosum: a clinical entity. *Acta Chir. Belg.* 2007 Jul-Aug;107(4):424-8.

[24] Tripathi AK, Erdmann MW. Bird-bite infection and pyoderma gangrenosum: a rare combination? *J. Plast Reconstr. Aesthet Surg.* 2008 Nov;61(11):1409-11.

[25] Lindberg-Larsen R, Fogh K. Traumatic pyoderma gangrenosum of the face: pathergy development after bike accident. *Dermatology.* 2009;218(3):272-4.

[26] Kühn C, Vente C, Dörner J, et al. Pyoderma gangrenosum: and important differential diagnosis from wound infection. Case report of a life threatening course. Anaesthesist. 2000 Sep;49(9):829-33.

[27] Torzecka JD, Kondras K, Dziankowska-Bartkowiak B, et al. Pyoderma gangrenosum--an interdisciplinary diagnostic problem *Pol. Merkur Lekarski.* 2008 Jan;24(139):34-7.

[28] Lê Huu S, Spertini F, Roggero P, et al. Pyoderma gangrenosum: a rare pathology or an omitted diagnosis? *Ann. Chir. Plast. Esthet.* 2009 Feb;54(1):82-7.

[29] Patel BB, Holland NW Atypical pyoderma gangrenosum presenting as paronychia. *Am. J. Med.* 2009 Jul;122(7):e5-6.

[30] Ho KK, Otridge BW, Vandenberg E, Powell FC. Pyoderma gangrenosum, polycythemia rubra vera, and the development of leukemia. *J. Am. Acad. Dermatol.* 1992 Nov;27(5 Pt 2):804-8.

[31] Bertram-Callens A, Machet L, Vaillant L, et al. Buccal and ocular localizations of pyoderma gangrenosum in Vaquez's disease. *Ann. Dermatol Venereol.* 1991;118(9):611-4.

[32] Gupta AK, Skinner AR. A review of the use of infliximab to manage cutaneous dermatoses. *J. Cutan. Med. Surg.* 2004 Mar-Apr;8(2):77-89. Epub 2004 May 4.

[33] Trent JT, Kerdel FA. Tumor necrosis factor alpha inhibitors for the treatment of dermatologic diseases. *Dermatol Nurs.* 2005 Apr;17(2):97-107.

[34] Roy DB, Conte ET, Cohen DJ. The treatment of pyoderma gangrenosum using etanercept. *J. Am. Acad. Dermatol.* 2006 Mar;54(3 Suppl 2):S128-34.

[35] Brooklyn TN, Dunnill MG, Shetty A, et al. Infliximab for the treatment of pyoderma gangrenosum: a randomised, double blind, placebo controlled trial. *Gut.* 2006 Apr;55(4):505-9.

[36] Fonder MA, Cummins DL, Ehst BD, et al. Adalimumab therapy for recalcitrant pyoderma gangrenosum. *J. Burns Wounds.* 2006 Nov 20;5:e8.

[37] Heffernan MP, Anadkat MJ, Smith DI. Adalimumab treatment for pyoderma gangrenosum. *Arch. Dermatol.* 2007 Mar;143(3):306-8.

[38] Suchak R, Macedo C, Glover M, Lawlor F. Intravenous immunoglobulin is effective as a sole immunomodulatory agent in pyoderma gangrenosum unresponsive to systemic corticosteroids. *Clin. Ex. Dermatol.* 2007 Mar;32(2):205-7.

[39] Hewitt D, Tait C. Use of infliximab in pyoderma gangrenosum. *Australas J. Dermatol.* 2007 May;48(2):95-8.

[40] Dini V, Romanelli M, Bertone M, et al. Improvement of idiopathic pyoderma gangrenosum during treatment with anti-tumor necrosis factor alfa monoclonal antibody. *Int. J. Low Extrem. Wounds.* 2007 Jun;6(2):108-13.

[41] Charles CA, Leon A, Banta MR, Kirsner RS. Etanercept for the treatment of refractory pyoderma gangrenosum: a brief series. *Int. J. Dermatol.* 2007 Oct;46(10):1095-9.

[42] Pomerantz RG, Husni ME, Mody E, Qureshi AA. Adalimumab for treatment of pyoderma gangrenosum. *Br. J. Dermatol.* 2007 Dec;157(6):1274-5.

[43] Jacob SE, Weisman RS, Kerdel FA. Pyoderma gangrenosum--rebel without a cure? *Int. J. Dermatol.* 2008 Feb;47(2):192-4.

[44] Adişen E, Oztaş M, Gürer MA. Treatment of idiopathic pyoderma gangrenosum with infliximab: induction dosing regimen or on-demand therapy? *Dermatology.* 2008;216(2):163-5.

[45] Azizan NZ, Gangaram HB, Hussein SH. A novel therapy for the treatment of pyoderma gangrenosum. *Med. J. Malaysia.* 2008 Mar;63(1):51-4.

[46] Rogge FJ, Pacifico M, Kang N. Treatment of pyoderma gangrenosum with the anti-TNFalpha drug - Etanercept. *J. Plast Reconstr Aesthet Surg.* 2008;61(4):431-3.

[47] Zold E, Nagy A, Devenyi K, et al. Successful use of adalimumab for treating fistulizing Crohn's disease with pyoderma gangrenosum: Two birds with one stone. *World J. Gastroenterol.* 2009 May 14;15(18):2293-5.

[48] Cohen PR. Neutrophilic dermatoses: a review of current treatment options. *Am. J. Clin. Dermatol.* 2009;10(5):301-12.

[49] Akhras V, Sarkany R, Walsh S, et al. Superficial granulomatous pyoderma treated preoperatively with infliximab. *Clin. Exp. Dermatol.* 2009 Jul;34(5):e183-5.

[50] Cummins DL, Anhalt GJ, Monahan T, Meyerle JH. Treatment of pyoderma gangrenosum with intravenous immunoglobulin. *Br. J. Dermatol.* 2007 Dec;157(6):1235-9.

[51] de Zwaan SE, Iland HJ, Damian DL. Treatment of refractory pyoderma gangrenosum with intravenous immunoglobulin. *Australas J. Dermatol.* 2009 Feb;50(1):56-9.

[52] Okuma K, Mitsuishi K, Hasegawa T, et al. A case report of steroid and immunosuppressant-resistant pyoderma gangrenosum successfully treated by granulocytapheresis. *Ther. Apher Dial.* 2007 Oct;11(5):387-90.

[53] Tutrone WD, Green K, Weinberg JM, et al. Pyoderma gangrenosum: dermatologic application of hyperbaric oxygen therapy. *J. Drugs Dermatol.* 2007 Dec;6(12):1214-9.
[54] Bellini V, Simonetti S, Lisi P. Successful treatment of severe pyoderma gangrenosum with pimecrolimus cream 1%. *J. Eur. Acad. Dermatol Venereol.* 2008 Jan;22(1):113-5.
[55] Proudfoot LE, Singh S, Staughton RC. Superficial pyoderma gangrenosum responding to treatment with isotretinoin. *Br. J. Dermatol.* 2008 Dec;159(6):1377-8.
[56] Parren LJ, Nellen RG, van Marion AM, et al. Penile pyoderma gangrenosum: successful treatment with colchicine. *Int. J. Dermatol.* 2008 Nov;47 Suppl 1:7-9.
[57] Herfarth AE, Martínez-Ramos D, Nomdedeu-Guinot J, Salvador-Sanchis JL. Hyperbaric oxygen therapy as an adjuvant treatment for pyoderma gangrenosum. *Cir. Esp.* 2009 May;85(5):319-21.
[58] Lucchina S, Parvex SL, Biegger P, Fusetti C. FK-506 ointment: an effective adjuvant therapy to treat a dramatic case of pyoderma gangrenosum of unilateral hand. *Chin J. Traumatol.* 2009 Jun;12(3):181-3.
[59] Eaton PA, Callen JP. Mycophenolate mofetil as therapy for pyoderma gangrenosum. *Arch. Dermatol.* 2009 Jul;145(7):781-5.

In: Hyperthermia: Recognition, Prevention and Treatment ISBN 978-1-61942-275-9
Editor: Xing-Hua Gao and Hong-Duo Chen, 223-240

Chapter 9

HYPERTHERMIA THERAPY IN MUCO-CUTANEOUS INFECTION BY HUMAN PAPILLOMAVIRUS: EFFICACY AND MECHANISMS OF ACTION

Xing-Hua Gao, Wei Huo, Yuxiao Hong, and Hong-Duo Chen
Department of Dermatology,
No. 1 Hospital of China Medical University, PRC

INTRODUCTION

Human papillomavirus (HPV) infection of the skin or mucosa leads to various clinical forms of warts, or even carcinomas. Most of the warts will regress in a matter of years, while some are recalcitrant. Warts are disfiguring or debilitating.

There is quite an array of physical, chemical, pharmaceutical or biological therapies for treating warts. Most of the currently applied methods have certain limitations, such as low cure rates, ablation related sufferings, long healing time. Local heat therapy, or local hyperthermia within or several degrees above fever range, has been anecdotally practiced in treating warts, while high degree evidence based studies are not abundant. In the present review, we collected and analyzed literatures on hyperthermia treatment of warts, in the aim to best evaluate its efficacy. The natural history of wart suggests a systemic immune response to resolve the HPV infection. HPV generally infects epidermal compartment.

It is thus reasonable to postulate that two major epidermal components, keratinocytes--the body of HPV infection, and Langerhans cells-- professional antigen presenting cells, take principal roles in initiating and regulating the immune responses against HPV. Here we reviewed the effect of hyperthermia on immune related responses in keratinocytes and Langerhans cells, at both cellular and molecular levels. Last but not the least, the effect of hyperthermia on the biology of HPV itself is also discussed.

HUMAN PAPILLOMAVIRUS

HPV is a ubiquitous family of viruses with more than 120 genotypes. Some HPV types are considered to be potentially oncogenic. Recent epidemiological data showed that HPV infection is the most common sexually transmitted disease with the highest economic, psychosocial, and disease burden. While, other HPV-associated diseases is also very common in the world such as skin warts, cervical cancers and anogenital cancer, buschke-Lowenstein tumor, bowenoid papulosis, laryngeal papilloma, etc [1, 2].

HPV generally infects the basal cells of the skin and of mucosal epithelia such as the mouth, the upper respiratory tract, the lower genital tract and the anal canal and its life cycle is linked to the progressive differentiation of epithelial cells. It is unclear whether HPV has the ability to penetrate intact skin and mucosa. Minor wounds certainly facilitate access of HPV to the basal layer of the epidermis, as evidenced by the predilection of warts for trauma sites[3].

HPV subtypes have been divided into three categories (low risk, intermediate risk and high risk) based on their likelihood of inducing dyplasia and carcinoma. Condyloma acuminate is usually caused by HPV6/11, common warts by HPV 1/2/4/7/27/29/40/54/57, flat warts by HPV 3/10/27/28/41, palms and soles warts by HPV 1/4, squamous intraepithelial neoplasia by HPV31/33/45/51/52/56/58/59, cervical dysplasia and anogenital cancers mainly by HPV16/18, Querat's erythroplasia by HPV16/18, bowenoid papulosis by HPV16/18/39/42, laryngeal papilloma by HPV6/11/18[4, 5]. Moreover, patients can be infected simultaneously with multiple HPV types[6].

In some people with HPV-associated diseases, their immune systems will eradicate the virus after a variable period of time. In regression of HPV disease, the immune system plays a central role via both cellular and humoral immunity. However, cellular immune response appears to be the prime means of repelling HPV infection [7]. Unfortunately, that immune response is type-specific [8]. It has been estimated that approximately 60% of common warts resolve within 2 years, although in the following 2 to 10 years the natural clearance rate is less than 10% [5, 9]. Genital warts may regress spontaneously as many as 30 % over 4 months [10]. Even the high-risk types of HPV 16, is cleared by 80% of women within 2 years, with a median time to clearance of 17 months [11]. Other types of HPV-associated diseases also have a self-limiting clinical course [12, 13].

Although spontaneous regression often occurs without seeming to trouble individuals at all, some patients with immune suppression (e.g., diabetes mellitus, HIV infection, pregnancy, post-organ transplantation and the use of immunosuppressant agents) are troubled by clinical infection lasting many years [14-19].

Conventional therapies of HPV-associated diseases are mainly nonspecific. Current treatments physically remove the visible lesions rather than clear the infection. Criteria for patient selection, therapeutic response and follow-up period are varied greatly in published trials. And the evaluation of treatments for warts is imprecise because it is hard to distinguish between relapse and reinfection. Many treatment options, including destructive therapy, virucidal therapy, antimitotic therapy, and immunotherapy, or a combination of these, have been commonly used with varying efficacies [20, 21]. Treatment of choice depends on the number, size, area covered by the warts and length of time the warts have been present [22]. Moreover, side effects, cost, cosmetic results, effectiveness and convenience also define the choice of therapy. However, there is little certainty that any treatment is more effective than another. And it is unclear if combinations of therapies are more effective than monotherapy.

Hyperthermia with temperatures ranging from 39°C to 48°C has been successfully used in the treatment of some neoplasms and HPV infections in the recent medical literature [3][23-26]. Therapeutic hyperthermia can be achieved by a variety of methods such as laser (Carbon Dioxide (CO2) Laser, Erbium: Yttrium/Aluminum/Garnet (Er:YAG) Laser, Neodymium:YAG (Nd:YAG) Laser and Pulsed Dye Laser), infrared, electrosurgery, radiofrequency, ultrasonic, microwave, warm water soaks, heating pads, and pocket warmers. Although these therapeutic modalities have been proven safe and effective, each has its benefits and drawbacks. However, hyperthermia is a more potent therapy to HPV-associated disease as compared to other modalities [27].

Hyperthermia: Possible Mechanism of Action

The biological role local hyperthermia plays in the HPV-associated diseases has not yet been fully understood. Although local hyperthermia is not hot enough to kill the cells directly, it can establish a specific immune response against HPV-infected keratinocytes to clear warts. In many treatments with hyperthermia, patients with multiple warts experienced almost simultaneous clearance of targeted warts as well as the remaining untargeted warts, suggesting that hyperthermia plays an indirect yet facilitating role in the specific immune response against HPV infection.

Absorption of heat energy by tissue is dependent on temperature and length of time applied (D=tRT-43)[28]. Fever-range (39-40°C) temperature can modulate the activities of immune cells (APCs, T cells and NK cells). Heat shock temperature (41-43°C) can increased the immunogenicity of cells. In these situations, hyperthermia can modify the immune cells to make the endogenous specific immune response more easily induced and sustained rather than kill the cells directly [29]. Hyperthermia at 43°C is regarded as the breakpoint above which there are more cells undergoing apoptosis [28, 30], a condition that would facilitate the establishment of a specific immune response[31].

Antigen presenting cells, particularly Langerhans cells (LCs), can capture these antigens and migrate to lymph nodes. Here they present these antigens to T cells to induce an immune response that is effective against virus cell antigens. Recent studies have shown that loss of LCs in skin enhances contact dermatitis and prevents chemical-mediated skin cancer [32]

Thus, hyperthermia-mediated LC "loss" may increase the immune- response necessary to kill virus-infected keratinocytes. Li et al. [33] recently noted that local hyperthermia could promote migrational maturation of LCs in both normal and HPV-infected skin and that the effect was stronger at a local hyperthermia temperature of 44°C than at 42°C; the results suggested that hyperthermia might augment the antigen-presenting capability of LCs. Ostberg et al. [34] showed that there was a decrease in the density of LCs in the mouse epidermis after whole body hyperthermia, possibly correlating to the migration of LCs to draining lymph nodes[34]. Hong et al. have also observed that hyperthermia at 41°C reduced the numbers of mouse epidermal LCs, which minimized on day 3 and then gradually returned to the original levels on day 7. Yoshioka et al.[35] observed that homeostasis of LCs in the epidermis was reached on day 14 after local hyperthermia at 43°C. The cellular immune response is responsible for the elimination of HPV-infected keratinocytes.

A study [36] under an organotypic culture condition, showed that the mRNA expressions of CCL-20 and IL-1alpha in CA specimen were significantly higher than those in normal skin. Local hyperthermia at 42°C and 45°C significantly decreased the mRNA levels of CCL-20 and IL-1alpha, as compared with the control groups. The decrease of CCL-20 was well correlated with that of IL-1alpha. The expression of TNF-alpha in CA remained unchanged in spite of the temperature variation. Local hyperthermia at 45°C concomitantly increased the mRNA expression of CCL-20 and IL-1alpha in normal skin. Hyperthermia decreases the expression of CCL-20 with concomitant decrease in IL-1alpha, and reduce the number of Langerhans cells in HPV infected skin. CCL-20 (also known as MIP-3alpha) is the most potent chemokine for recruitment of LCs precursors into the skin. CCL-20 expression can be increased by TNF-alpha and IL-1alpha. The effects of local hyperthermia on the mRNA expressions of CCL-20, TNF-alpha, IL-1alpha have been investigated in both CA and normal skin.

HPV infects the germinal keratinocytes. Infection and maturation of HPV is dependent on the differentiation of keratinocytes. The more the differentiation processes in the affected keratinocytes progresses, the more the expression of the viral proteins is strengthened, and they consequently form the viral particle. The survival of HPV infected keratinocytes at different stages might involve different machineries [37]

In vitro studies showed that HPV E5, E6 proteins could respectively impair apoptosis of infected keratinocytes by TRAIL and FasL-mediated pathways, rendering host cells live long enough for virus replication[38-41].

The hyperthermia could induce apoptosis in normal and HPV infected skin, with stronger potency in the latter case. Although the apoptosis of keratinocytes by hyperthermia might not kill the virus directly, the dissolution of keratinocytes, together with the release of HPV components, might promote specific immunity against HPV [42]. Wang et al.[43]noticed that the apoptotic index was correlated with hyperthermia temperatures applied, in that heating samples at 45°C caused more apoptotic cells than that heated at 42°C. There were more apoptotic keratinocytes in HPV infected lesions than in normal skin, a result suggesting that HPV infected tissue was more susceptible to heat induced apoptosis. Keratinocytes, the bulk component of outer skin, undergoes programmed cell death to maintain homeostasis and this process could be interrupted by some pathological conditions or external stimuli [44]. There was increased transcriptional and translational expression of Fas and Bax, but not BCL-2, in accordance with temperatures applied. Likewise, the temperature dependant increase in transcriptional and translational expression of Fas and Bax in HPV infected lesions, similar to that in normal skin, but with much higher magnitude. In addition, a temperature dependant decrease in the expression of BCL-2 in CA, but not in normal skin.

These results suggested the involvement of both intrinsic and extrinsic pathways in hyperthermia induced apoptosis in normal and HPV infected human skin and, BCL-2 was additionally involved in the apoptosis of HPV infected skin. A temperature dependant increased transcriptional expression of DR4 and DR5 in CA, but not in normal skin. Unexpectedly, we could hardly detect the expression of corresponding proteins in CA tissue.

Thus it is plausible that HPV infected keratinocytes developed machineries to counteract with a potential DR4 and DR5 mediated apoptosis by hyperthermia, as in the case of adenovirus infected cells [45].

Hyperthermia took effect partly by inducing the expression of local endogenous IFN and partly by subsequent IFN-induced antiviral activity via Jak–STATs signalling pathway in CA[46].

Zhu et al.[46] found that hyperthermia at 42°C and 45°C could induce a significant increase in the transcriptional expression of interferon(IFN)-α, IFN-ß and IFN-ɤ, in a temperature-dependent manner in CA, but not in normal skin. Accordingly, local hyperthermia could enhance the expression of 2'-5' oligoadenylate synthase and double-stranded RNA (dsRNA)-dependent protein kinase, two antiviral enzymes downstream of the IFN-dependant pathway.

Hyperthermia led to an increase in IFN-α/ß receptor transcripts, and an increase in the levels in phospho-Stat1 and phospho-Stat2 in CA, though it had no influence on the levels of Jak1, Tyk2, Stat1 and Stat2 transcriptional expression.

Immunotherapy acts in association with hyperthermia to help to control HPV-associated diseases recurrence. And they may be able to achieve a greatest effect as compared to other therapies. Furthermore, this effect may be a more potent treatment to HPV-associated disease than other modalities[27]. It is to be hoped that hyperthermia in a modified pattern will transform the management of HPV in the future.

Efficacy of Hyperthermia on HPV Infection

The laser (Carbon Dioxide (CO2) Laser, Erbium:Yttrium /Aluminum /Garnet (Er:YAG) Laser, Neodymium:YAG(Nd:YAG)Laser, and Pulsed Dye Laser)has numerous applications in dermatology and dermatologic surgery. Laser is an excellent modality for treating extensive or recurrent warts without bleeding. During laser treatments, target structures absorb monochromatic coherent light. Light energy gets converted to thermal energy. Target structure is excised by coagulation or blasting. The advantages of laser therapy are precision which results in sparing of normal tissue, probable elimination of the infective agent. However, disadvantages of laser are more complex and costly and may require local anesthesia for larger lesions and pediatric patients. Future studies examining optimal laser fluence and pulse duration depending on the location of the warts would expand our knowledge on how to use laser therapy more effectively.

CO2 Laser

CO2 laser therapy is widely used in HPV-infection diseases. The CO2 laser, which creates thermal injury absorption of the infrared light 10,600 nm wavelength by water in tissue, has be used to excise tissue with a combination of vaporization and resection of the wart root until a healthy tissue is obtained[47]. Several studies have examined its effectiveness[48] and CO2 lasers can be used to treat extensive and thick warts without bleeding. However, disadvantage of this treatment in patients include postoperative pain, prolonged healing time, and scar formation. The recurrence rates of condyloma acuminate treated with CO2 laser are high[49]. In using CO2 laser, surgical mask and evacuating equipment are need to protect operators from potential inhalation exposure to HPV DNA releasing during laser vaporization [50].

The carbon dioxide laser probable eliminate the infective lesions, partly because of deeper penetration. However, CO2 laser ablation would not be enough to remove the HPV DNA or viral particles. After CO2 laser therapy, these HPV DNA or viral particles might be retransmitted to the ambient defective epidermis, resulting in recurrent warts. The remaining HPV DNA or viral particles would be involved in recurrence[51]. A study of 25 women with subclinical papillomavirus infection of the lower genital tract show that 88% of patients treated with CO2 laser were documented histologic persistence of subclinical papillomavirus infection. This suggest that even extended carbon dioxide laser can not effectively eradicate subclinical papillomavirus infection of the female lower genital tract[51]. However, in a study of 35 patients with plantar warts, PCR assay was used to show that 31(88.6%)of 35 warts treated with a combination of CO2 laser therapy and artificial dermis application were cleared of HPV DNA in the upper epidermis of the postoperative site. These results suggest that recurrence would not occurred in the patients without HPV DNA in the future [52].

There are no new randomized, controlled trials comparing the benefit of carbon dioxide laser therapy versus placebo. Previous studies of laser treatment to external genital warts described a 23% to 52% cure rate with 3-18 months follow up, while recurrence rates ranged from 60% to 77%[53-55]. Studies dealing with CO2 laser surgery of common warts reported on cure rates of 47–80%[56-60].

Further more, CO2 laser is also used in the treatment of periungual and subungual warts [61], immunosuppressed patients with recalcitrant warts, pregnant women with CA, cervical intraepithelial neoplasia, laryngeal papilloma [62], recurrent squamous papillomata of the conjunctiva[63], etc. A retrospective study of 40 patients with subungual and periungual viral warts treated with CO2 laser vapourisation results a cure rate of 57.4% (39/68) over 10 months follow-up period, while recurrences rate is 36.8% (25/68) [57]. A study dealing with CO2 laser surgery of warts in immunosuppressed persons, a majority of them renal allograft recipients, reports 6/13 patients showed full remission[64]. The results are remarkably in line with immunocompetent persons dealed with CO2 laser surgery[56-60]. A retrospective long-term study of CO2 laser surgery for extensive, anogenital condylomata acuminata shows similar clearance and recurrence rates. The cure rate in HIV-positive patients was 58%(11/19) versus 71%(32/45) in HIV-negative patients[65]. A study of 32 pregnant women with CA, using CO2 laser therapy and 85% trichloroacetic acid showed 97% (31/32) of women were controlled with this combination therapy[66]. Another case-control study of 536 pregnant women with clinical HPV infection, using CO2 laser therapy during pregnancy or after delivery showed recurrence rates were higher in the women treated in postpartum than in the group treated during gestation after 2 years follow-up. Clinical HPV infections treated during the second trimester of pregnancy showed a sensitive decrease in recurrence-rate of infection [67]. A study of 32 women of cervical intraepithelial neoplasia with one or more oncogenic types, using CO2 laser showed that no recurrences of HPV or dysplasia were seen after five years follow-up. A miniconization procedure with CO2 laser for treatment of cervical intraepithelial neoplasia proved useful also for simultaneous therapy of concomitant HPV infection of the uterine cervix[68].

However, some studies hold opposite views. A study of 62 patients with refractory genital warts, using CO2 laser therapy showed an immediate cure rate of 27% (17/62) two weeks after operation. It is concluded that a single CO2 laser treatment often is an insufficient procedure[69, 70].

Pulsed Dye Laser (PDL)

Pulsed dye laser, emitting in the yellow light spectrum at 585 nm[71, 72], have been used effectively in the treatment of warts with the dilated capillaries in the warts as the target tissue without damaging the surrounding skin[71, 73-77]. The use of the PDL depends upon the energy absorption within the capillary loops of the wart and hence localized tissue necrosis by abolishing nutritional blood flow. When used against skin, this light is best absorbed by hemoglobin and oxyhemoglobin since hemoglobin in blood has strong absorption peaks at wavelengths ranging from 585 to 595 nm[77].

The mechanism of action of the PDL is not clear, but thermal damage, removal of nutritional blood flow, and cellular immune response are believed to contribute to warts regression [78]. In a study using PDL in a manual mode suggests that the mechanism of this laser treatment is thought to be vascular instead of thermal[79]. As the vessels burst, purpura develops within minutes after the therapy in the treated areas. PDL causes minimal postoperative pain and completely heals in 2-4 weeks.

PDL has been used for facial warts and perianal warts in children [74, 79, 80]for less pain and scarring than CO2 laser treatment. It is recommended as a simple, safe, cost and time saving alternative [81] treatment option for viral warts with a completely remission in 48% to 100% of cases [74-76, 82-94]. However it is not listed in genital warts treatment guidelines and therefore PDL does not exist as a widely accepted therapy option for HPV-associated diseases [95, 96].

Er: YAG Laser

The Er:YAG laser is widely used in dermatology and aesthetic surgery since it allows precise ablation in a thin upper layer of tissue avoiding strong inflammation with minimal thermal damage. It emits an infrared light with the wavelength of 2,940 nm that is absorbed 12 to 18 times more efficiently by water-containing superficial cutaneous tissues than is the 10,600 nm wavelengths emitted by the CO2 laser[97].

The advantages of Er:YAG laser are a smaller zone of thermal damage, precise thermal ablation, and minimal scar. However, postoperative healing time is 7 to 10 days, and erythema that occurred after the treatment required 2 months to subside[97]. A potential safety feature of this laser is that HPV DNA has not been detected in the laser plume[98].

The use of Er:YAG lasers in the treatment of warts has been reported in the literature, showing variable response rates and relapse rates[99]. Warts in a variety of locations have been successfully eliminated in 72.5% of patients after treatment of Er:YAG, with a 6-24% relapse rate within 3-12 months after treatment[100-102]. Approximately 13.5% of patients are non-responders[103]. The combination of both Er:YAG laser and topical 0.5% podophyllotoxin solution in palmoplantar warts was investigated to show a 88.6% complete response and 5.7% recurrence rates after 3 months[103].

Nd: YAG Laser

The Nd:YAG laser have gained widespread acceptance in the treatment of HPV efflorescences. The Nd:YAG lasers emit a light spectrum at 1,064 nm, still in the infrared range. The 1,064 nm long-pulsed Nd:YAG laser effect depends on the absorption of red structures such as cutaneous blood vessels. As a result, blood vessels heat up rapidly and burst. Depth of coagulation can be controlled by laser power, spot size and exposure time[104]. The warts were irradiated until the surface temperature was 40°C to 42°C during the Nd:YAG laser treatment[26, 105].

However, the mechanism of action of Nd:YAG laser is not clear. Light microscopic evaluation of treated areas by Nd:YAG laser at 7 days after treatment in a study showed separation of the dermoepidermal junction, epidermal necrosis, and red blood cell extravasation. Moreover, destroyed blood vessels in the dermis were surrounded by a dense inflammatory infiltrate. This destruction may remove the nutrient supply to the wart or destroy the rapidly dividing epidermal cells that contain HPV[104].

The advantage of Nd:YAG Laser Treatment is excellent cosmetic results, high cure rates, and low recurrences. Side effects including transient numbness, purpura, hemorrhagic bullae, hyperpigmentation, and hypopigmentation are generally mild and don't prevent normal activity[106]. The treatment success ratio with 1064 nm Nd:YAG laser is quite high in comparison with the other treatment methods and its side effects are not too much, except for pain. We suggest that Nd:YAG laser is the most efficient and time-saving method in patients with needle fear, unwilling to receive longstanding local treatment, and in children.

In 1999, El-Tonsy MH et al.[26]employed the in situ hybridization to detect HPV DNA in 100% (25/25) warts of untreated lesions in plantar warts leions and in 0% of Nd:YAG laser treated lesions. From this results obtained, it can be concluded that Nd:YAG laser hyperthermia eradicates HPV DNA efficiently.

The Nd:YAG Laser is one of the most effective methods for treatment of acute condylomata. In 2010, a prospective descriptive study with 198 patients was planned to determine the efficiency of Nd:YAG laser in different types of verruca, such as periungual warts, facial - plana warts, palmar warts, plantar warts, and genital warts. At clinical recovery evaluation, the recovery rate was 97% (range: 75-100%).

In addition Nd:YAG Laser has been used in the gynecology, obstetrics, and urology. A study of seven girls (children) with CA, using Nd:YAG laser showed 85.7%(6/7) cure rate after the first treatment[107]. And a study of nineteen pregnant women between 10 to 38 gestation weeks with genital acute condyloma shows that Nd:YAG laser is effective without complications or recurrence after one year follow-up[108].

What's more, Nd:YAG laser can excise urethral condylomata that is difficult to eradicate by other methods. However, it may require several treatments and recurrence is frequently observed in patients with extensive lesions. Previous studies of Nd:YAG laser treatment to urethral condylomata described a 30% to 88.9% cure rate with 3-16 months follow-up, while recurrence rates ranged from 0% to 35.2%[109-111].

Nd:YAG Laser has also been found effectively in the treatment of children with recurrent respiratory papillomatosis (RRP). During 1-year follow-up visits, all Nd:YAG laser light-treated patients, showed a regression of the disease[112]. Furthermore, we have been treating successfully cervical neoplasias using Nd:YAG lasers regardless of margin positivity and

HPV-DNA status. Involved margins and HPV infection did not influence the cure of early neoplasia of the uterine cervix achieved by the laser conization procedure[113].

Electrosurgery

Electrosurgery is widely used to destroy HPV affected lesions, especially in the treatment of extensive anogenital warts. Thermal coagulation and electrocautery are two forms of electrosurgery. In the direct-current form of electrocautery, electricity flows only through the instrument producing heat that is applied to the lesion. While, in the alternating-current form of electrosurgery electricity flows from the instrument through the patient to a grounding plate. The alternating current systems produce cutting, coagulation, or a blend of both. Fine-wire loops can be used for excision [114, 115]

Electrosurgery appears to be as effective as CO2 laser surgery[116] or cryotherapy[117, 118] for treating anogenital condylomas[116]. There are no recent studies of electrosurgical methods for genital wart treatment. Old studies showed that complete clearance rates of warts reached 38% -57%[116, 119] after one treatment session and 64%-78% after repeated sessions[53, 116]. After 3 months follow-up, 14%-24% patients had recurrences[116, 120]. Schoenfeld et al. studied 28 women and showed 86% lesions treated with the loop electrosurgical excision procedure (LEEP) and 75% lesions treated with a laser showed no HPV DNA within a 20-mm circumference of the treated lesion[121].

However, electrosurgery is considered as a second-line treatment for some severe discomforts and delayed complications occurred in a few of patients such as scarring and vitiligo [116].

Radiofrequency (RF)

Radiofrequency inducing thermal destruction of the targeted tissue by an ultrahigh frequency current waveform at a frequency of 0.5 MHz, has been widely used in the treatment of warts.

RF heat generators contain a bipolar probe that comes in contact with the infected skin and generates heat via tissue resistance to current flow. As the electrode tip is not heated, there is minimal thermal damage to the surrounding tissues, producing good esthetic results. Localized heating with radiofrequency heat generators as well as surgical excision with radiofrequency electrosurgical knives have been used with moderate success [78].

The advantages of RF are negligible intra-operative bleeding, short operative time, rapid healing[122]. Drawbacks include the cost of machinery and local anesthetic administration is adopted for the intense heat that can be generated by such devices[123]. Complications are uncommon and mainly occur due to an improper technique [124].

A placebo controlled clinical trial with 29 common warts, using RF(50°C) showed the cure rate is 86%(24/29) after one to four treatments, versus 41%(7/17) in the control group. No recurrence was noted after a average of 15.6 weeks follow-up [123].

In recent clinical trials, RF has be used effectively in HPV correlated extra-cervical genital lesions and esophageal squamous papillomatosis (ESP). A study of 78 patients with HPV correlated extra-cervical genital lesions, using radiofrequency electrosurgical knife, showed that 10.3% (7/68) patients with subclinical condylomatosis presented histologically documented recidivation during the course of follow-up[125]. Balloon-based radiofrequency ablation was successfully used to treat ESP with dysplasia besides type 4 ESP with high-grade dysplasia [126]

Ultrasound Hyperthermia

Ultrasound thermotherapy of warts was not widely used today. There were several early reports about ultrasound thermotherapy of warts with variable response and recurrence rates [127-135]. Nowadays a prospective study of 11 patients with anal and perianal CA, using a ultrasound-driven harmonic scalpel blade operating at a vibration frequency of 55.5 kHz and within a temperature range of 65-120°C, showed all CA were excised in a single-step procedure with complete clearing without injury of subepidermal skin. No side effect or recurrence was observed during 4-26 months follow-up [136].

Infrared Coagulation (IRC)

The Infrared coagulation is a useful therapeutic modality for common warts but rarely used today. The infrared coagulator is an instrument that delivers a narrow beam of noncoherent infrared light to the skin surface. The instrument allows adjustable tissue necrosis without tissue adhesion [137]. In comparison to electrocoagulation, IRC produces similar cure rates, with lower side effects. The overall cure rates at the end of the 6-11 months follow-up were 66.7%-70%[138] with a 10.8%-30% recurrence rates [137-139].

Gao etal.[140] devised a hyperthermia device with an infrared emitting source without direct contact with skin and found that local hyperthermia at a mean temperature of 45.3°C attained a 65.3% cure rate(15/23) among patients with plantar warts. Further more, a randomized, patient-blinded, placebo-controlled trial was conducted with the same hyperthermia device, and found that 53.57% of patients (15/28) in the treatment group (44°C) and 11.54% of patients(3/26) in the control group were cured by the end of 3 months follow-up[141].

Microwave Treatment

A vitro study of 12 patients with excised warts, showed that microwave produce more HPV DNA damage than CO2 laser [142]. In this trial DNA was extracted from tissue by proteolytic digestion and amplified by the polymerase chain reaction(PCR). HPV DNA was amplified in 100%, 83.3% and 50% of untreated specimens, specimens treated with CO2 laser and microwave, respectively[142].

Hot Water

In the old study, hyperthermia therapy was described effectively in treating warts by hot water at temperature level of 40-45°C[143]. A study of 15 patients with palmoplantar verruca, using local water baths in a temperature level of 45-48°C showed that thirteen of 15 patients' warts regressed or disappeared after 2 to 6 weeks of once or twice weekly immersion of involved skin for 30 minutes to 1.5 hours[144]. Moreover, water baths at temperatures of 45-50°C also have been effectively used to deal with warts in a patient with HIV infection [145].

Exothermic Patches

A self-administered, self-adhesive disposable exothermic patche that produces heat through oxidation upon exposure to air has been effectively used in the treatment of warts[146]. Exothermic patche is used to provide continuous, long-lasting, and reproducible heating of the skin surface for at least 2 hours at a temperature level of 42-43°C, without burning or injuring the normal skin. Dvoretzky etal.[146] reported their experience in using the exothermic patch to effectively deal two patients with recalcitrant warts. No recurrence was observed by the end of 8 months follow-up. However, the regression of warts cannot be attributed entirely to thermotherapy because 17% salicylic acid was also employed in one of the patients.

REFERENCES

[1] Weaver, B.A., Epidemiology and natural history of genital human papillomavirus infection. *J. Am. Osteopath Assoc*, 2006. 106(3 Suppl 1): p. S2-8.

[2] Chaturvedi, A.K., et al., Human Papillomavirus Infection with Multiple Types: Pattern of Coinfection and Risk of Cervical Disease. *J. Infect. Dis.* 203(7): p. 910-920.

[3] Beutner, K.R. and S. Tyring, Human papillomavirus and human disease. *Am. J. Med.*, 1997. 102(5A): p. 9-15.

[4] Leman, J.A. and E.C. Benton, Verrucas. Guidelines for management. *Am. J. Clin. Dermatol*, 2000. 1(3): p. 143-9.

[5] Grussendorf-Conen, E.I., et al., Topical 5% imiquimod long-term treatment of cutaneous warts resistant to standard therapy modalities. *Dermatology*, 2002. 205(2): p. 139-45.

[6] Brown, D.R., et al., Detection of multiple human papillomavirus types in Condylomata acuminata lesions from otherwise healthy and immunosuppressed patients. *J. Clin. Microbiol*, 1999. 37(10): p. 3316-22.

[7] Coleman, N., et al., Immunological events in regressing genital warts. *Am. J. Clin. Pathol*, 1994. 102(6): p. 768-74.

[8] Dillner, J., The serological response to papillomaviruses. *Semin Cancer Biol*, 1999. 9(6): p. 423-30.

[9] Kyriakis, K., et al., Lifetime prevalence fluctuations of common and plane viral warts. *J. Eur. Acad. Dermatol. Venereol.*, 2007. 21(2): p. 260-2.

[10] Ho, G.Y., et al., Natural history of cervicovaginal papillomavirus infection in young women. *N. Engl. J. Med.,* 1998. 338(7): p. 423-8.
[11] Xi, L.F., et al., Acquisition and natural history of human papillomavirus type 16 variant infection among a cohort of female university students. *Cancer Epidemiol Biomarkers Prev*, 2002. 11(4): p. 343-51.
[12] Suligoi, B., M.C. Salfa, and L. Mariani, [Epidemiology and management of patients with ano-genital warts in Italy]. *Ig. Sanita Pubbl.* 66(6): p. 733-56.
[13] Mayeaux, E.J., Jr. and C. Dunton, Modern management of external genital warts. *J. Low Genit Tract Dis*., 2008. 12(3): p. 185-92.
[14] Firnhaber, C., et al., Human papillomavirus types in HIV seropositive men with penile warts in Johannesburg, South Africa. *Int. J. STD AIDS.* 22(2): p. 107-9.
[15] Jones, L., Sexually-transmitted infections for HIV-positive people. Signs and symptoms of common STDs, warts and all. *Posit Aware*, 2005. 16(3): p. 35-6, 42.
[16] Walzman, M., Successful treatment of profuse recalcitrant extra-genital warts in an HIV-positive patient using 5% imiquimod cream. *Int. J. STD AIDS,* 2009. 20(9): p. 657-8.
[17] Yong, M., et al., Diabetes and genital warts: an unhappy coalition. *Int. J. STD AIDS.* 21(7): p. 457-9.
[18] Palefsky, J., Biology of HPV in HIV infection. *Adv. Dent Res.,* 2006. 19(1): p. 99-105.
[19] Rozmus-Warcholinska, W., et al., [Genital warts associated with HPV infection during II and III trimester of pregnancy--a case report and analysis of treatment options]. *Ginekol Pol*, 2007. 78(11): p. 888-91.
[20] Simonart, T. and V.D. Maertelaer, Systemic treatments for cutaneous warts: A systematic review. *J. Dermatolog Treat.*
[21] Lipke, M.M., An armamentarium of wart treatments. *Clin. Med. Res,* 2006. 4(4): p. 273-93.
[22] Wilson, J.D., C.B. Brown, and P.P. Walker, Factors involved in clearance of genital warts. *Int. J. STD AIDS*, 2001. 12(12): p. 789-92.
[23] Hurwitz, M.D., et al., Hyperthermia combined with radiation for the treatment of locally advanced prostate cancer: long-term results from Dana-Farber Cancer Institute study 94-153. *Cancer.* 117(3): p. 510-6.
[24] Fotopoulou, C., et al., Regional abdominal hyperthermia combined with systemic chemotherapy for the treatment of patients with ovarian cancer relapse: Results of a pilot study. *Int. J. Hyperthermia*. 26(2): p. 118-26.
[25] Hildebrandt, B., et al., Current status of radiant whole-body hyperthermia at temperatures >41.5 degrees C and practical guidelines for the treatment of adults. The German 'Interdisciplinary Working Group on Hyperthermia'. *Int. J. Hyperthermia*, 2005. 21(2): p. 169-83.
[26] El-Tonsy, M.H., et al., Density of viral particles in pre and post Nd: YAG laser hyperthermia therapy and cryotherapy in plantar warts. *Int. J. Dermatol,* 1999. 38(5): p. 393-8.
[27] Fox, P.A. and M.Y. Tung, Human papillomavirus: burden of illness and treatment cost considerations. *Am. J. Clin. Dermatol*, 2005. 6(6): p. 365-81.
[28] Hildebrandt, B., et al., The cellular and molecular basis of hyperthermia. *Crit. Rev. Oncol. Hematol.,* 2002. 43(1): p. 33-56.

[29] Zhang, H.G., et al., Hyperthermia on immune regulation: a temperature's story. *Cancer Lett*, 2008. 271(2): p. 191-204.

[30] Wang, X., et al., Local hyperthermia induces apoptosis of keratinocytes in both normal skin and condyloma acuminata via different pathways. *Apoptosis,* 2009. 14(5): p. 721-8.

[31] Russo, V., et al., Dendritic cells acquire the MAGE-3 human tumor antigen from apoptotic cells and induce a class I-restricted T cell response. *Proc. Natl. Acad. Sci. USA*, 2000. 97(5): p. 2185-90.

[32] Kaplan, D.H., et al., Epidermal langerhans cell-deficient mice develop enhanced contact hypersensitivity. *Immunity,* 2005. 23(6): p. 611-20.

[33] Li, X.D., et al., [Effects of local hyperthermia on maturation of Langerhans cells in HPV-infected skin]. *Zhonghua Shi Yan He Lin Chuang Bing Du Xue Za Zhi*. 24(2): p. 122-4.

[34] Kida, Y., et al., Increased liver temperature efficiently augments human cellular immune response: T-cell activation and possible monocyte translocation. Cancer Immunol Immunother, 2006. 55(12): p. 1459-69.

[35] Yoshioka, A., et al., Suppression of contact sensitivity by local hyperthermia treatment due to reduced Langerhans cell population in mice. Br J Dermatol, 1989. 120(4): p. 493-501.

[36] Wang, X., et al., Local hyperthermia decreases the expression of CCL-20 in condyloma acuminatum. *Virol J*. 7: p. 301.

[37] Moody, C.A., et al., Human papillomaviruses activate caspases upon epithelial differentiation to induce viral genome amplification. *Proc. Natl. Acad. Sci. USA*, 2007. 104(49): p. 19541-6.

[38] Garnett, T.O., M. Filippova, and P.J. Duerksen-Hughes, Accelerated degradation of FADD and procaspase 8 in cells expressing human papilloma virus 16 E6 impairs TRAIL-mediated apoptosis. *Cell Death Differ,* 2006. 13(11): p. 1915-26.

[39] Filippova, M., L. Parkhurst, and P.J. Duerksen-Hughes, The human papillomavirus 16 E6 protein binds to Fas-associated death domain and protects cells from Fas-triggered apoptosis. *J. Biol. Chem.,* 2004. 279(24): p. 25729-44.

[40] Kabsch, K., et al., The HPV-16 E5 protein inhibits TRAIL- and FasL-mediated apoptosis in human keratinocyte raft cultures. *Intervirology,* 2004. 47(1): p. 48-56.

[41] Kabsch, K. and A. Alonso, The human papillomavirus type 16 E5 protein impairs TRAIL- and FasL-mediated apoptosis in HaCaT cells by different mechanisms. *J. Virol*, 2002. 76(23): p. 12162-72.

[42] Albert, M.L., B. Sauter, and N. Bhardwaj, Dendritic cells acquire antigen from apoptotic cells and induce class I-restricted CTLs. *Nature,* 1998. 392(6671): p. 86-9.

[43] Ostberg, J.R., E. Kabingu, and E.A. Repasky, Thermal regulation of dendritic cell activation and migration from skin explants. *Int. J. Hyperthermia,* 2003. 19(5): p. 520-33.

[44] Boehm, I., Apoptosis in physiological and pathological skin: implications for therapy. *Curr. Mol. Med.*, 2006. 6(4): p. 375-94.

[45] Tollefson, A.E., et al., Forced degradation of Fas inhibits apoptosis in adenovirus-infected cells. Nature, 1998. 392(6677): p. 726-30.

[46] Zhu, L.L., et al., Local hyperthermia could induce antiviral activity by endogenous interferon-dependent pathway in condyloma acuminata. *Antiviral Res*. 88(2): p. 187-92.

[47] Serour, F. and E. Somekh, Successful treatment of recalcitrant warts in pediatric patients with carbon dioxide laser. *Eur. J. Pediatr. Surg*, 2003. 13(4): p. 219-23.
[48] Ting, P.T. and M.T. Dytoc, Therapy of external anogenital warts and molluscum contagiosum: a literature review. *Dermatol. Ther*, 2004. 17(1): p. 68-101.
[49] Padilla-Ailhaud, A., Carbon dioxide laser vaporization of condyloma acuminata. *J. Low Genit. Tract. Dis.*, 2006. 10(4): p. 238-41.
[50] Weyandt, G.H., et al., Low risk of contamination with human papilloma virus during treatment of condylomata acuminata with multilayer argon plasma coagulation and CO laser ablation. *Arch. Dermatol. Res*. 303(2): p. 141-4.
[51] Riva, J.M., et al., Extended carbon dioxide laser vaporization in the treatment of subclinical papillomavirus infection of the lower genital tract. *Obstet. Gynecol.*, 1989. 73(1): p. 25-30.
[52] Mitsuishi, T., et al., Combination of carbon dioxide laser therapy and artificial dermis application in plantar warts: human papillomavirus DNA analysis after treatment. *Dermatol. Surg*. 36(9): p. 1401-5.
[53] Wiley, D.J., et al., External genital warts: diagnosis, treatment, and prevention. *Clin. Infect. Dis*, 2002. 35(Suppl 2): p. S210-24.
[54] Petersen, C.S., et al., Systemic interferon alpha-2b increases the cure rate in laser treated patients with multiple persistent genital warts: a placebo-controlled study. *Genitourin Med.,* 1991. 67(2): p. 99-102.
[55] Davis, B.E. and M.J. Noble, Initial experience with combined interferon-alpha 2B and carbon dioxide laser for the treatment of condylomata acuminata. *J. Urol.,* 1992. 147(3): p. 627-9.
[56] Landsman, M.J., J.E. Mancuso, and S.P. Abramow, Carbon dioxide laser treatment of pedal verrucae. *Clin. Podiatr. Med. Surg.,* 1992. 9(3): p. 659-69.
[57] Lim, J.T. and C.L. Goh, Carbon dioxide laser treatment of periungual and subungual viral warts. *Australas J. Dermatol*, 1992. 33(2): p. 87-91.
[58] Logan, R.A. and C.B. Zachary, Outcome of carbon dioxide laser therapy for persistent cutaneous viral warts. *Br. J. Dermatol*, 1989. 121(1): p. 99-105.
[59] Sloan, K., H. Haberman, and C.W. Lynde, Carbon dioxide laser-treatment of resistant verrucae vulgaris: retrospective analysis. J Cutan Med Surg, 1998. 2(3): p. 142-5.
[60] Street, M.L. and R.K. Roenigk, Recalcitrant periungual verrucae: the role of carbon dioxide laser vaporization. *J. Am. Acad. Dermatol*, 1990. 23(1): p. 115-20.
[61] Sterling, J.C., S. Handfield-Jones, and P.M. Hudson, Guidelines for the management of cutaneous warts. *Br. J. Dermatol.,* 2001. 144(1): p. 4-11.
[62] Kunachak, S., P. Sithisarn, and B. Kulapaditharom, Are laryngeal papilloma virus-infected cells viable in the plume derived from a continuous mode carbon dioxide laser, and are they infectious? A preliminary report on one laser mode. *J. Laryngol Otol.,* 1996. 110(11): p. 1031-3.
[63] Bosniak, S.L., N.L. Novick, and M.E. Sachs, Treatment of recurrent squamous papillomata of the conjunctiva by carbon dioxide laser vaporization. *Ophthalmology,* 1986. 93(8): p. 1078-82.
[64] Lauchli, S., et al., CO2 laser treatment of warts in immunosuppressed patients. Dermatology, 2003. 206(2): p. 148-52.

[65] Carrozza, P.M., et al., CO(2) laser surgery for extensive, cauliflower-like anogenital condylomata acuminata: retrospective long-term study on 19 HIV-positive and 45 HIV-negative men. *Dermatology,* 2002. 205(3): p. 255-9.
[66] Schwartz, D.B., et al., Genital condylomas in pregnancy: use of trichloroacetic acid and laser therapy. *Am. J. Obstet. Gynecol*, 1988. 158(6 Pt 1): p. 1407-16.
[67] Frega, A., et al., Regression rate of clinical HPV infection of the lower genital tract during pregnancy after laser CO2 surgery. *Clin. Exp. Obstet. Gynecol.,* 2006. 33(2): p. 93-5.
[68] Bekassy, Z., et al., Carbon dioxide laser miniconization for treatment of human papillomavirus infection associated with cervical intraepithelial neoplasia. *Acta Obstet. Gynecol. Scand*, 1995. 74(10): p. 822-6.
[69] Larsen, J. and C.S. Petersen, The patient with refractory genital warts in the STD-clinic. Treatment failure with CO2-laser. *Dan. Med. Bull.,* 1990. 37(2): p. 194-5.
[70] Scheinfeld, N.S. and D. Lehman, Condyloma with lichen nitidus. *Skinmed,* 2005. 4(3): p. 177-8.
[71] Kopera, D., Verrucae vulgares: flashlamp-pumped pulsed dye laser treatment in 134 patients. *Int. J. Dermatol*, 2003. 42(11): p. 905-8.
[72] Sexually transmitted diseases treatment guidelines 2002. Centers for Disease Control and Prevention. *MMWR Recomm. Rep*, 2002. 51(RR-6): p. 1-78.
[73] Badawi, A., et al., Treatment of genital warts in males by pulsed dye laser. *J. Cosmet Laser Ther*, 2006. 8(2): p. 92-5.
[74] Robson, K.J., et al., Pulsed-dye laser versus conventional therapy in the treatment of warts: a prospective randomized trial. *J. Am. Acad. Dermatol,* 2000. 43(2 Pt 1): p. 275-80.
[75] Kenton-Smith, J. and S.T. Tan, Pulsed dye laser therapy for viral warts. *Br. J. Plast. Surg.,* 1999. 52(7): p. 554-8.
[76] Ross, B.S., et al., Pulsed dye laser treatment of warts: an update. *Dermatol. Surg.*, 1999. 25(5): p. 377-80.
[77] Weiss, R.A. and M.A. Weiss, Early clinical results with a multiple synchronized pulse 1064 NM laser for leg telangiectasias and reticular veins. *Dermatol. Surg.,* 1999. 25(5): p. 399-402.
[78] Tosti, A. and B.M. Piraccini, Warts of the nail unit: surgical and nonsurgical approaches. *Dermatol. Surg.,* 2001. 27(3): p. 235-9.
[79] Tuncel, A., et al., Treatment of anogenital warts by pulsed dye laser. *Dermatol. Surg.,* 2002. 28(4): p. 350-2.
[80] Vargas, H., et al., The treatment of facial verrucae with the pulsed dye laser. *Laryngoscope*, 2002. 112(9): p. 1573-6.
[81] Alam, M. and M. Stiller, Direct medical costs for surgical and medical treatment of condylomata acuminata. *Arch. Dermatol.*, 2001. 137(3): p. 337-41.
[82] Tan, O.T., R.M. Hurwitz, and T.J. Stafford, Pulsed dye laser treatment of recalcitrant verrucae: a preliminary report. *Lasers Surg. Med.,* 1993. 13(1): p. 127-37.
[83] Kauvar, A.N., D.H. McDaniel, and R.G. Geronemus, Pulsed dye laser treatment of warts. *Arch. Fam. Med.,* 1995. 4(12): p. 1035-40.
[84] Huilgol, S.C., R.J. Barlow, and A.C. Markey, Failure of pulsed dye laser therapy for resistant verrucae. *Clin. Exp. Dermatol*, 1996. 21(2): p. 93-5.

[85] Fernandez-Guarino, M., A. Harto, and P. Jaen, Treatment of recalcitrant viral warts with pulsed dye laser MAL-PDT. *J. Dermatolog Treat.*

[86] Sethuraman, G., et al., Effectiveness of pulsed dye laser in the treatment of recalcitrant warts in children. *Dermatol. Surg*. 36(1): p. 58-65.

[87] Park, H.S. and W.S. Choi, Pulsed dye laser treatment for viral warts: a study of 120 patients. *J. Dermatol,* 2008. 35(8): p. 491-8.

[88] Schellhaas, U., et al., Pulsed dye laser treatment is effective in the treatment of recalcitrant viral warts. *Dermatol. Surg.,* 2008. 34(1): p. 67-72.

[89] Chen, T. and I.J. Frieden, Development of extensive flat warts after pulsed dye laser treatment of a port-wine stain. *Dermatol. Surg*, 2007. 33(6): p. 734-5.

[90] Park, H.S., et al., Pulsed dye laser therapy for pediatric warts. *Pediatr. Dermatol,* 2007. 24(2): p. 177-81.

[91] Komericki, P., M. Akkilic, and D. Kopera, Pulsed dye laser treatment of genital warts. *Lasers Surg. Med.*, 2006. 38(4): p. 273-6.

[92] Smucler, R. and E. Jatsova, Comparative study of aminolevulic acid photodynamic therapy plus pulsed dye laser versus pulsed dye laser alone in treatment of viral warts. *Photomed Laser Surg.*, 2005. 23(2): p. 202-5.

[93] Wu, C., et al., Efficacy of pulsed-dye laser for viral warts--an internal audit. *Ir. Med. J.,* 2003. 96(3): p. 80, 82-3.

[94] Togsverd-Bo, K., et al., Paring and intense pulsed light versus paring alone for recalcitrant hand and foot warts: a randomized clinical trial with blinded outcome evaluation. Lasers Surg Med. 42(2): p. 179-84.

[95] von Krogh, G., et al., European guideline for the management of anogenital warts. *Int. J. STD AIDS*, 2001. 12 Suppl 3: p. 40-7.

[96] National guideline for the management of anogenital warts. Clinical Effectiveness Group (Association of Genitourinary Medicine and the Medical Society for the Study of Venereal Diseases). *Sex Transm. Infect,* 1999. 75 Suppl 1: p. S71-5.

[97] Kaufmann, R. and R. Hibst, Pulsed 2.94-microns erbium-YAG laser skin ablation--experimental results and first clinical application. *Clin. Exp. Dermatol.,* 1990. 15(5): p. 389-93.

[98] Hughes, P.S. and A.P. Hughes, Absence of human papillomavirus DNA in the plume of erbium:YAG laser-treated warts. *J. Am. Acad Dermatol.,* 1998. 38(3): p. 426-8.

[99] Hruza, G.J., Laser treatment of epidermal and dermal lesions. Dermatol Clin, 2002. 20(1): p. 147-64.

[100] Park, J.H., et al., Er:YAG laser treatment of verrucous epidermal nevi. *Dermatol. Surg.,* 2004. 30(3): p. 378-81.

[101] Trelles, M.A., I. Allones, and E. Mayo, Er:YAG laser ablation of plantar verrucae with red LED therapy-assisted healing. *Photomed Laser Surg*, 2006. 24(4): p. 494-8.

[102] Wollina, U., H. Konrad, and T. Karamfilov, Treatment of common warts and actinic keratoses by Er:YAG laser. *J. Cutan Laser Ther.*, 2001. 3(2): p. 63-6.

[103] Wollina, U., Er:YAG laser followed by topical podophyllotoxin for hard-to-treat palmoplantar warts. *J. Cosmet Laser Ther.,* 2003. 5(1): p. 35-7.

[104] Han, T.Y., et al., Long-pulsed Nd:YAG laser treatment of warts: report on a series of 369 cases. *J. Korean Med. Sci.,* 2009. 24(5): p. 889-93.

[105] Pfau, A., et al., Nd:YAG laser hyperthermia in the treatment of recalcitrant verrucae vulgares (Regensburg's technique). Acta Derm Venereol, 1994. 74(3): p. 212-4.

[106] Schneede, P. and R. Muschter, [Laser applications in condylomata acuminata]. *Urologe A,* 1994. 33(4): p. 299-302.
[107] Vasileva, P., V. Ignatov, and E. Kiriazov, [Laser therapy as a method for treating condylomata at a prepubescent age]. *Akush Ginekol* (Sofiia), 1994. 33(2): p. 23-4.
[108] Buzalov, S. and E. Khristakieva, [The treatment of neglected cases of condylomata acuminata in pregnant women with the Nd:Yag laser]. *Akush Ginekol* (Sofiia), 1995. 34(2): p. 38-9.
[109] Nathan, M., Outpatient treatment of intrameatal warts in a genitourinary medicine department. *Int. J. STD AIDS*, 1994. 5(3): p. 218-20.
[110] Huguet Perez, J., et al., [Urethral condyloma in the male: experience with 48 cases]. *Arch. Esp. Urol.*, 1996. 49(7): p. 675-80.
[111] Volz, L.R., V.L. Carpiniello, and T.R. Malloy, Laser treatment of urethral condyloma: a five-year experience. *Urology*, 1994. 43(1): p. 81-3.
[112] Janda, P., et al., Preliminary report of endolaryngeal and endotracheal laser surgery of juvenile-onset recurrent respiratory papillomatosis by Nd:YAG laser and a new fiber guidance instrument. *Otolaryngol Head Neck Surg.,* 2004. 131(1): p. 44-9.
[113] Izumi, T., et al., Margin clearance and HPV infection do not influence the cure rates of early neoplasia of the uterine cervix by laser conization. *Eur. J. Gynaecol Oncol.*, 2000. 21(3): p. 251-4.
[114] Beutner, K.R., et al., Genital warts and their treatment. *Clin. Infect. Dis.,* 1999. 28 Suppl 1: p. S37-56.
[115] Scheinfeld, N. and D.S. Lehman, An evidence-based review of medical and surgical treatments of genital warts. *Dermatol. Online* J., 2006. 12(3): p. 5.
[116] Ferenczy, A., et al., Treating vaginal and external anogenital condylomas with electrosurgery vs CO2 laser ablation. *J. Gynecol. Surg.,* 1995. 11(1): p. 41-50.
[117] Stone, K.M., et al., Treatment of external genital warts: a randomised clinical trial comparing podophyllin, cryotherapy, and electrodesiccation. *Genitourin Med.,* 1990. 66(1): p. 16-9.
[118] Simmons, P.D., F. Langlet, and R.N. Thin, Cryotherapy versus electrocautery in the treatment of genital warts. *Br. J. Vener Dis.,* 1981. 57(4): p. 273-4.
[119] Armstrong, D.K., et al., Combined therapy trial with interferon alpha-2a and ablative therapy in the treatment of anogenital warts. *Genitourin Med.,* 1996. 72(2): p. 103-7.
[120] Challenor, R. and I. Alexander, A five-year audit of the treatment of extensive anogenital warts by day case electrosurgery under general anaesthesia. *Int. J. STD AIDS*, 2002. 13(11): p. 786-9.
[121] Schoenfeld, A., et al., Laser versus loop electrosurgical excision in vulvar condyloma for eradication of subclinical reservoir demonstrated by assay for 2'5' oligosynthetase human papillomavirus. *Gynecol Obstet Invest,* 1995. 40(1): p. 46-51.
[122] Gupta, P.J., Radio surgery: a new tool in proctology practice. *Bratisl Lek Listy*, 2004. 105(7-8): p. 270-6.
[123] Willard, R.J., et al., Cutaneous leishmaniasis in soldiers from Fort Campbell, Kentucky returning from Operation Iraqi Freedom highlights diagnostic and therapeutic options. *J. Am. Acad. Dermatol,* 2005. 52(6): p. 977-87.
[124] Thappa, D.M., First impact factor of Indian Journal of Dermatology, Venereology and Leprology. *Indian J. Dermatol. Venereol. Leprol.* 77(1): p. 1-2.

[125] Zaza, M., et al., [The use of electrosurgery in the treatment of extra-cervical genital condylomatosis]. *Minerva Ginecol*, 1998. 50(9): p. 367-71.

[126] Kibria, R., et al., Esophageal squamous papillomatosis with dysplasia. Is there a role of balloon-based radiofrequency ablation therapy? *Acta Gastroenterol Belg.,* 2009. 72(3): p. 373-6.

[127] Kent, H., Plantar wart treatment with ultrasound. *Arch Phys. Med. Rehabil,* 1959. 40(1): p. 15-8.

[128] Rowe, R.J. and J.M. Gray, Ultrasound therapy of plantar warts. *Arch. Dermatol,* 1960. 82: p. 1008-9.

[129] Braatz, J.H., B.R. McAlistar, and M.D. Broaddus, Ultrasound and plantar warts: a double blind study. *Mil. Med.,* 1974. 139(3): p. 199-201.

[130] Vaughn, D.T., Direct method versus underwater method in the treatment of plantar warts with ultrasound. A comparative study. *Phys. Ther,* 1973. 53(4): p. 396-7.

[131] Griffin, J.E. and J.W. Gersten, What is the best method in the use of ultrasound for the treatment of plantar warts? *Phys. Ther,* 1971. 51(1): p. 82-3.

[132] Cohen, H.J., Warts and ultrasound therapy. *Arch. Dermatol,* 1969. 100(4): p. 489.

[133] Kent, H., Warts and ultrasound. Arch Dermatol, 1969. 100(1): p. 79-81.

[134] Rowe, R.J. and J.M. Gray, Ultrasound Treatment of Plantar Warts. *Arch. Phys. Med. Rehabil.,* 1965. 46: p. 273-4.

[135] Cherup, N., J. Urben, and L.F. Bender, The Treatment of Plantar Warts with Ultrasound. *Arch. Phys. Med. Rehabil,* 1963. 44: p. 602-4.

[136] Colombo-Benkmann, M., et al., Ultrasonic technology: a new treatment option for anal condylomata acuminata. *Dis. Colon Rectum*, 2008. 51(11): p. 1681-5.

[137] Bekassy, Z. and L. Westrom, Infrared coagulation in the treatment of condyloma acuminata in the female genital tract. *Sex Transm Dis.,* 1987. 14(4): p. 209-12.

[138] Piskin, S., T. Aksoz, and A. Gorgulu, The treatment of common warts with infrared coagulation. *J. Dermatol*, 2004. 31(12): p. 989-92.

[139] Halasz, C.L., Treatment of common warts using the infrared coagulator. *J. Dermatol. Surg. Oncol.,* 1994. 20(4): p. 252-6.

[140] Gao, X.H., et al., Non-ablative controlled local hyperthermia for common warts. *Chin. Med. J. (Engl),* 2009. 122(17): p. 2061-3.

[141] Huo, W., et al., Local hyperthermia at 44 degrees C for the treatment of plantar warts: a randomized, patient-blinded, placebo-controlled trial. *J. Infect Dis.* 201(8): p. 1169-72.

[142] Li, H.X., W.Y. Zhu, and M.Y. Xia, Detection with the polymerase chain reaction of human papillomavirus DNA in condylomata acuminata treated with CO2 laser and microwave. *Int. J. Dermatol*, 1995. 34(3): p. 209-11.

[143] .Badgwell Doherty, C., S.D. Doherty, and T. Rosen, Thermotherapy in dermatologic infections. *J. Am. Acad. Dermatol.* 62(6): p. 909-27; quiz 928.

[144] Locricchio, J., Jr. and J.R. Haserick, Hot-water treatment for warts. *Cleve Clin. Q,* 1962. 29: p. 156-61.

[145] Kang, S. and T.B. Fitzpatrick, Debilitating verruca vulgaris in a patient infected with the human immunodeficiency virus. Dramatic improvement with hyperthermia therapy. *Arch. Dermatol*, 1994. 130(3): p. 294-6.

[146] Dvoretzky, I., Hyperthermia therapy for warts utilizing a self-administered exothermic patch. Review of two cases. *Dermatol. Surg,* 1996. 22(12): p. 1035-8; discussion 1038-9.

In: Hyperthermia: Recognition, Prevention and Treatment ISBN 978-1-61942-275-9
Editor: Xing-Hua Gao and Hong-Duo Chen, 241-254

Chapter 10

HIGHLY FOCALIZED THERMOTHERAPY – A NEW TECHNIQUE FOR SOLID TUMOURS TREATMENT

A. Portela[1], M. Vasconcelos[1], F. Gartner,[2] and J. Cavalheiro[3]
[1]FMD, Universidade do Porto, Porto, Portugal
[2]IPATIMUP, ICBAS, Universidade do Porto, Porto, Portugal
[3]INEB/FEUP, Universidade do Porto, Porto, Portugal

ABSTRACT

Ferrimagnetic materials, exposed to an electromagnetic field, are largely utilized in the experimental treatment of tumours by hyperthermia. Hyperthermia is a widely studied method in the cancer treatment, operating in the tumour cells, tumour neovascularization, and it may induce an immunologic and apoptotic response in the whole organism. The methods currently available to produce hyperthermia are generally limited by the inability to selectively target the tumour cells, with the subsequent risk of affecting adjacent healthy tissues. To overcome this limitation and in the context of the magnetically mediated hyperthermia, the Highly Focalized Thermotherapy (HFT) technique was developed. A new material, a ferrimagnetic cement (FC), composed by a silicate cement and magnetite was created to be used with this technique. FC is injected within the tumour and then exposed to an external high frequency magnetic field. The data obtained in the in vitro and the in vivo studies allows us to conclude that FC can provide a minimally invasive technique to treat solid tumours with HFT.

INTRODUCTION

The hyperthermia, in the oncologic treatment, is a widely studied method and refers to the application of heat (42-46ºC) as a vital adjunct in regional tumour control due to its selective effect on cancer cells, tumour neovascularization, immunologic and apoptotic response induction in the whole organism and to its potentiation of chemotherapy and radiation therapy [1-4]. Clinical hyperthermia can be divided into three almost separate domains: whole body hyperthermia, regional hyperthermia and local hyperthermia (including superficial local and interstitial local hyperthermia). The clinical application of heat can be induced by different methodologies like ultrasounds, microwaves, perfusion methods and electromagnetic fields [3-5]. One of those approaches, the Magnetically Mediated Hyperthermia (MMH), was first described by Gilchrist et al. in 1957 [6] and it has the potential to address the targeting and tissue heat distribution shortcomings of the other hyperthermia modalities.

The MMH technique consists of localizing magnetic particles or seeds within the tumour tissue and then applying an external alternating magnetic field to cause them to heat by hysteresis loss, Neel relaxation or induced eddy currents. This heat then conducts into the surrounding cancerous tissue [7-9].

Currently, There seems to be a renewed interest in this tumour treatment, due to new developments including improved heating techniques to uniform the heat only in the tumour region without damaging normal tissue.

Many investigation teams are working on the development of new strategies to improve the MMH techniques according to the tumour type and location. In this context, a new technique was developed, the Highly Focalized Thermotherapy (HFT).

Highly Focalized Thermotherapy

This technique consists in the direct injection of a magnetic material into the tumour mass, its immobilization in the injected site and the subsequent exposition to an external high frequency magnetic field that will heat the magnetic particles and subsequently, the injected tumour.

The HFT technique was primarily described by Cavalheiro et al. in 1996 [10]. The first material used for this propose was a bioceramic, a ferrimagnetic ceramic, which works as a vehicle and support of ferrimagnetic particles. Successful in vitro studies were performed to assess the biocompatibility of this material, which was later used in vivo to treat solid tumours. Treatments were effective in small tumours, with a significant reduction of the tumour, but the injection of larger quantities in larger tumours was lethal to some animals, due to of the material toxicity [11-12].

In an attempt to overcome this problem, the same author developed a new material, a ferrimagnetic cement (FC), to be used in the treatment of solid tumours using the previously described methodology.

Ferrimagnetic Cement

The composition of this new material, the ferrimagnetic cement, is similar to the silicon–calcium cements but with a large amount of ferrimagnetic oxide. FC is a fine powder structure, which has a chemical composition in weight percent, based on oxides, corresponding to $10SiO2$, $2Al2O3$, $52Fe2O3$, $0.6MgO$, $33CaO$, $(SO3 + K2O)$ R.

In previous studies [13], it was determined that the FC paste obtained by the powder/water combination ratio (3:2) flows, enabling the product to be injected within a reasonable period of time (5 minutes) and with an adequate setting time.

Over the past few years, many in vitro and in vivo studies have been performed to establish the bioactivity of bioceramic materials, such as calcium phosphate, bioglass, and the silicon–calcium cements (the mineral trioxide aggregate, the Portland-like cements). These materials are indicated to be used in orthopaedics and in dentistry due to their osteoconductivity and bone replacement capability [14-16].

The magnetic particles applied to hyperthermia, must satisfy some properties like non-toxicity, biocompatibility, injectability, high-level accumulation in the target tumour and effective absorption of the energy of alternating magnetic fields.

Based on the principle that the physical structure of a material and the properties of its surface is an indicative of a material's biocompatibility [17], in vitro studies were performed through the FC samples immersion in a simulated body fluid - SBF (where the calcium was substituted by strontium - SBF Sr), and the interaction between FC and the simulated physiological environment was determined using SEM/EDS and XRD. Significant alterations were detected in the FC samples' surface morphology and composition before and after immersion in SBF Sr. The porous surface observed in the control samples, not immersed in SBF Sr (Figure 1), was covered by a layer of precipitates (Figure 2) with the continuous formation, that tends to stabilize 48h after the immersion. The EDS spectrum of the control surface (Figure 3) shows peaks of Fe, Ca and Si, as well as small percentages of Al and Mg, which are components of FC.

The layer which covers the initial surface (Figure 4) is composed of Ca, Mg and Sr associated to high peaks of C and P. Crossing the EDS results with XRD analysis (Figure 5) proves to be mainly crystalline SrCO3 in association with amorphous calcium compounds, because it was not possible to identify crystalline phases of phosphorus or carbonate calcium compounds.

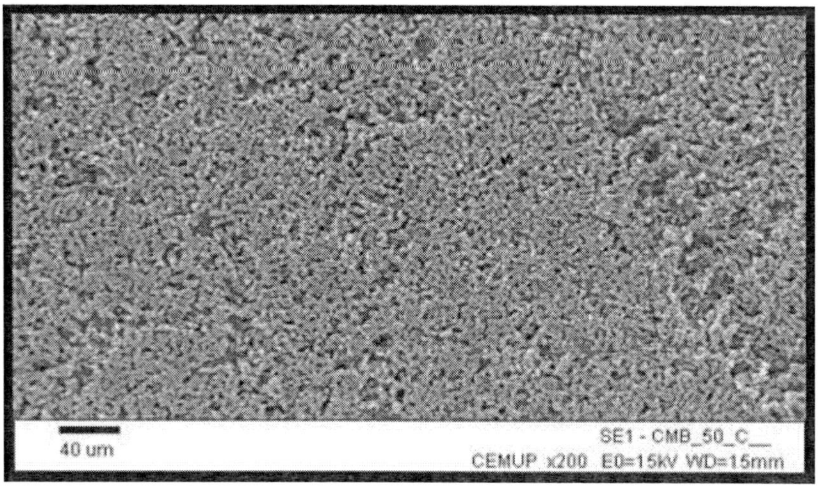

Figure 1. FC control samples after cement preparation. It was observed a very porous surface (200x magnification).

Figure 2. FC samples immersed for 43 hours in SBF Sr. The initial FC surface is completely covered with precipitates (200x magnification).

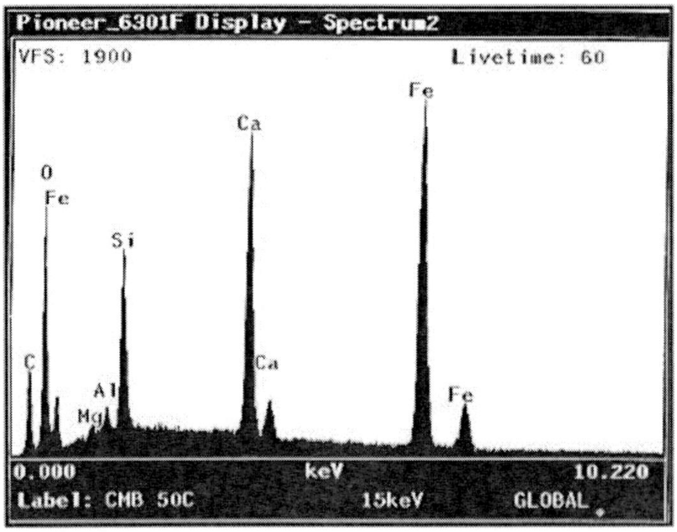

Figure 3. Initial surface spectrum, where peaks of Fe, Ca and Si can be seen, as well as small percentages of Al and Mg, which are components of FC.

Due to the inexistence of Ca in the special SBF Sr medium, the existent Ca of the precipitate layer can only comes from the FC. Part of the cement will dissolve and the Ca will reprecipitate on the surface. Therefore, there was an ionic exchange between the initial cement sample and the surrounding environment resulting in the formation of the layer of precipitates.

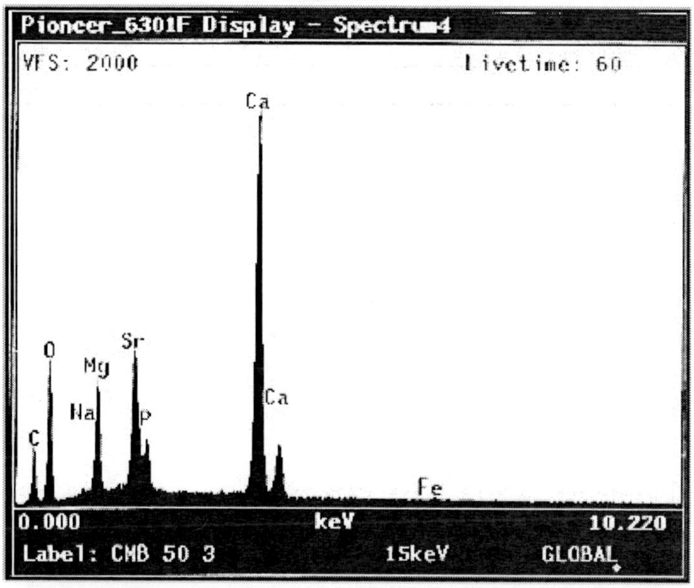

Figure 4. Spectrum of the sample surface that is covered with Ca, Sr, Mg, C and P, resulting from interaction with the synthetic plasma.

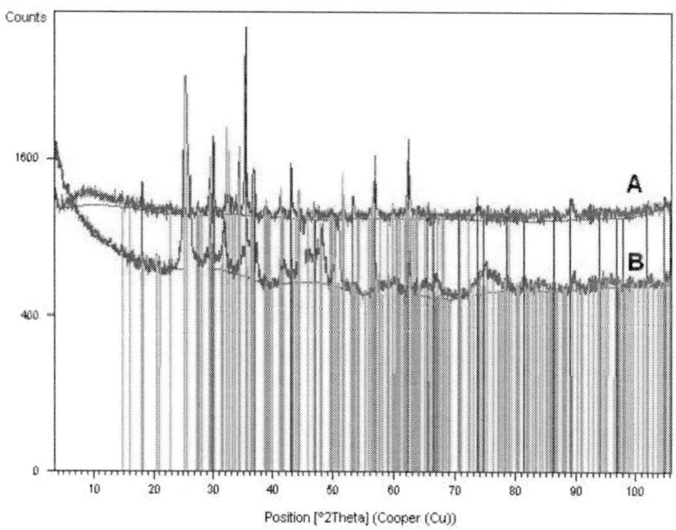

Figure 5. X-ray diffraction patterns of control samples (A) allow the identification of peaks of the two major compounds of the cement: magnetite (2θ=35.48; 62.56) and calcium silicate Ca3SiO5 (2θ=29.46; 32.26; 34.40). In the samples immersed 8 days in SBF Sr (B) the surface is covered with large crystalline precipitates. The main peaks belong to crystalline SrCO3 (2θ=25.42; 25.84; 36.64).

Some authors report the formation of that layer on the surface of different materials during immersion in SBF and describe the bioactive behaviours as an ion exchange mechanism between the biomaterial and the physiological solution [18-21].

Although in vitro studies provide information about certain elements (the cellular, molecular interaction with biomaterials), they cannot replace in vivo studies, owing to the complexity of the biological medium [22,23]. So, in vivo studies of FC were also performed [13]. They consisted in the intramuscular implantation of FC samples and a control material in Wistar rats. The biological response was assessed through the histological evaluation, by haematoxylin-eosin staining (HandE), in accordance with the ISO 10993-6:1994 [22]. The histological data regarding the obtained samples revealed a transient local inflammatory reaction, similar to the control material, only slightly more abundant during the first weeks, with a gradual decrease over the implantation time (Figures 6 and 7).

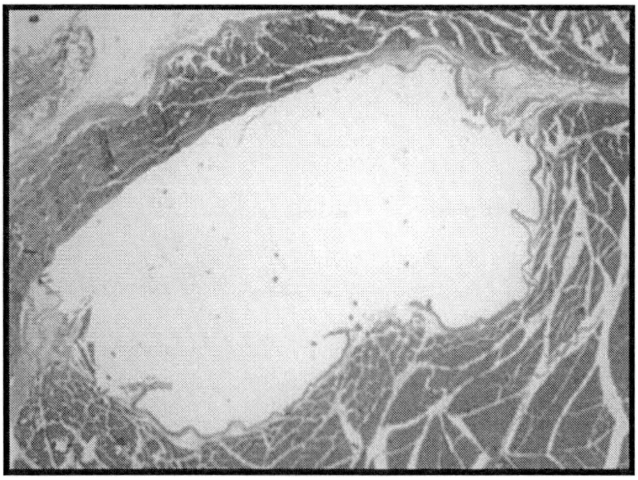

Figure 6. Samples of the control material, nine weeks after implantation (50x magnification). It can be observed a cystic lesion surrounding the area of the control material with few lymphocytes in a thin layer of inflammatory cells.

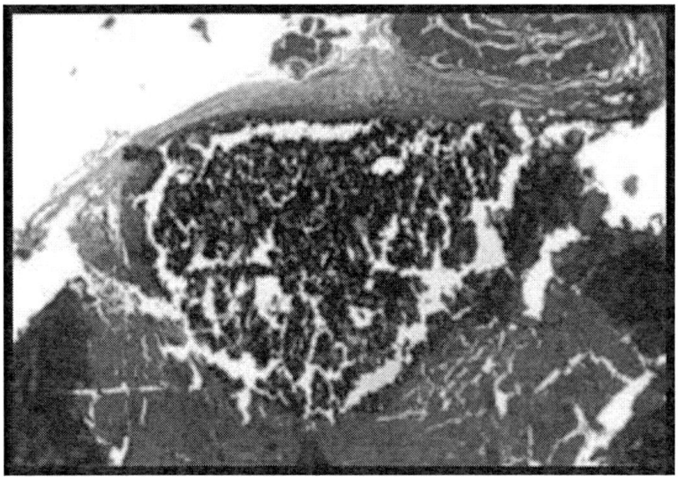

Figure 7. Image of the intramuscular FC implantation area, after nine weeks (50x magnification). The FC is surrounded by muscle tissue in the lower portion of the sample and by an inflammatory infiltrate in the upper portion. The inflammatory cells are only few macrophages and lymphocytes.

FC demonstrates to be biocompatible allowing the remaining in the organism during the treatment period without side effects.

Since FC will be used in the treatment of solid tumours by the HFT methodology, it's necessary that the FC has the capability to generate heat when exposed to a magnetic field. The in vitro evaluation of the specific heating power of FC, exposed to a magnetic field (10 kHz) in a vertical coil (diameter 11 cm, 12 turns), using an induction system High Frequency Electronic Furnace K10/RV (CALAMARI and Milan, Italy), demonstrate a temperature increase in the FC sample environment (maximum heating power of 2.11 W g-1).

In Vivo HFT Application in a Melanoma Model

The following step was accomplished to evaluate the FC performance within the tumour and the HFT technique effectiveness in the tumour treatment. The animal model used for this assessment was the C57BL6 inbreed mouse strain and the tumour was established by the subcutaneous injection of mouse B16F10 melanoma cells. After the tumour model achieved a size of 10mm in diameter, the FC was directly injected within the tumour. Tacking into account that FC is a radiopaque material, its location and distribution in the tumour can be assessed through radiographic image.

To allow repeated treatments, dispensing a repeated injection, it's necessary that FC stays located in the injected site, the tumour. The FC stabilization was confirmed by radiographic image and SEM/EDS (Figures 8, 9 and 10).

The injected FC surface is composed of Ca, N associated with P and a high peak of C while the inside is composed by Ca, Fe, Si, Mg, the main compounds of FC, associated to P and a high peak of C, confirming the results obtained in the previous in vitro studies [13]: the FC in contact with body fluids become incapsulated in a calcium phosphate layer.

Animals were then exposed to an external high frequency magnetic field (10 kHz) in a vertical coil (diameter 11 cm, 12 turns), using an induction system High Frequency Electronic Furnace K10/RV (CALAMARI and Milan, Italy) (Figure 11) for 3 periods of 30 minutes, with a 48hours interval, avoiding heat shock protein-induced thermoresistance [24]. The repeated exposition was based on hyperthermia studies, which confirm the effectiveness of multiple exposures to achieve tumour regression [25,26].

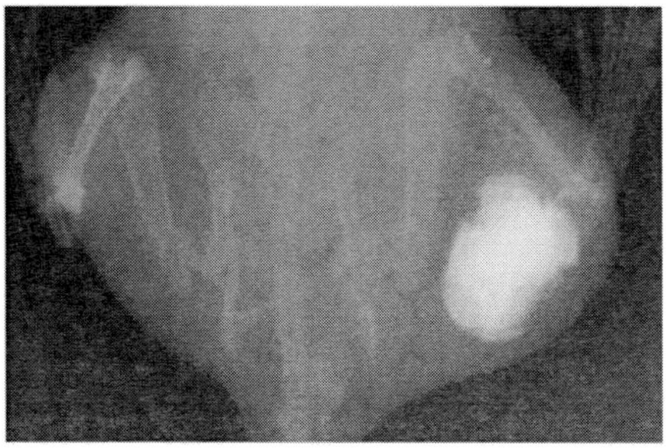

Figure 8. Rx image of FC injected within the tumour.

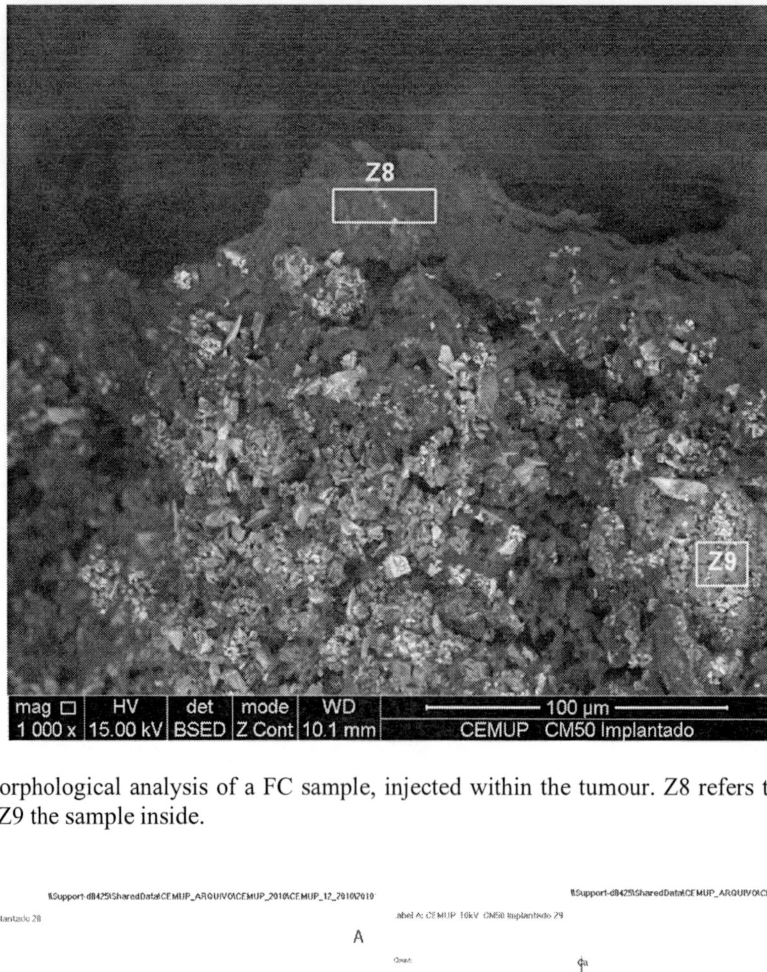

Figure 9. Morphological analysis of a FC sample, injected within the tumour. Z8 refers to the sample surface and Z9 the sample inside.

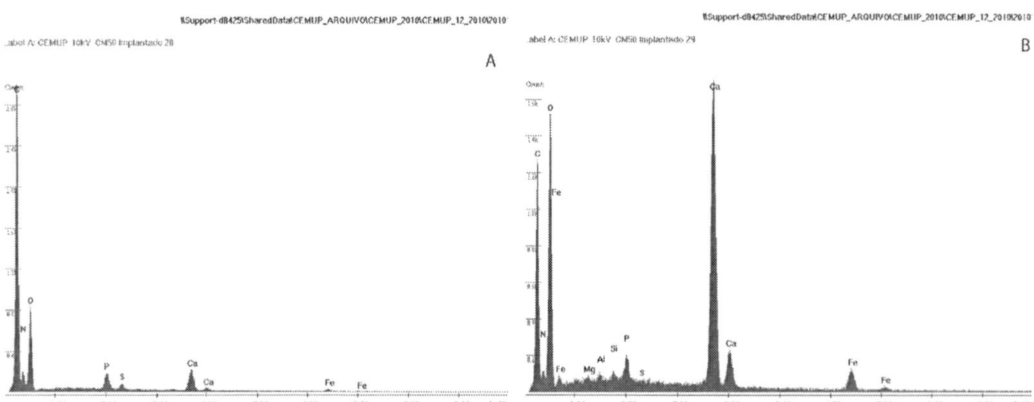

Figure 10. EDS analysis of the injected FC sample. The surface (A-Z8) is composed of Ca, N associated with P and a high peak of C. The inside (B-Z9) is composed by Ca, Fe, Si, Mg associated to P and a high peak of C.

Figure 11. Animal exposed to the electromagnetic field.

The animal's whole body temperature was monitored using a thermographic camera (ThermaCAM TM Researcher Professional 2.9; FLIR Systems, 2007) and when the initial tumour temperature increased to 42°C, it was maintained during all the treatment period (30 minutes), by controlling the magnetic field intensity. In contrast, the average of the body temperature showed only a limited increase (38°C). It indicates that with HFT treatment it was feasible to heat only the tumour and not damage healthy tissues (Figure 12).

The macroscopic observation and measurement of the tumours dimensions (determined by applying a formula described in the literature) [27], was realised every 2 days; the first corresponds to the day when FC was injected, followed by the 1°st, 2°nd and 3°rd treatments, and the final measurement corresponds to the day 16 after the last treatment. It was determined that the time courses of HFT treated tumour growth, compared to the control group, significantly decreased (Figure 13). Even though all the mice examined were not completely cured by the HFT treatment, a significant prolongation of survival time was observed compared to the control group.

A comprehensive understanding of the mechanisms through which HFT kills tumour cells is extremely important. It would provide useful information to, if necessary, improve the methodology.

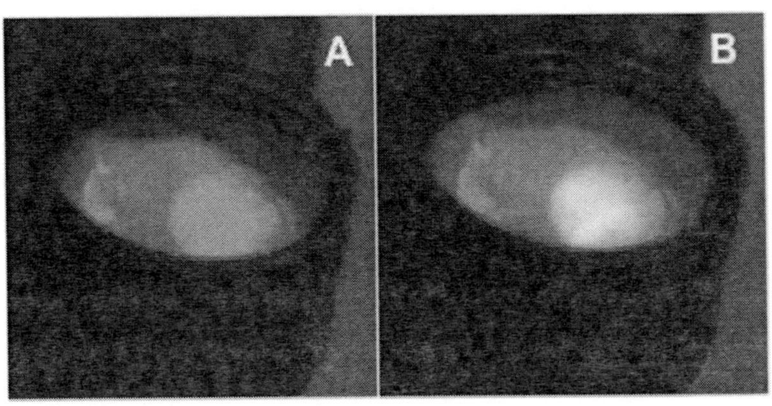

Figure 12. Thermographic camera image of the animal injected with FC. A) Before the magnetic field activation. B) One minute after the magnetic field activation.

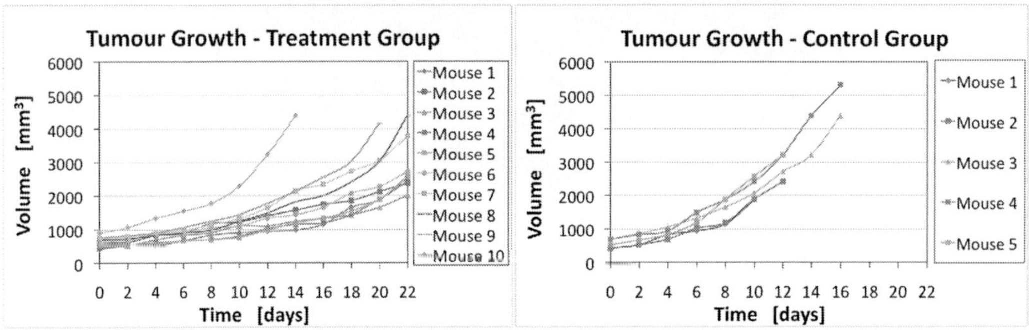

Figure 13. Time courses of tumour growth after 3 expositions to the electromagnetic field, treatment group (n=10) and of normal tumour growth, without any intervention, control group (n=5). Each line represents to the tumour growth in a single mouse. The end of the line indicates the mouse death.

The programmed cell death is a basic biological phenomenon that plays an important role during development and preservation of tissue homeostasis and removal of damaged cells.

According to the morphology of programmed cell death, it can be classified as type I, apoptotic cell death, type II autophagy and type III, programmed necrosis [28, 29]. The histological evaluation of the treated tumour samples at the end of treatment period was assessed by routine HandE staining. This observation revealed severe necrosis with haemorrhage (coagulate necrosis) associated to inflammatory infiltrate and many apoptotic cells with condensed nuclei surrounding FC. The tumour damage was extensive (Figure 14).

It was also important to examine whether the HFT treatment induce apoptosis. Apoptotic cell death acts as part of a quality control and repair mechanism by elimination of unwanted, genetically damaged, or senescent cells, and as such it is critically important for the development of organisms. It is now clear that caspase activation (cleavage of procaspase to active caspase) is a hallmark of almost all apoptotic systems. Caspase 3 is a central effector caspase in many cells and mediates the cleavage of itself, other downstream caspases, and other caspase substrates [30].

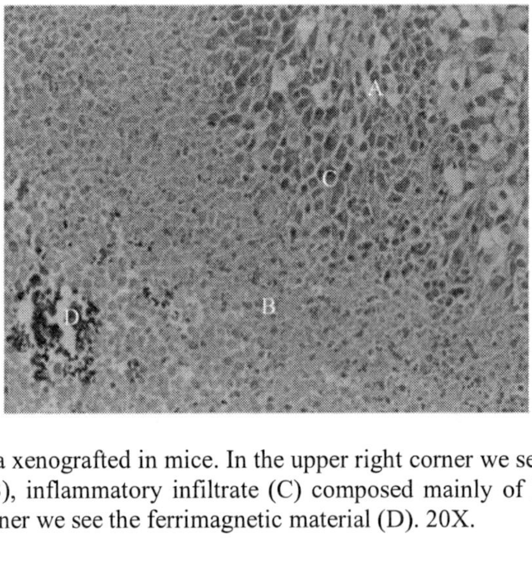

Figure 14. BL6 melanoma xenografted in mice. In the upper right corner we see the neoplastic cells (A) with areas of necrosis (B), inflammatory infiltrate (C) composed mainly of lymphocytes and plasma cells. In the lower left corner we see the ferrimagnetic material (D). 20X.

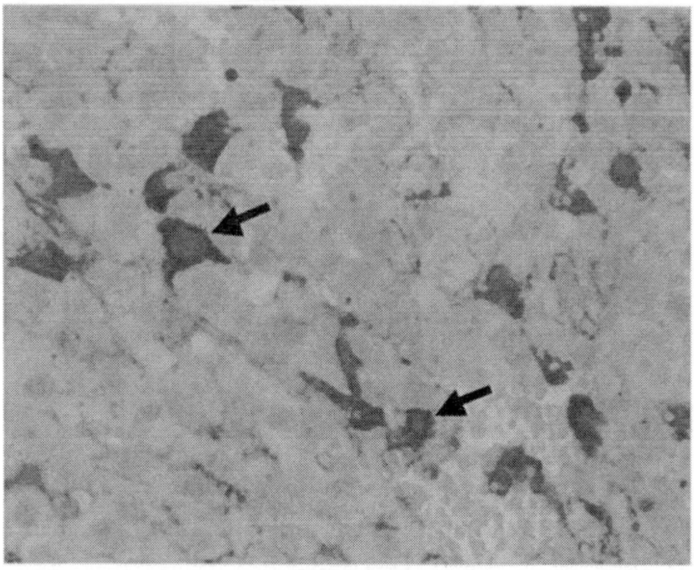

Figure 15. Immunohistochemistry (IHC) for caspase-3 (Cell Signaling Technology) in a BL6 melanoma xenografted in mice. Moderate cytoplasmic immunostaining in most of the tumour cells and strong immunostaining in few scattered cells (arrow). 40X

For the immunohistochemical studies, thin serial sections were obtained from selected tumour and used for immunohistochemistry (IHC) using a polyclonal antibody against Caspase-3 (Cell Signaling Technology, Inc.). It was observed a moderate cytoplasmic immunostaining in most of the tumour cells and strong immunostaining in few scattered cells (Figure 15).

It is well established that tumour cells have a wide range of sensitivity to hyperthermia [1-4], but it is a temperature- and time-dependent treatment [31] and it also activates the process of apoptosis in tumours [32-35]. In the histological evaluation of melanoma mouse model treated with HFT it was observed many dead cells and it was clarified the mechanism through which the apoptotic pathway was induced. Since the HFT technique with FC works, we can now propose some improvements in the treatment application, like the temperature increase, the time of exposition, the repeated treatments and to experiment this methodology in other tumours.

CONCLUSION

The HFT, using the FC, is a promising technique for the treatment of solid tumours. The FC paste can be directly injected within the tumour, it has the capability to generate heat when placed within a magnetic field and its stabilization within the tumour allows repeated treatments without repeated FC injection, providing a minimally invasive technique. HFT can be an effective therapy for the treatment of melanoma, since this treatment can kill the tumour cells not only by heat but also by inducing an immune response.

Acknowledgement

This work was supported by FCT and FEDER/ Compete program: PTDC/SAU-BEB/69497/2006.

Authors would like to thank to the Laboratory of Pathology from ICBAS- Instituto de Ciências Biomédicas Abel Salazar (Universidade do Porto) and to the FEUP-Faculdade de Engenharia da Universidade do Porto which provided the Thermographic camera.

References

[1] Lagendijk, JJW. Hyperthermia treatment planning. *Phys. Med. Biol.* 2000;45:61-76.

[2] Moyer, HR; Delman, KA. The role of hyperthermia in optimizing tumor response to regional therapy. *Int. J. Hyperthermia* 2008;24(3):251-261.

[3] Falk, MH; Issels, RD. Hyperthermia in oncology. *Int. J. Hyperthermia* 2001;17(1):1-18.

[4] Hildebrandt, B; Wust, P; Ahlers, O; Dieing, A; Sreenivasa, G; Kerner, T; Felix, R; Riess, H. The cellular and the molecular basis of hyperthermia. *Crit. Rev. in Oncol. Hematol* 2002;43:33-56.

[5] van der Zee, J. Heating the patient: a promising approach? *Ann. Oncol.* 2002;13:1173-1184.

[6] Gilchrist, TK; Medal, R; Shorey, WD; Hanselman, RC; Parrot, JC; Taylor, CB. Selective inductive heating of lymph nodes. *Ann. Surg.* 1957;146(4):596-606.

[7] Moroz, P; Jones, SK; Gray, BN. Magnetically mediated hyperthermia: current status and future directions. *Int. J. Hyperthermia* 2002;18(4), 267-284.

[8] Shinkai, M. Functional magnetic particles for medical application. *J. Biosci. Bioeng* 2002;94(6):606-613.

[9] Bahadur, D; Giri, J. Biomaterials and magnetism. Sadhana 2003;28:639-656.

[10] Cavalheiro, J; Vasconcelos, M; Afonso, A; Branco, R. Ferrimagnetic and biocompatible cement for tumours thermotherapy. *Bioceramics* 1996;9:243-246.

[11] Almeida, T. Estudo in vivo de um Cerâmico Ferrimagnético com potencial em Oncoterapia. Tese de Mestrado em Engenharia Biomédica pela Faculdade de Engenharia da Universidade do Porto. 2001.

[12] Almeida, T; Soares, ME; Cavalheiro, J; Bastos, ML. Silicon and iron levels in tissues of animals treated with a "ferrimagnetic ceramic" with oncotherapeutic potential (anti-tumor) value. *J. Trace Elem. Med. Biol.* 2002;16:255-259.

[13] Portela, A; Vasconcelos, M; Branco, R; Cavalheiro, J; Gartner, F; Faria, M. An in vitro and in vivo investigation of the biological behaviour of a ferrimagnetic cement for Highly Focalised Thermotherapy. *J. Mat. Sci. Mat. Med.* 2010;21:2413-2423.

[14] Xin, R; Leng, Y; Chen, J; Zhang, Q. A comparative study of calcium phosphate formation on bioceramics in vitro and in vivo. Biomaterials 2005;26(33):6477-6486.

[15] Burguera, EF; Xu, HH; Takagi, S; Chow, LC. High early strength calcium phosphate bone cement: effects of dicalcium phosphate dihydrate and absorbable fibers. *J. Biomed. Mater. Res.* A 2005; 75(4):966-975.

[16] Islam, I; Chng, HK; Yap, AU. X-ray diffraction analysis of mineral trioxide aggregate and Portland cement. *Int. Endod. J.* 2006;39(3): 220-225.

[17] Ratner, BD; Hoffman, A; Schoen, FJ; Lemons, JL. *Biomaterials science: an introduction to materials in medicine.* 2nd ed. London: Academic Press; 1996.

[18] Kokubo, T; Miyaji, F; Min-Kim, H; Nakamura, T. Spontaneous formation of bonelike apatite layer on chemically treated titanium metals. *J. Am. Ceram. Soc.* 1996;79(4):1127-1129.

[19] Takadama, H; Kim, M; Kokubo, T; Nakamura, J. An X-ray photoelectron spectroscopy study of the process of apatite formation on bioactive titanium metal. *J. Biomed.Mater. Res.* 2001;55(2):185-193.

[20] Liang, F; Zhou, L; Wang, K. Apatite formation on porous titanium by alkali and heat-treatment. *Surf. Coat Technol* 2003;165:133-139.

[21] Spriano, S; Bronzoni, M; Vernè, E; Maina, G; Bergo, V; Windler, M. Characterization of surface modified Ti–6Al–7Nb alloy. *J. Mater. Sci. Mater. Med.* 2005;16(4):301-312.

[22] International standard. Biological evaluation of medical devices - Part 6: Test for local effects after implantation. ISO 10993-6. 1994.

[23] Kirkpatrick, CJ; Krump-Konvalinkova, V; Unger, RE; Bittinger, F; Otto, M; Peters, K. Tissue response and biomaterial integration: the efficacy of in vitro methods. *Biomol. Eng.* 2002;19:211-217.

[24] Tucker, RD; Loening, SA; Landas, S; Paulus, JA; Ren, ZY; Park, JB. The effect of interstitial hyperthermia on the Dunning prostate tumor model. *J. Urol.* 1992;147:1129-1133.

[25] Yanase, M; Shinkai, M; Honda, H; Wakabayashi, T; Yoshida, J; Kobayashi, T. Intracellular hyperthermia for cancer using magnetite cationic liposomes: an in vivo study. *Jpn J. Cancer Res.* 1998;89:463-469.

[26] Saito, H; Mitobe, K; Ito, A; Sugawara, Y; Maruyama, K; Minamiya, Y; Motoyama, S; Yoshimura, N; Ogawa, J. Self-regulating hyperthermia induced using thermosensitive ferromagnetic material with a low Curie temperature. *Cancer Sci.* 2008;99(4):805-809.

[27] Buhtoiarov, IN; Lum, HE; Berke, G; Sondel, PM; Rakhmilevich, AL. Synergistic activation of macrophages via CD40 and TLR9 results in T cells independent anti-tumor effects. *J. Immunol.* 2006;176:308-318.

[28] Gozuacik, D; Kimchi, A. Autophagy and cell death. *Curr. Top. Dev. Biol.* 2007;78:217–45.

[29] Clarke, PGH. Apoptosis: from morphological types of cell death to interacting pathways. *Trends Pharmacol. Sci.* 2002;23:308–309.

[30] Green, DR. Apoptotic pathways: paper wraps stone blunts scissors. *Cell* 2000;102:1–4.

[31] Shellman, YG; Howe, WR; Miller, LA; Goldstein, NB; Pacheco, TR; Mahajan, RL; LaRue, SM; Norris, DA. Hyperthermia induces endoplasmic reticulum-mediated apoptosis in melanoma and non-melanoma skin cancer cells. *J. Inv. Derm.* 2008;128:949-956.

[32] Dyson, JED. Kinetic and physical studies of cell death induced by chemotherapeutic agents or hyperthermia. *Cell Tissue Kinet* 1986;19:311-324.

[33] Barry, MA; Behnke, CA; Eastman, A. Activation of programmed cell death (apoptosis) by cisplatin, other anticancer drugs, toxins and hyperthermia. *Biochem. Pharmacol.* 1990;40:2353-23621.

[34] Takano, YS; Harmon, BV; Kerr, JFR. Apoptosis induced by mild hyperthermia in human and murine tumor cell lines: a study using electron microscopy and DNA gel electrophoresis. *J. Pathol.* 1991;163:329-336.

[35] Harmon, BV; Corder, AM; Collins, RJ; Gobe, GC, Allen, J, Allan, DJ; Kerr, JFR. Cell death induced in a murine mastocytoma by 42-47ªC in vitro: evidence that the form of death changes from apoptosis to necrosis above a critical heat load. *Int. J. Radiat. Biol.* 1990;58:845-858.

INDEX

A

B

C

D

E

F

G

H

I

K

L

M

N

O

P

Q

R

S

V

W

Y